Mechanical Behavior of High-Strength Low-Alloy Steels

Mechanical Behavior of High-Strength Low-Alloy Steels

Special Issue Editors

Ricardo Branco
Filippo Berto

MDPI • Basel • Beijing • Wuhan • Barcelona • Belgrade

MDPI

Special Issue Editors
Ricardo Branco
University of Coimbra
Portugal

Filippo Berto
Norwegian University of Science and Technology
Norway

Editorial Office
MDPI
St. Alban-Anlage 66
Basel, Switzerland

This is a reprint of articles from the Special Issue published online in the open access journal *Metals* (ISSN 2075-4701) from 2017 to 2018 (available at: http://www.mdpi.com/journal/metals/special_issues/low_alloy_steels)

For citation purposes, cite each article independently as indicated on the article page online and as indicated below:

LastName, A.A.; LastName, B.B.; LastName, C.C. Article Title. *Journal Name* **Year**, *Article Number*, Page Range.

ISBN 978-3-03897-204-4 (Pbk)
ISBN 978-3-03897-205-1 (PDF)

Cover image courtesy of Ricardo Branco.

Contents

About the Special Issue Editors

Ricardo Branco ompleted his PhD degree in Mechanical Engineering from the University of Coimbra. He is currently Assistant Professor at the Department of Mechanical Engineering of the University of Coimbra. His research interests include the mechanical behavior of materials, fatigue and fracture of engineering materials, multiaxial fatigue life prediction models, low-cycle fatigue of high-strength steels, and the numerical modelling of fatigue crack growth.

Filippo Berto ompleted his PhD degree in Mechanical Engineering from the University of Florence. He is currently Chair of Fatigue and Fracture at the Department of Mechanical Engineering of the Norwegian University of Science and Technology. His research interests include local approaches for fatigue assessment, the fatigue and fracture of advanced materials, multiaxial fatigue, the structural integrity of additive, materials and the numerical modeling of fatigue crack growth.

metals

MDPI

Editorial

Mechanical Behavior of High-Strength, Low-Alloy Steels

Ricardo Branco [1,*] and Filippo Berto [2,*]

[1] CEMMPRE, Department of Mechanical Engineering, University of Coimbra, 3030-788 Coimbra, Portugal
[2] Department of Industrial and Mechanical Engineering, Norwegian University of Science and Technology, 7491 Trondheim, Norway
* Correspondence: ricardo.branco@dem.uc.pt (R.B.); filippo.berto@ntnu.no (F.B.)

Received: 25 July 2018; Accepted: 2 August 2018; Published: 6 August 2018

1. Introduction and Scope

High-strength low-alloy steels are designed to provide specific desirable combinations of properties, such as strength, toughness, formability, weldability, and corrosion resistance. These features make them ideal for critical applications under severe service conditions and in aggressive environments, namely rail and road vehicles, passenger car components, construction machinery, industrial equipment, offshore structures, gas pipelines, and bridges, among others. This Special Issue aims to address the mechanical behavior of high-strength low-alloy steels from different perspectives, namely in terms of mechanical deformation, damage, and failure. It gathers scientific contributions from authors working in various fields, for instance processing techniques, the modeling of the mechanical behavior, the characterization of material microstructure, the influence of environmental parameters, temperature dependence, as well as advanced applications.

2. Contributions

The modern automotive and rail industries are facing significant challenges in terms of weight reduction for fuel economy as well as in terms of mechanical properties to develop safer and more durable products. In this Special Issue, attention is paid to the relation of the final mechanical properties and the microstructural features. The paper by Branco et al. [1] provides an insight into the role of the bainitic morphologies in fatigue cyclic plastic properties; the paper by Liang et al. [2] studies the effect of phase fraction on deformation and fracture behavior in low-carbon ferrite-martensite steels; and the paper by Evin et al. [3] addresses the microstructure characteristics of the mechanical properties of dual-phase, high-strength steels.

Due to the incentives for the widespread use of natural gas as a source of clean energy, high-strength low-alloy steels have been one of the prime choices for high-pressure gas pipeline networks. In this context, knowledge of the microstructural and mechanical properties of such materials, as well as the development of accurate design and inspection methodologies is of major importance. Lavigne et al. [4] present a comprehensive study of the microstructural and mechanical characterization of API 5L X52 steel; Silva et al. [5] discuss the effect of precipitation and grain size on the tensile strain-hardening exponent of API X80 steel; Liu et al. [6] analyze the local buckling behavior and plastic deformation capacity of high-strength X80 steel pipelines subjected to strike–slip fault displacements; and Vilkys et al. [7] study the influence of mechanical surface defects on the safe operation of gas pipelines on the basis of fragments collected from operating parts.

The effect of aggressive environments on the design of engineering structures is also particularly important. This is well documented in the paper written by Krivy et al. [8], which examines the dependence between the deposition of chlorides and the corrosion layers of two steel bridges; or in the paper written by Cho et al. [9], which correlates hydrogen-induced corrosion cracking with

Metals **2018**, *8*, 610

maintenance interventions in the wires of cable suspension bridges; or in the paper written by Cabrini et al. [10], which evaluates the critical ranges of pH for the initiation of stress corrosion cracking in high-strength steel bars for pre-stressed concrete structures.

Finally, another important topic is the fracture behavior of high-strength steels. Zhang et al. [11] investigate the brittle fracture of large press die holders following a systematic approach that encompasses chemical composition analysis, mechanical property testing, and microstructure examination of the critical area; Riyanta et al. [12] determine the fracture resistance of AISI 304 welding by Charpy impact testing, relating the values of energy absorption to the presence of chromium interstitial solute and chromium carbide precipitation. Last, but not least, Suárez et al. [13] review the recent experimental and numerical advances dealing with fracture mechanics in steels with a primary focus on flat-fracture surfaces.

References

1. Branco, R.; Berto, F.; Zhang, F.; Long, X.; Costa, J.D. Comparative Study of the Uniaxial Cyclic Behaviour of Carbide-Bearing and Carbide-Free Bainitic Steels. *Metals* **2018**, *8*, 422. [CrossRef]
2. Liang, J.; Zhao, Z.; Wu, H.; Peng, C.; Sun, B.; Guo, B.; Liang, J.; Tang, D. Mechanical Behavior of Two Ferrite–Martensite Dual-Phase Steels over a Broad Range of Strain Rates. *Metals* **2018**, *8*, 236. [CrossRef]
3. Evin, E.; Kepič, J.; Buriková, K.; Tomáš, M. The Prediction of the Mechanical Properties for Dual-Phase High Strength Steel Grades Based on Microstructure Characteristics. *Metals* **2018**, *8*, 242. [CrossRef]
4. Lavigne, O.; Kotousov, A.; Luzin, V. Microstructural, Mechanical, Texture and Residual Stress Characterizations of X52 Pipeline Steel. *Metals* **2017**, *7*, 306. [CrossRef]
5. Silva, R.A.; Pinto, A.L.; Kuznetsov, A.; Bott, I.S. Precipitation and Grain Size Effects on the Tensile Strain-Hardening Exponents of an API X80 Steel Pipe after High-Frequency Hot-Induction Bending. *Metals* **2018**, *8*, 168. [CrossRef]
6. Liu, X.; Zhang, H.; Wang, B.; Xia, M.; Wu, K.; Zheng, Q.; Han, Y. Local Buckling Behavior and Plastic Deformation Capacity of High-Strength Pipe at Strike-Slip Fault Crossing. *Metals* **2018**, *8*, 22. [CrossRef]
7. Vilkys, T.; Rudzinskas, V.; Prentkovskis, O.; Tretjakovas, J.; Višniakov, N.; Maruschak, P. Evaluation of Failure Pressure for Gas Pipelines with Combined Defects. *Metals* **2018**, *8*, 346. [CrossRef]
8. Krivy, V.; Kubzova, M.; Kreislova, K.; Urban, V. Characterization of Corrosion Products on Weathering Steel Bridges Influenced by Chloride Deposition. *Metals* **2017**, *7*, 336. [CrossRef]
9. Cho, T.; Delgado-Hernandez, D.J.; Lee, K.; Son, B.; Kim, T. Bayesian Correlation Prediction Model between Hydrogen-Induced Cracking in Structural Members. *Metals* **2017**, *7*, 205. [CrossRef]
10. Cabrini, M.; Lorenzi, S.; Pastore, T.; Bucella, D.P. Effect of Hot Mill Scale on Hydrogen Embrittlement of High Strength Steels for Pre-Stressed Concrete Structures. *Metals* **2018**, *8*, 158. [CrossRef]
11. Zhang, W.; Wang, H.; Zhang, J.; Dai, W.; Huang, Y. Brittle Fracture Behaviors of Large Die Holders Used in Hot Die Forging. *Metals* **2017**, *7*, 198. [CrossRef]
12. Riyanta, B.; Wardana, I.N.G.; Irawan, Y.S.; Choiron, M.A. AISI 304 Welding Fracture Resistance by a Charpy Impact Test with a High Speed Sampling Rate. *Metals* **2017**, *7*, 543. [CrossRef]
13. Suárez, F.; Gálvez, J.C.; Cendón, D.A.; Atienza, J.M. Distinct Fracture Patterns in Construction Steels for Reinforced Concrete under Quasistatic Loading—A Review. *Metals* **2018**, *8*, 171. [CrossRef]

metals

MDPI

Article

Comparative Study of the Uniaxial Cyclic Behaviour of Carbide-Bearing and Carbide-Free Bainitic Steels

Ricardo Branco [1,*], Filippo Berto [2], Fucheng Zhang [3,4], Xiaoyan Long [3] and José Domingos Costa [1]

[1] CEMMPRE, Department of Mechanical Engineering, University of Coimbra, Rua Luís Reis Santos, 3030-788 Coimbra, Portugal; jose.domingos@dem.uc.pt
[2] Department of Mechanical and Industrial Engineering, Norwegian University of Science and Technology (NTNU), Richard Birkelands vei 2b, 7491 Trondheim, Norway; filippo.berto@ntnu.no
[3] State Key Laboratory of Metastable Materials Science and Technology, Yanshan University, Qinhuangdao 066004, China; zfc@ysu.edu.cn (F.Z.); longxiaoyanlxy@163.com (X.L.)
[4] National Engineering Research Center for Equipment and Technology of Cold Strip Rolling, Yanshan University, Qinhuangdao 066004, China
* Correspondence: ricardo.branco@dem.uc.pt; Tel.: +351-239-790-700

Received: 21 May 2018; Accepted: 1 June 2018; Published: 5 June 2018

Abstract: Bainitic steels play an important role in the modern automotive and rail industries because of their balanced properties. Understanding the relationship between the bainitic microstructure features and the fatigue performance is a fundamental ingredient in developing safer and durable products. However, so far this relationship is not sufficiently clear. Therefore, there is the need to strengthen the knowledge within this field. The present paper aims at comparing the uniaxial cyclic behaviour of carbide-bearing and carbide-free bainitic steels. To meet this goal, fully-reversed strain-controlled tests at various strain amplitudes were performed. After the final failure, fracture surfaces were observed by transmission electron microscopy to relate the bainitic morphology to the fatigue performance. The main findings of this work show that the carbide-free lower bainite has superior fatigue performance compared to the carbide-bearing lower bainite. This is explained by the presence of stable carbides and thick bainite ferrite plates.

Keywords: low-cycle fatigue; cyclic behaviour; strain energy density; bainite; carbide

1. Introduction

Bainitic steels play an important role in the modern automotive and rail industries because of their balanced properties in terms of strength, fatigue and fracture characteristics, wear, elongation, machinability, and production costs [1–3]. In this context, understanding the relationship between the bainitic morphology and the mechanical performance of the produced steels is pivotal to meet these requirements and, ultimately, to develop safer and more durable products.

Mechanical performance is directly related to the bainitic morphology and the chemical composition. In the above-mentioned industries, critical components are usually subjected to time-varying loading histories and therefore, superior cyclic mechanical properties are of major engineering significance. According to Georgiyev et al. [4], the highest performance with respect to crack resistance in medium-carbon steels of similar strength is obtained from carbide-free lower bainite microstructures. The main outcome of a recent study published by Long et al. [5] also attests to the improved performance of the carbide-free lower bainitic steels when compared with carbide-bearing lower bainitic steels in a low-cycle fatigue regime.

Nevertheless, systematic studies dealing with fatigue behaviour of carbide-free and carbide-bearing lower bainitic steels are quite scarce [6,7]. Therefore, there is the need to strengthen the research in this field. From an engineering point of view, as is well-known, fatigue design is usually carried out using stress-based, strain-based, or energy-based relationships [8–10]. In the modern fatigue life prediction models, cyclic plasticity plays a major role and is considered to be the main cause of damage [11,12]. An accurate knowledge of cyclic plastic behaviour is a fundamental ingredient to obtain accurate lifetime predictions, as well as to develop feasible elastic-plastic numerical models [13,14].

The present paper aims at comparing the uniaxial cyclic behaviour of carbide-free and carbide-bearing bainitic steels under fully-reversed strain-controlled conditions. In this ambit, low-cycle fatigue tests at room temperature in standard cylindrical specimens will be performed at various strain amplitudes. Then, the cyclic stress-strain response, the shapes of the hysteresis loops, the fatigue-strength and fatigue-ductility properties, and the plastic strain energy densities of both bainitic steels will be assessed and evaluated. Moreover, before fatigue testing, the microstructures will be analysed by scanning electron microscopy (SEM) and transmission electron microscopy (TEM). After the total failure of the specimens, fracture morphologies will be observed by SEM to relate the phase effects to the cyclic behaviour.

2. Experimental Procedure

Low-cycle fatigue tests, summarised in Table 1, were performed in a 100 kN MTS servo-hydraulic testing machine (MTS, Eden Prairie, MN, USA) under fully-reversed strain-controlled conditions, for strain amplitudes between 0.6% and 1.0%. This was done using sinusoidal waveforms and a constant strain rate of 6×10^{-3} s^{-1}. Each individual test was initiated in tension, and failure was defined as a 25% load drop relative to the maximum load. Hysteresis loops were collected from a uniaxial extensometer. The samples, with a gauge length of 10 mm and a gauge diameter of 5 mm, were fabricated from two bainitic steels [5], termed here carbide-bearing lower bainite and carbide-free lower bainite, whose chemical compositions and mechanical properties are listed in Tables 2 and 3, respectively. Both chemical compositions are virtually the same, except for the content of Si and Al. In the former steel, those elements were alloyed to introduce carbide-free bainite, while in the latter, the elements were not alloyed to introduce carbide-bearing bainite. The contents of S, P, and N are far below 0.01% which means that their effects on the steel performance can be neglected [5]. The steels were synthesised via vacuum smelting and forging, with a forging ratio equal to 6.

Microstructures were observed using a Hitachi H-800 TEM (Hitachi, Tokyo, Japan) operated at 200 kV and a SU-5000 Hitachi thermal-emission SEM (Hitachi, Tokyo, Japan). Before examination, the samples were thinned to perforation on a TenuPol-5 twinjet unit with an electrolyte composed of 7% perchloric and 93% glacial acetic acids. Electropolishing was performed at a temperature of 25 °C and a voltage of 29 V. After the fatigue tests, fracture surfaces were observed by the TEM to characterise the surface morphologies and to identify the main fatigue damage mechanisms.

Table 1. Low-cycle fatigue test program.

Specimen Reference	Total Strain Amplitude, $\Delta\varepsilon/2$ (%)	Elastic Strain Amplitude, $\Delta\varepsilon_e/2$ (%)	Plastic Strain Amplitude, $\Delta\varepsilon_p/2$ (%)	Stress Amplitude, $\Delta\sigma/2$ (MPa)	Plastic strain Energy Density, ΔW_P (MJ/m^3)	Number of Cycles to Failure, N_f
Carbide-bearing lower bainite						
CB-0.6	0.5985	0.4724	0.1261	929.29	3.962	3572
CB-0.7	0.6980	0.4837	0.2143	951.42	6.896	2357
CB-0.8	0.7980	0.4963	0.3017	976.17	9.961	1069
CB-1.0	0.9990	0.5287	0.4703	1040.01	16.541	514
Carbide-free lower bainite						
CF-0.6	0.5995	0.5440	0.0555	1070.00	2.010	6305
CF-0.7	0.6990	0.5906	0.1084	1162.43	4.259	4144
CF-0.8	0.7965	0.6249	0.1716	1229.76	7.138	2003
CF-1.0	0.9950	0.6664	0.3285	1311.62	14.573	783

Table 2. Chemical composition in weight percentage.

Material	C	Si	Mn	Cr	Ni	Mo	Al	S	P	N
Carbide-free lower bainite	0.34	1.48	1.52	1.15	0.93	0.4	0.71	0.003	0.006	0.002
Carbide-bearing lower bainite	0.34	0.01	1.61	1.24	0.96	0.45	0.04	0.002	0.005	0.003

Table 3. Main mechanical properties.

Property	Carbide-Bearing Lower Bainite	Carbide-Free Lower Bainite
Yield strength, σ_{YS} (MPa)	1033	1080
Tensile strength, σ_{UTS} (MPa)	1390	1498
Young's modulus, E (GPa)	198.3	197.4
Total elongation, δ_t (%)	12.5	16.0
HRC	43.6	46.0

3. Results and Discussion

3.1. Microstructure

The microstructures of both carbide-bearing and carbide-free lower bainitic steels that were observed via the SEM and TEM microscopes, are exhibited in Figure 1a–d, respectively. As can be seen in Figure 1a–c, the former is mainly formed by bainitic ferrite (BF) and carbides, with a volume fraction of 5.4%, and was distributed within the bainitic ferrite or between the bainitic ferrite along a certain direction. Whereas, the latter is essentially formed by bainitic ferrite (BF) and retained austenite (RA) of varying sizes, with a volume fraction of 9.9%, and alternating between small sizes (RA1) and large sizes (RA2).

The carbide-bearing lower bainite, when compared with the carbide-free lower bainite, contains thicker bainitic ferrite plates (314 ± 34 nm compared to 133 ± 18 nm) and lower dislocation densities (3.3×10^{15} m^{-2} compared to 4.6×10^{15} m^{-2}). This can be explained by the addition of Si and Al elements in the carbide-free lower bainite, which results in a smaller C-diffusion; inhibition of the precipitation of carbides; and higher distribution of carbon atoms in the retained austenite. Moreover, Al can also increase both the transformation driving force and the nucleation density, giving rise to finer bainitic ferrite plates.

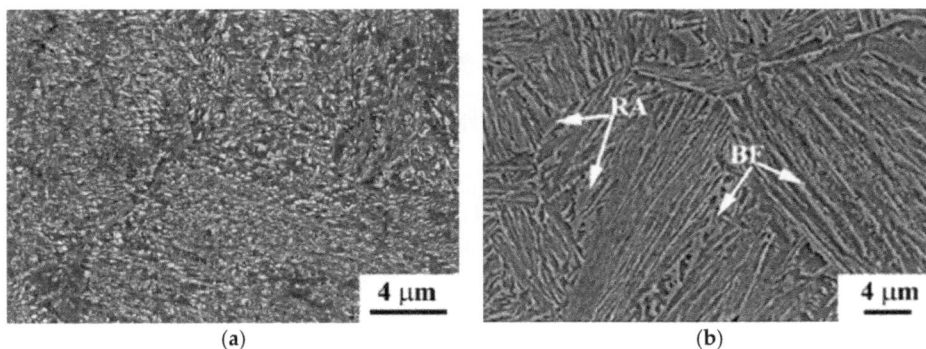

4 μm   4 μm

(a)   (b)

Figure 1. *Cont.*

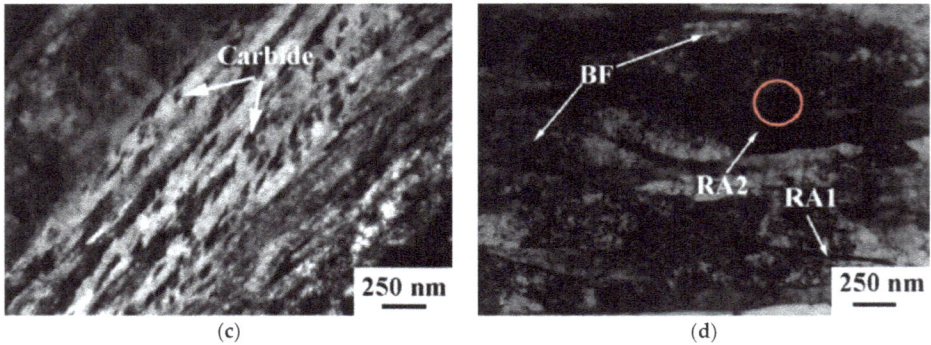

Figure 1. SEM and TEM micrographs [5]: (**a,c**) refer to the carbide-bearing lower bainite; (**b,d**) refer to the carbide-free lower bainite (RA: Retained austenite; BF: Bainitic ferrite). Reproduced from [5], with permission from publisher Elsevier, 2018.

3.2. Cyclic Stress-Strain Deformation Behaviour

Figure 2 plots the peak stress against the normalised fatigue life (N/N_f) for the carbide-bearing and the carbide-free bainitic steels at different strain amplitudes, under fully-reversed strain-controlled conditions. Both steels, irrespective of the strain amplitude, undergo an initial cyclic hardening with growing intensity in the early cycles, followed by a progressive reduction of peak stress, which is more pronounced for the carbide-bearing bainite. In the second stage, peak stress variations tend to be tenuous, and the material response is close to a saturated state, particularly for the lower strain amplitudes. After this period, in the final stage, the peak stress drops more steeply, leading to total failure.

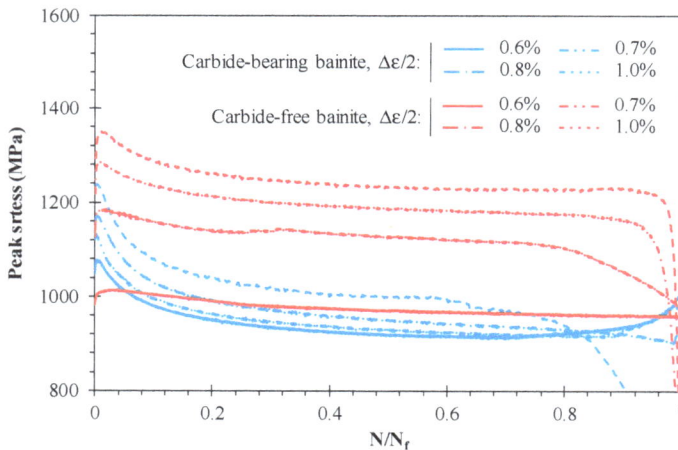

Figure 2. Variation of peak stress with the normalised fatigue life for both carbide-bearing and carbide-free bainitic steels at different strain amplitudes under fully-reversed strain-controlled conditions.

The maximum stress occurs in the early cycles of the tests. The life ratios of these values, designated here by N_p/N_f, are represented in Figure 3a for the tested steels. As can be seen in the figure, although some scatter is observed, there is a clear trend for each case. The maximum stress amplitudes are attained faster for the carbide-bearing lower bainite than for the carbide-free lower bainite, and the N_p/N_f values are, on average, equal to 0.33% (see the dashed line) and 0.98% (see the dash-dotted line) of the life ratio, respectively.

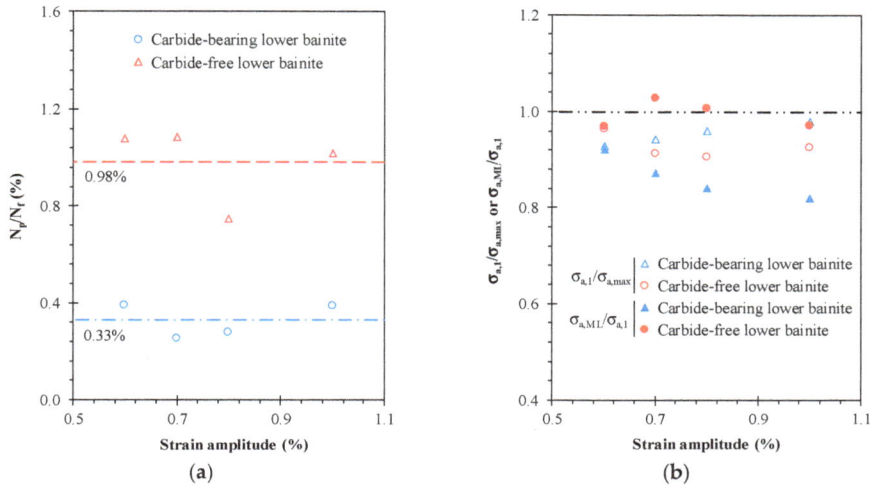

Figure 3. Variation of the: (**a**) N_p/N_f ratio with the strain amplitude; (**b**) $\sigma_{a,1}/\sigma_{a,max}$ and the $\sigma_{a,ML}/\sigma_{a,1}$ ratios with the strain amplitude.

Figure 3b shows, for various strain amplitude levels, the relation between the stress amplitude of the first cycle ($\sigma_{a,1}$) and the maximum stress amplitude ($\sigma_{a,max}$). Both steels have similar relations ($\sigma_{a,1}/\sigma_{a,max}$) slightly below the unity (i.e., 0.93 for the carbide-bearing lower bainite and 0.97 for the carbide-free lower bainite), which denotes cyclic hardening behaviour. As far as what can be inferred from the figure, the above-mentioned relations reach minimum values: (i) at the lowest strain amplitudes for the carbide-bearing lower bainite; and (ii) at intermediate strain amplitudes for the carbide-free lower bainite. With regard to the relations between the stress amplitude of the mid-life cycle ($\sigma_{a,ML}$) and the stress amplitude of the first cycle ($\sigma_{a,1}$), as displayed in Figure 3b, both steels behave differently, i.e., the carbide-bearing bainitic steel exhibits a cyclic softening behaviour with $\sigma_{a,ML}/\sigma_{a,1}$ becoming increasingly lower as the strain amplitude increases, while the other shows a mixed cyclic hardening-softening response that is not particularly intense, with $\sigma_{a,ML}/\sigma_{a,1}$ values quite close to 1.

The study of the cyclic stress-strain response for both tested materials, was performed on the basis of the data being collected for the mid-life cycle [15–17]. Total plastic and elastic strain amplitudes, stress amplitudes, and plastic strain energy densities of the selected hysteresis loops are listed in Table 1. Figure 4a plots the mid-life stress-strain circuits of the carbide-bearing and carbide-free bainitic steels in relative coordinates, with the lower tips tied together at different strain amplitudes. When first looking, it can be concluded that the upper branches are not perfectly coincident, and that we are therefore in the presence of non-Masing type materials. A more in-depth analysis for the carbide-bearing bainitic steel is provided in Figure 4b, which compares perfect Masing-type circuits (dashed lines) with those that were obtained in the experiments (full lines) in relative coordinates, with the upper branches overlapped. Not surprisingly, it is possible to distinguish a reduction of the linear region where, the higher the strain amplitudes, the bigger the differences. The comparison of the mid-life circuits shows that the areas—or in other words, the plastic strain energy densities—are larger for the carbide-bearing lower bainite than for the carbide-free lower bainite (see Table 1) at similar strain amplitude levels.

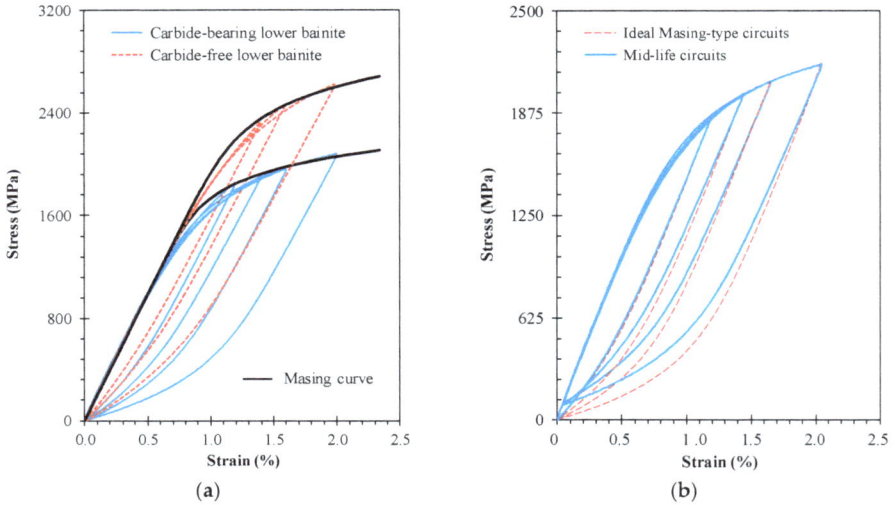

Figure 4. Mid-life circuits at different strain amplitudes with: (**a**) lower tips tied together for the carbide-bearing and carbide-free bainitic steels; (**b**) upper branches overlapped and ideal Masing-type circuits for the carbide-bearing bainitic steel.

The cyclic stress-strain curves, obtained from the mid-life hysteresis circuits, can be seen in Figure 5a. The constants k' and n' of Equation (1), which respectively represent the cyclic hardening coefficient and the cyclic hardening exponent, were determined using the least square method, and are listed in Table 4. Monotonic stress-strain curves are also plotted for comparison purposes. Both steels behave differently. The carbide-free lower bainite exhibits a strain-hardening response in the entire range, as the experimental cyclic data are above the monotonic curve. On the contrary, the carbide-bearing lower bainite is characterized by a strain-softening behaviour. The degree of strain-hardening (DH) and the degree of strain-softening (DS) are presented in Figure 5b. These variables were accounted for from two different approaches: the first was given by the difference between the stress amplitude of the first and the mid-life circuits (circles and rectangles); and the second was given by the difference between the cyclic and the monotonic curves (dashed lines). The insights drawn from the two approaches are similar: DS increases with the strain amplitude for the carbide-bearing lower bainite and DH decreases with the strain amplitude for the carbide-free lower bainite. Furthermore, the experimental results are close to those collected from the fitted cyclic curves.

$$\frac{\Delta\varepsilon}{2} = \frac{\Delta\sigma}{2E} + \left(\frac{\Delta\sigma}{2k'}\right)^{1/n'} \tag{1}$$

Figure 6 displays the stress amplitude against the plastic strain amplitude for both of the bainitic steels. These two variables (see the dash-dotted lines) can be related by a power law. Figure 6 also displays the relationship between the stress amplitude and the elastic strain amplitude, which is defined on the basis of the unloading moduli obtained in the experimental tests (see the dashed lines). Similarly, these two variables can also be related via a power law. The k'' and n'' constants were determined using the least square method and are summarised in Table 4. The variations of the unloading moduli with the elastic strain amplitude (which are evident when compared with the solid lines—both overlapped—that were obtained from the values of the Young's moduli, given in Table 3) indicate a non-linear behavior, not only in the plastic regime but also in the elastic regime.

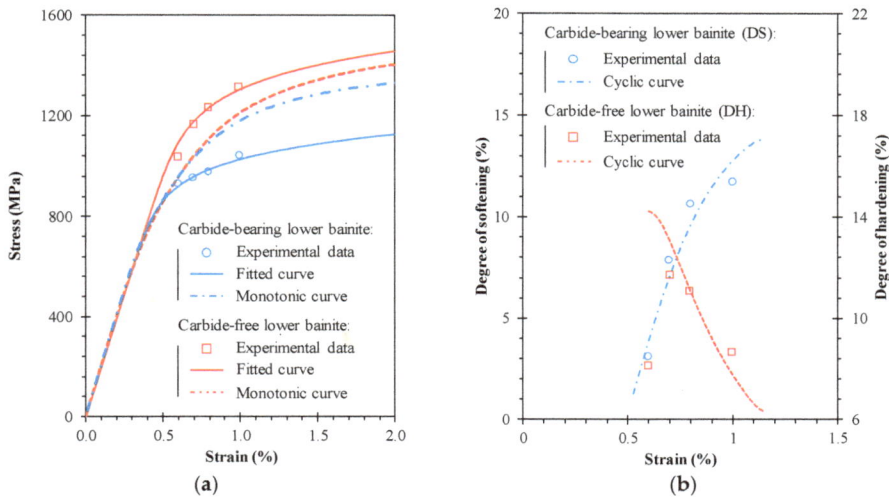

Figure 5. (**a**) Cyclic stress-strain curve; (**b**) variation of the degree of softening and the degree of hardening for the carbide-bearing and carbide-free bainitic steels.

Table 4. Summary of the cyclic mechanical properties.

Material	k' (MPa)	n'	k'' (MPa)	n''
Carbide-bearing lower bainite	1093.60	0.08374	19,823.03	0.57854
Carbide-free lower bainite	1427.49	0.08421	42,027.73	0.70602

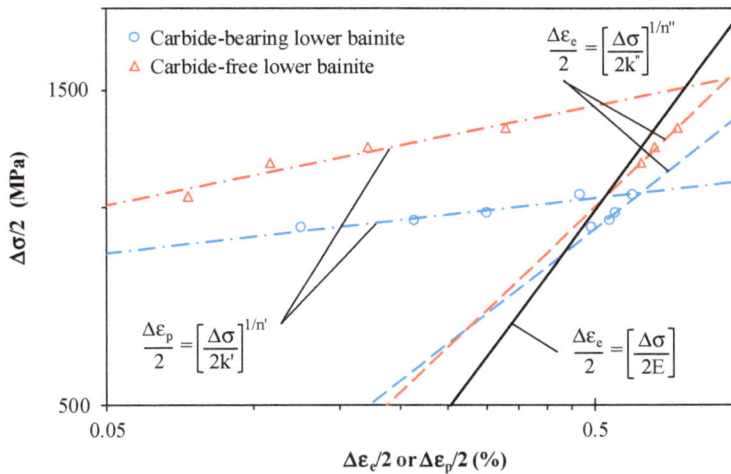

Figure 6. Stress amplitude versus elastic strain amplitudes on the basis of both of the unloading Young's moduli that were evaluated in the tests (dashed lines) and the Young's modulus that was determined from the monotonic tensile test (solid lines).

The stress-life relations, expressed in terms of the number of reversals to failure versus stress amplitude of the mid-life circuits, can be written in the following form:

$$\frac{\Delta\sigma}{2} = \sigma_f' \left(2N_f\right)^b \tag{2}$$

where σ_f' is the fatigue strength coefficient, and b is the fatigue strength exponent. The constants, determined via the least square method, are reported in Table 5. In both cases, but particularly for the carbide-bearing lower bainite, a significant correlation between the experiments and the proposed functions was obtained [5]. As already noted by Long et al. [5], the carbide-free bainite can deal with higher stress amplitudes for a similar number of cycles, particularly for lower fatigue lives. As the fatigue life increases, the differences tend to be attenuated.

The total strain amplitude, defined as the sum of the elastic and plastic parts, can be related to the fatigue life from the following equation:

$$\frac{\Delta\varepsilon}{2} = \frac{\sigma_f'}{E} \left(2N_f\right)^b + \varepsilon_f' \left(2N_f\right)^c \tag{3}$$

where σ_f' is the fatigue strength coefficient, b is the fatigue strength exponent, ε_f' is the fatigue ductility coefficient, and c is the fatigue ductility exponent. Figure 7 presents the strain-life relations of the carbide-bearing and the carbide-free steels that were obtained from the experiments. The constants, fitted using the least square method, are reported in Table 5, where the experimental data that was collected in the tests were omitted for the sake of clarity. Overall, the fatigue resistance of the carbide-bearing lower bainite is smaller than that of the carbide-free lower bainite. As suggested by Long et al. [5], fatigue durability is negatively affected by the stable carbides and thick bainite ferrite plates and, on the contrary, the existence of fine bainite ferrite plates and metastable retained austenite positively affects the fatigue performance. The difference between the two tested steels is the greatest for lower lives, and tends to disappear as the strain amplitude decreases. For lives greater than 10^4, the total strain versus life curves tend to be overlapped. Regarding the transition lives, represented by $2N_T$ in Figure 7, the outcomes are also notoriously different: $2N_T$ of the carbide-bearing lower bainite is two times higher than that of the carbide-free lower bainite.

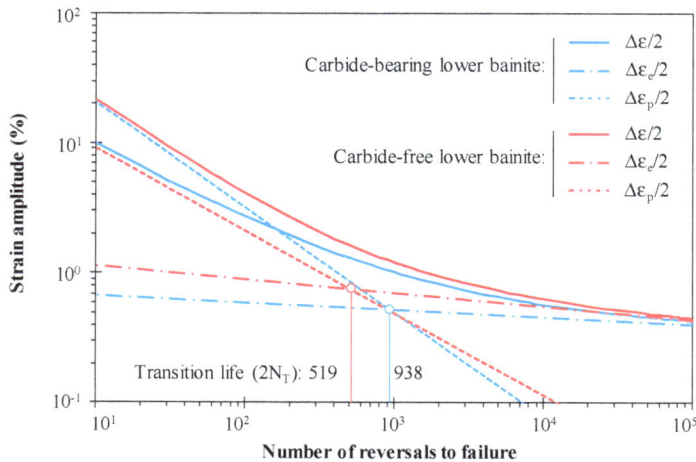

Figure 7. Strain-life relationships accounted for in terms of total strain, elastic strain, and plastic strain components for the carbide-bearing and carbide-free bainitic steels.

Table 5. Summary of the fatigue strength and fatigue ductility properties.

Material	σ_f' (MPa)	b	ε_f'	c
Carbide-bearing lower bainite	1513.44	−0.05522	0.39141	1.30301
Carbide-free lower bainite	2601.08	−0.09171	−0.63354	−0.80511

Figure 8 plots the plastic strain energy density that was evaluated from the mid-life hysteresis loops against the number of cycles to failure for the carbide-bearing and carbide-free bainitic steels. In a log-log scale, the relationship between these variables can be described by a straight line, i.e.,

$$\Delta W_p = \kappa_p (2N_f)^{\alpha_p} \tag{4}$$

where κ_p and α_p are two unknowns determined from the experimental data. The constants were calculated via the least square method and are summarised in Table 6. Not surprisingly, there is a strong correlation between these two variables. This demonstrates the adequacy of such a variable to account for the fatigue damage based on the energy dissipated [18,19].

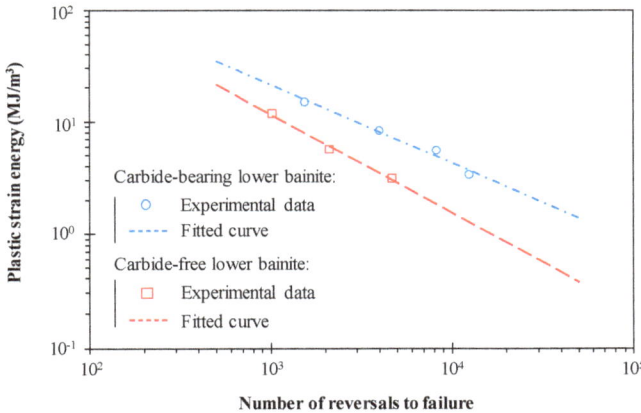

Figure 8. Plastic strain energy density versus fatigue life for the carbide-bearing and carbide-free bainitic steels.

The plastic strain energies measured in the experiments are compared in Figure 9 with those of Masing-type materials, defined as follows:

$$\Delta W_{pM} = \frac{1 - n'}{1 + n'} \Delta\sigma \, \Delta\varepsilon_P \tag{5}$$

where $\Delta\varepsilon_p$ is the plastic strain range, $\Delta\sigma$ is the stress range, and n' is the cyclic hardening exponent. The experimentally measured values (ΔW_p) are relatively far from those of the Masing-type materials (ΔW_{pM}) for both the carbide-bearing and carbide-free bainitic steels, and the differences increase with the strain amplitude, irrespective of the tested steel. This outcome is in line with the conclusions drawn from Figure 4.

Table 6. Summary of the energy-life properties.

Material	κ_p	α_p
Carbide-bearing lower bainite	2506.0	−0.69293
Carbide-free lower bainite	4773.1	−0.87335

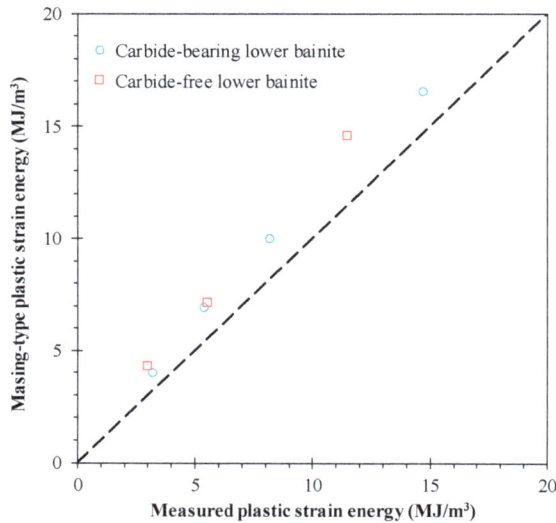

Figure 9. Comparison of the experimentally measured plastic strain energy densities (ΔW_p) and the ideal Masing-type plastic strain energy densities (ΔW_{pM}).

3.3. Analysis of Fracture Surfaces

Figure 10 shows the TEM micrographs of the carbide-free and carbide-bearing bainitic steels at strain amplitudes of 1%. The morphologies of bainite ferrite and metastable austenite, as already highlighted by Long et al. [5], are closely related to the fatigue life. Carbide-free lower-bainite steel presents higher fatigue lives under strain-controlled conditions due to the presence of fine bainitic plates which reduce its susceptibility to crack nucleation. When subjected to fatigue loading histories, metastable retained austenite gives rise to martensite, particularly in the presence of large sizes of retained austenite, as exhibited in Figure 10a. This strain-induced martensitic transformation is beneficial to increase the fatigue life in carbide-free lower-bainite steels, as it absorbs the energy required to crack propagation. In the presence of a cyclic stress state, it can lead to a tensile stress relaxation, as the tensile stress relaxation introduces compressive stresses which are likely to promote the crack closing. Regarding the carbide-bearing lower bainitic steel, it behaves differently. As documented in Figure 10b, only the occurrence of deformation is observed. As is well-known, the secondary carbide phase of the carbide-bearing lower bainitic steel is very prone to the formation of micropores which act as local stress raisers and contribute to higher rates of crack formation and, consequently, lower fatigue lives.

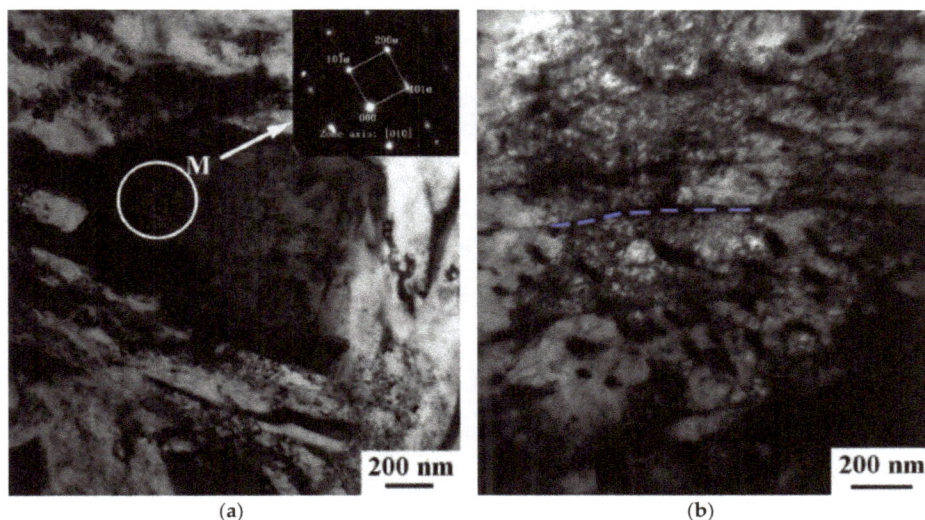

Figure 10. TEM images of low-cycle fatigue tests ($\Delta\varepsilon/2 = 1.0\%$) for the [5]: (**a**) carbide-free lower-bainite; (**b**) carbide-bearing lower-bainite. Reproduced from [5], with permission from publisher Elsevier, 2018.

4. Conclusions

This paper aimed at comparing the cyclic behaviour of carbide-bearing and carbide-free bainitic steels. In order to meet this goal, strain-controlled tests in low-cycle fatigue regime, under fully-reversed conditions at various strain amplitudes, were performed. In addition, microstructures were observed by SEM and TEM before testing, and the morphologies of the fracture surfaces were observed by TEM after total fatigue failure. The following conclusions can be drawn:

- The microstructure of the carbide-free lower bainite is formed by bainitic ferrite and retained austenite of varying size, while the microstructure of the carbide-bearing lower bainite is formed by bainitic ferrite and carbides that are distributed within the bainitic ferrite or between the bainitic ferrite;

- The cyclic stress response of the tested steels comprises three different stages: an initial cyclic hardening behaviour with a strong effect in the early cycles, followed by a smoother decrease of the stress amplitude; a second stage, close to a saturated response, with tenuous cyclic stress variations; and a final stage of a marked reduction of stress amplitude until fatigue failure occurs;

- The maximum stress amplitudes, under strain-controlled conditions, are observed in the early cycles of the tests for life ratios lower than 1%, irrespective of the strain amplitude. Regarding the ratios of the initial stress amplitude to the maximum stress amplitude, it can be noted that the responses of both steels are similar, with ratios close to 1—or more precisely, 0.93 and 0.97 for the carbide-bearing lower bainite and carbide-free lower bainite, respectively;

- The shapes of the mdi-life hysteresis loops at similar strain amplitudes are quite different. The area—i.e., the plastic strain energy density—is higher for the carbide-bearing lower bainite. However, in both cases, we are in the presence of non-Masing type materials which are associated with the changes in the linear region of the stable circuits;

- Based on the differences between the monotonic curve and the cyclic stress-strain curve that were evaluated from the mid-life circuits, it was possible to identify two antagonistic responses: the carbide-bearing lower bainite exhibited a strain-softening behaviour, while the carbide-free lower bainite was characterised by a strain-hardening behaviour. The degree of softening was in the range of 3–15%, and the degree of hardening varied within the range of 3–8%;

- The fatigue resistance of the carbide-bearing lower bainite is lower than that of the fatigue resistance of the carbide-free lower bainite. At low strain amplitudes, the differences are more relevant. Nevertheless, for lives greater than 10^4 reversals, the fatigue responses tend to be similar. With regard to the transition lives, the main outcomes are markedly different: the transition life of the carbide-bearing lower bainite is almost twice the value of the carbide-free lower bainite;

- Fracture surface morphologies of the carbide-free lower bainite exhibited fine bainitic plates which are associated with a reduced susceptibility to crack nucleation, which contributes to higher fatigue resistance, which is verified in the experiments.

Author Contributions: F.Z. and X.L. conceived and designed the experiments; F.Z. and X.L. performed the experiments; R.B., F.B. and J.D.C. analysed the data; R.B. wrote the paper.

Acknowledgments: The authors would like to acknowledge the sponsoring under the project No. 016713 (PTDC/EMS-PRO/1356/2014) financed by Project 3599: Promover a Produção Científica e Desenvolvimento Tecnológico e a Constituição de Redes Temáticas (3599-PPCDT) and FEDER funds.

Conflicts of Interest: The authors declare no conflict of interest.

References

1. Caballero, F.G.; Santofimia, M.J.; García-Mateo, C.; Chao, J.; Andrés, C.G. Theoretical design and advanced microstructure in super high strength steels. *Mater. Des.* **2009**, *30*, 2077–2083. [CrossRef]
2. Zhang, F.C.; Yang, Z.N.; Kang, J. Research progress of bainitic steel used for railway crossing. *J. Yanshan Univ.* **2013**, *37*, 1–7.
3. Aglan, H.A.; Liu, Z.Y.; Hassan, M.F.; Fateh, M. Mechanical and fracture behavior of bainitic rail steel. *J. Mater. Process. Technol.* **2004**, *151*, 268–274. [CrossRef]
4. Georgiyev, M.N.; YuKaletin, A.; Siminov, Y.N.; Schastlivtsev, V.M. Influence of stability of retained austenite on crack resistance of engineering steel. *Phys. Met. Metall.* **1999**, *69*, 110–118.
5. Long, X.; Zhang, F.; Yang, Z.; Lv, B. Study on microstructures and properties of carbide-free and carbide-bearing bainitic steels. *Mater. Sci. Eng. A* **2018**, *715*, 10–16. [CrossRef]
6. Zhang, F.C.; Long, X.Y.; Kang, J.; Cao, D.; Lv, B. Cyclic deformation behaviors of a high strength carbide-free bainitic steel. *Mater. Des.* **2016**, *94*, 1–8. [CrossRef]
7. Long, X.Y.; Zhang, F.C.; Zhang, C.Y. Effect of Mn content on low-cycle fatigue behaviors of low-carbon bainitic steel. *Mater. Sci. Eng. A* **2017**, *697*, 111–118. [CrossRef]
8. Golos, K.M.; Debski, D.K.; Debski, M.A. A stress-based fatigue criterion to assess high-cycle fatigue under in-phase multiaxial loading conditions. *Theor. Appl. Fract. Mech.* **2014**, *73*, 3–8. [CrossRef]
9. Lu, C.; Melendez, J.; Martínez-Esnaola, J.M. A universally applicable multiaxial fatigue criterion in 2D cyclic loading. *Int. J. Fatigue* **2017**, *110*, 95–104. [CrossRef]
10. Branco, R.; Costa, J.D.; Berto, F.; Antunes, F.V. Effect of loading orientation on fatigue behaviour in severely notched round bars under non-zero mean stress bending-torsion. *Theor. Appl. Fract. Mech.* **2017**, *92*, 185–197. [CrossRef]
11. Wu, S.C.; Zhang, S.Q.; Xu, Z.W.; Kang, G.Z.; Cai, L.X. Cyclic plastic strain based damage tolerance for railway axles in China. *Int. J. Fatigue* **2016**, *93*, 64–70. [CrossRef]
12. Madrigal, C.; Navarro, A.; Chaves, V. Numerical implementation of a multiaxial cyclic plasticity model for the local strain method in low cycle fatigue. *Theor. Appl. Fract. Mech.* **2015**, *80*, 111–119. [CrossRef]
13. Branco, R.; Prates, P.; Costa, J.D.; Berto, F.; Kotousov, A. New methodology of fatigue life evaluation for multiaxially loaded notched components based on two uniaxial strain-controlled tests. *Int. J. Fatigue* **2018**, *111*, 308–320. [CrossRef]
14. Firat, M. A numerical analysis of combined bending-torsion fatigue of SAE notched shaft. *Finite Elem. Anal. Des.* **2012**, *54*, 16–27. [CrossRef]
15. Branco, R.; Costa, J.D.; Antunes, F.V.; Perdigão, S. Monotonic and cyclic behaviour of DIN 34CrNiMo6 martensitic steel. *Metals* **2016**, *6*, 98. [CrossRef]

16. Branco, R.; Costa, J.D.; Berto, F.; Razavi, S.M.J.; Ferreira, J.A.M.; Capela, C.; Santos, L.; Antunes, F.V. Low-cycle fatigue behaviour of AISI 18Ni300 maraging steel produced by selective laser melting. *Metals* **2018**, *8*, 32. [CrossRef]

17. Morrow, J. Cyclic plastic strain energy and fatigue of metals. In *International Friction, Damping and Cyclic Plasticity*; ASTM STP 378; American Society for Testing and Materials: Philadelphia, PA, USA, 1965; pp. 45–87.

18. Song, W.; Liu, X.; Berto, F.; Razavi, S. Low-cycle fatigue behavior of 10CrNi3MoV high strength steel and its undermatched welds. *Materials* **2018**, *11*, 661. [CrossRef] [PubMed]

19. Branco, R.; Costa, J.D.; Berto, F.; Antunes, F.V. Fatigue life assessment of notched round bars under multiaxial loading based on the total strain energy density approach. *Theor. Appl. Fract. Mech.* **2018**, in press. [CrossRef]

metals

MDPI

Article

Mechanical Behavior of Two Ferrite–Martensite Dual-Phase Steels over a Broad Range of Strain Rates

Jiangtao Liang [1,2], Zhengzhi Zhao [1,*], Hong Wu [3], Chong Peng [3], Binhan Sun [2], Baoqi Guo [2], Juhua Liang [1] and Di Tang [1]

[1] Collaborative Innovation Center of steel Technology, University of Science and Technology Beijing, Beijing 100083, China; liangjtao@126.com (Ji.L.); liangjuhua0721@126.com (Ju.L.); tangdi@nercar.ustb.edu.cn (D.T.)

[2] Department of Materials Engineering, McGill University, 3610 University Street, Montreal, QC H3A 2B2, Canada; binhan.sun@mail.mcgill.ca (B.S.); baoqi.guo@mail.mcgill.ca (B.G.)

[3] Xin Yu Iron and Steel Co., Ltd., Xinyu 338001, Jiangxi, China; wuhong36@126.com (H.W.); 019pen@xinsteel.com.cn (C.P.)

[*] Correspondence: zhaozhzhi@ustb.edu.cn; Tel.: +86-136-6113-5191

Received: 27 February 2018; Accepted: 26 March 2018; Published: 3 April 2018

Abstract: The present study concerns the deformation and fracture behavior of two ferrite–martensite dual phase steels (FMDP660 and FMDP780) with different phase fractions subjected to different strain rate (0.001 s^{-1} to 1000 s^{-1}) tensile testing. For both steels, the yield strength (YS) monotonically increased with strain rates, whereas the values of ultimate tensile strength (UTS), uniform elongation (UE) and post-uniform elongation (PUE) were maintained stable at the low strain rate range (0.001–0.1 s^{-1}), followed by a significant increase with strain rate at high strain rate levels (0.1–1000 s^{-1}). The FMDP780 steel with a higher fraction of martensite possessed a stronger strain rate sensitivity of tensile strength and elongation (UE and PUE) values at the high strain rate stage, compared with the FMDP660 sample. The change of UTS and UE with different strain rates and phase fractions was highly related to the strain hardening behavior, which was controlled by the dislocation multiplication in ferrite, as validated by transmission electron microscopy (TEM). The fracture surface of the two steels was characterized by dimpled-type fracture associated with microvoid formation at the ferrite–martensite interfaces, regardless of the strain rates. The change of the dimple size and PUE value of the two steels with strain rates was attributed to the effect of adiabatic heating during tensile testing.

Keywords: Ferrite–martensite dual-phase (FMDP) steel; dynamic tensile testing; dislocation multiplication; adiabatic heating; failure mechanism

1. Introduction

The automotive industry is facing stringent regulations and significant challenges on weight reduction of road transportation vehicles for fuel economy, as well as improvement of crashworthiness performance to ensure vehicle safety. The response to these challenges requires the intensive use of advanced high strength steels (AHSS) in body-in-white assemblies. Ferrite–martensite dual-phase (FMDP) steels are one of the most widely used AHSS in the automotive industry [1–4]. The composite structure consisting of soft ferrite and hard martensite results in a great strength–ductility combination and interesting mechanical features, such as continuous yielding and initial high work hardening rate [5,6].

It is well documented that there is a large difference of the deformation behavior in automotive materials under collision and the quasi-static condition [7]. Understanding the dynamic deformation behavior of FMDP steels is thus very important in improving the safety of the vehicles.

Despite the large amounts of the study on the dynamic deformation behavior of DP steels [8–11], some inconsistencies still remain with respect to the effect of strain rate on the change of mechanical property values. For example, Yu et al. [12] has investigated the strain rate-dependent mechanical behavior of a DP600 steel, using the John–Cook (JC) rate-dependent constitutive model. It was found that the yield strength (YS) and ultimate tensile strength (UTS) of the steel increased with higher strain rates, whereas the uniform elongation (UE) and total elongation (TE) at the low strain rates were much larger compared with the values at high strain rate ranges. Song et al. [13] studied the deformation behavior of the DP1000 steel by the JC and Zerilli–Armstrong (ZA) model, and found that the ZA model can reflect the deformation behavior more accurately at strain rates from 10^{-4} to 2000 s^{-1}. Cadoni et al. [14] compared the dynamic mechanical behavior of the DP1200 and DP1400 steel, and observed that the two steels had good potential to absorb energy, and the DP1400 possessed a lower strain rate sensitivity. Wang et al. [15] investigated the high strain rate behavior of some DP steels and a martensitic (MS) steel. They found that the strain rate sensitivity of the UTS decreased with increasing martensite volume fractions, yet the fracture elongation (FE) demonstrated a reverse trend. Dong et al. [10] investigated the change of strength and ductility with increasing strain rate of DP600 steel and its welded joint. They found that the YS and UTS increase with the increase of strain rate, but the UE and TE present a fluctuation tendency. Singh et al. [7] researched the dynamic tensile behavior of MP800HY steel, they found the YS, UTS, UE and TE increased with increasing strain rate. Since most studies only focus on the change of the mechanical properties under various strain rates for one specific alloy composition and microstructure (i.e., phase fraction), such inconsistencies from the literature are most likely derived from the different alloy composition and microstructure selected in the literature [8–10]. Therefore, it is necessary to carry out systematic studies on the influence of the microstructure with respect to different phase fractions on the dynamic deformation behavior of FMDP steels.

In the present work, two types of ferrite–martensite dual-phase steels with different phase fractions were selected and subjected to tensile testing at different strain rates from 0.001–1000 s^{-1}. The effects of the phase fraction on the dynamic mechanical properties and energy absorption, the strain hardening behavior associated with dislocation motion, fracture mechanism, the adiabatic heating and fracture behavior were addressed in detail.

2. Experiments

2.1. Materials

Two commercial non-galvanized ferrite–martensite dual-phase (FMDP) steel sheets (FMDP660 and FMDP780) were selected; the chemical composition of the two steels is listed in Table 1. The thickness of the two tested steels is ~1.2 mm.

Table 1. Chemical compositions of the investigated steels (in wt %).

Steel	C	Si	Mn	P	S	Al	N	Cr	Mo	Nb	Ti	B	Fe
FMDP660	0.12	1.2	1.5	0.015	0.005	0.05	0.003	-	-	-	-	-	Balanced
FMDP780	0.08	0.1	2.0	0.015	0.015	0.6	-	0.25	0.2	0.018	0.012	0.0012	Balanced

2.2. Tensile Testing

Tensile tests at low strain rates were conducted on a material testing machine (CMT5105, SANS Testing Machine Co., Shenzhen, China), with a constant crosshead speed of 0.00002, 0.0002 and 0.002 m/s, corresponding to a nominal strain rate of 0.001, 0.01 and 0.1 s^{-1}. Dynamic tensile tests with an average strain rate of the order of 1, 10, 100 and 1000 s^{-1} were carried out on a servo-hydraulic high-speed tensile testing machine (Zwick HTM16020, Zwick Roell, Ulm, Germany). The strain rate was determined by $\dot{\varepsilon} = \frac{v}{L_c}$ (v: the set speed; L_c: the length of the parallel part, L_c = 20 mm). In the low strain rate stage, we used an extensometer to measure the strain, while the strain gauge

(HBM K-LY4, Hottinger Baldwin Measurement Co., Ltd, Suzhou, China) and the laser measurement were applied on measuring strain under high strain rate stage. Tensile specimens were directly machined from the steel sheets along the rolling direction. The dimension of the specimens tested at low and high strain rates is shown in Figure 1a,b, respectively. In order to avoid the size effect of the tensile specimen on the experimental results, the dimension of the gage part was maintained the same for all the samples. Three specimens were repeated for each set of conditions. Yield strength was determined according to the 0.2% offset strength, and the uniform elongation was obtained from the intersection point of the true stress-strain curve and the strain hardening rate curve ($\frac{d\sigma}{d\varepsilon}$). It can be seen that the strain rate has no significant change under the dynamic tensile test (1–1000 s^{-1}) (Figure 2), which can be considered as a constant value during the high strain rate stage. It is important to note that the strain rate sensitivity (m) is usually defined as $m = \frac{d\ln\sigma}{d\ln\varepsilon}$ [6]; however, in this paper we used the strain rate sensitivity term to describe the extent of the mechanical property variation with different strain rates. The microhardness measurements were carried out using a THV-1MD digital micro Vickers hardness tester (TEST-TECH Co., Ltd., Shanghai, China) with 50 g load and 10 s dwell time.

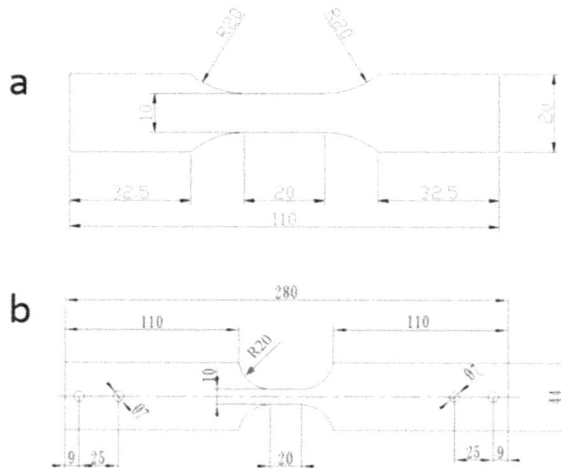

Figure 1. Tensile specimens at low strain rate (**a**) and high strain rate (**b**) with parallel region of 20 mm for the CMT5105 and the Zwick HTM 16020.

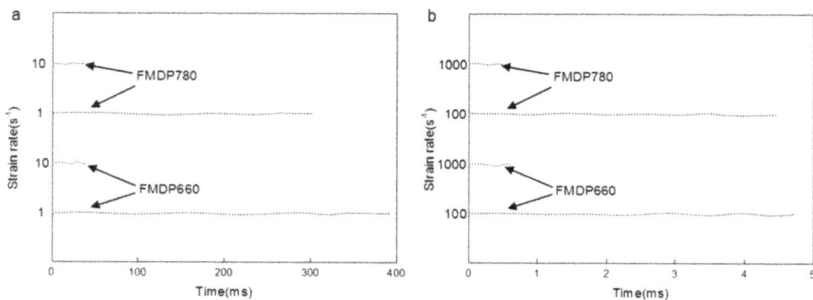

Figure 2. Values of strain rate (**a**) 1 s^{-1} and 10 s^{-1}; (**b**) 100 s^{-1} and 1000 s^{-1} of the two steels as a function of time.

2.3. Microstructural Characterization

The microstructure and the fracture surface were characterized using a field emission scanning electron microscopy (SEM, FEI Quanta 450, FEI, Hillsboro, OR, USA). A transmission electron microscopy (TEM, JEM2100, Japan Electronics Co., Ltd., Tokyo, Japan) was used to study the micromechanical details pertaining to dislocation motions in the deformed specimens. The samples for SEM were polished and etched using 4% nital (4% nitric acid and 96% alcohol), and TEM specimens were mechanically ground to a thickness of about 0.06 mm and subsequently electro-polished in a twin-jet polishing machine in a solution of 5% perchloric acid and 95% alcohol at about $-20\ ^{\circ}$C. The ferrite and martensite volume fractions were measured based on the ASTM E1245-03 standard using the image processing software ImageJ. Grain size was determined by the standard linear intercept method based on the SEM images.

3. Results and Discussion

3.1. Microstructure and Quasi-Static Mechanical Properties

Figure 3 shows the SEM micrographs of the FMDP660 and FMDP780 steel. For both steels, it is obvious that white martensite islands are imbedded in dark ferrite matrix and mainly distributed on the ferrite grain boundaries; the volume fraction of martensite is around 20% and 35% for the FMDP660 and FMDP780 steel, respectively. The grain size of the two steels is similar, which was determined to be 7.5 μm and 7.3 μm, respectively.

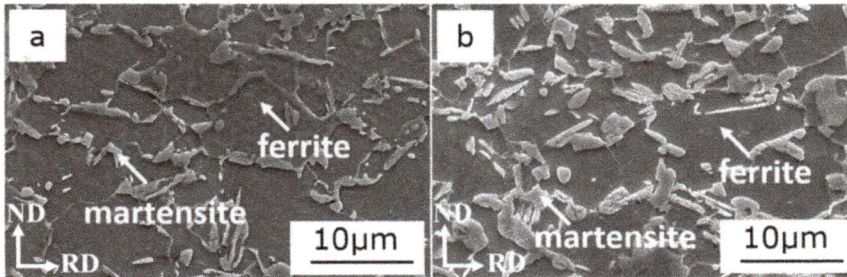

Figure 3. Scanning electron microscopy (SEM) micrographs of (**a**) FMDP660 and (**b**) FMDP780 steel of RD–ND plane. (RD represents rolling direction, ND represents normal direction)

The quasi-static engineering stress-stain curve ($0.001\ \mathrm{s}^{-1}$) of the two steels is present in Figure 4, with the strength/hardness and elongation values listed in Table 2. It is shown that both steels yield without an obvious yielding platform, namely, the two steels undergo continuous yielding, which is usually associated with the existing high mobile dislocation density in the microstructure [16–18]. The higher volume fraction of martensite in the FMDP780 steel leads to a higher hardness and strength and a lower elongation.

Table 2. Mechanical properties of FMDP660 and FMDP780 steel at the quasi-static condition, where YS, UTS, UE, TE are yield strength, ultimate tensile strength, uniform elongation, total elongation, respectively.

Tested Steel	YS/MPa	UTS/MPa	UE/%	TE/%	Microhardness/HV
FMDP660	425 ± 3	674 ± 4	19.25 ± 0.5	29.45 ± 0.5	211 ± 2
FMDP780	624 ± 5	792 ± 6	14.72 ± 0.5	23.43 ± 0.4	263 ± 3

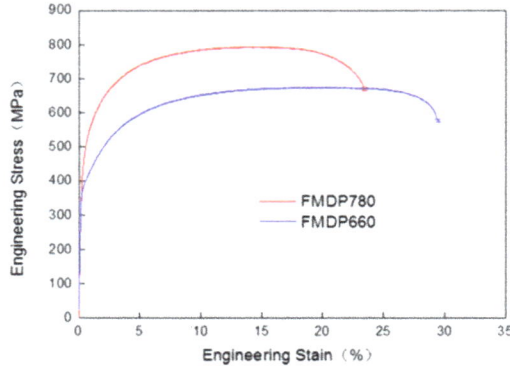

Figure 4. Quasi static engineering stress-strain curves of FMDP660 and FMDP780 steel.

3.2. Mechanical Properties at Different Strain Rates

Figure 5 shows the engineering stress–strain curve of the two steels at strain rates ranging from 0.001 to 1000 s^{-1}. Figure 6 plots the strain rate dependent tensile properties including YS, UTS, UE and PUE, with the values listed in Table 3. For all the investigated strain rates, the higher martensite fractioned sample, FMDP 780, possesses higher strength (YS and UTS) and lower elongation (UE and PUE) values compared with FMDP660. The effect of strain rate on the YS and UTS is similar for the two steels. Both steels show a monotonically increasing YS with higher strain rates, as demonstrated in Figures 5 and 6a. This is similar to the behavior of other AHSS, such as low Mn TRIP steels [19]. The UTS of the two steels, however, presents a slight increase at low strain rates (0.001–0.1 s^{-1}), followed by a significant rise at the higher strain rate range (0.1–1000 s^{-1}). Besides, we use the ($m = \frac{\mathrm{d}\ln\sigma}{\mathrm{d}\ln\dot{\varepsilon}}$) to assessment the strain rate sensitivity, where m is the strain rate sensitivity parameter, σ is the engineering stress and $\dot{\varepsilon}$ is the strain rate. The UTS strain rate sensitivity parameter is 0.02087 and 0.02177 for the FMDP660 and FMDP780. Obviously, the FMDP780 has a higher strain rate sensitivity. The reduction of the cross-section (Z) also keeps stable in the low strain rate stage (0.001–0.1 s^{-1}) and increases with higher strain rates during the high strain rate stage (1–1000 s^{-1}).

Figure 6c,d describes the trend of UE and PUE as a function of the strain rate for the two steels. At the low strain rate stage (0.001–0.1 s^{-1}), the values of UE and PUE of both steels are maintained relatively stable; they start to increase with higher strain rates when the strain rate achieves above 0.1 s^{-1}. The UE and PUE increment is larger for the FMDP780 steel at the high strain rate stage, that is, the FMDP780 steel possesses a higher strain rate sensitivity of ductility compared with the FMDP660 steel.

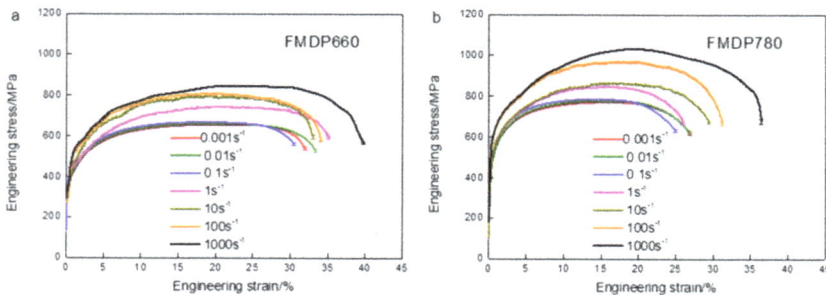

Figure 5. Engineering stress–strain curves of the FMDP660 (**a**) and FMDP780 (**b**) steel at strain rates from 0.001 to 1000 s^{-1}.

Table 3. Comparison of mechanical properties of FMDP660 and FMDP780 steels at strain rates from 0.001 to 1000 s^{-1}, where YS, UTS, UE, PUE and Z are, respectively, the yield strength, ultimate tensile strength, uniform elongation, post uniform elongation and the reduction of the cross-section area of the specimens.

Materials	$\dot{\varepsilon}$ (s^{-1})	YS (MPa)	UTS (MPa)	UE (%)	PUE (%)	Z (%)
FMDP660	0.001	392 ± 5	656 ± 6	21.09 ± 0.4	10.87 ± 0.6	31.96 ± 0.5
	0.01	399 ± 6	660 ± 7	19.84 ± 0.3	13.52 ± 0.4	33.36 ± 0.3
	0.1	433 ± 4	667 ± 5	18.65 ± 0.3	11.83 ± 0.5	30.48 ± 0.4
	1	450 ± 5	743 ± 8	21.11 ± 0.4	13.99 ± 0.6	35.1 ± 0.5
	10	470 ± 6	798 ± 8	20.76 ± 0.5	12.29 ± 0.6	33.05 ± 0.3
	100	523 ± 6	806 ± 3	19.42 ± 0.3	14.68 ± 0.5	34.1 ± 0.4
	1000	576 ± 7	847 ± 6	22.94 ± 0.4	16.76 ± 1.0	39.7 ± 0.9
FMDP780	0.001	509 ± 6	769 ± 7	16.05 ± 0.5	10.91 ± 0.7	26.96 ± 0.5
	0.01	520 ± 7	778 ± 10	15.63 ± 0.4	11.17 ± 0.5	26.8 ± 0.3
	0.1	547 ± 8	785 ± 6	14.22 ± 0.3	10.78 ± 0.5	25 ± 0.4
	1	581 ± 4	844 ± 7	16.12 ± 0.3	9.88 ± 0.5	26 ± 0.3
	10	586 ± 7	865 ± 6	16.91 ± 0.4	12.54 ± 0.5	29.45 ± 0.3
	100	663 ± 5	963 ± 8	18.67 ± 0.3	12.58 ± 0.4	31.25 ± 0.3
	1000	671 ± 8	1031 ± 10	19.67 ± 0.5	16.73 ± 0.7	36.4 ± 0.5

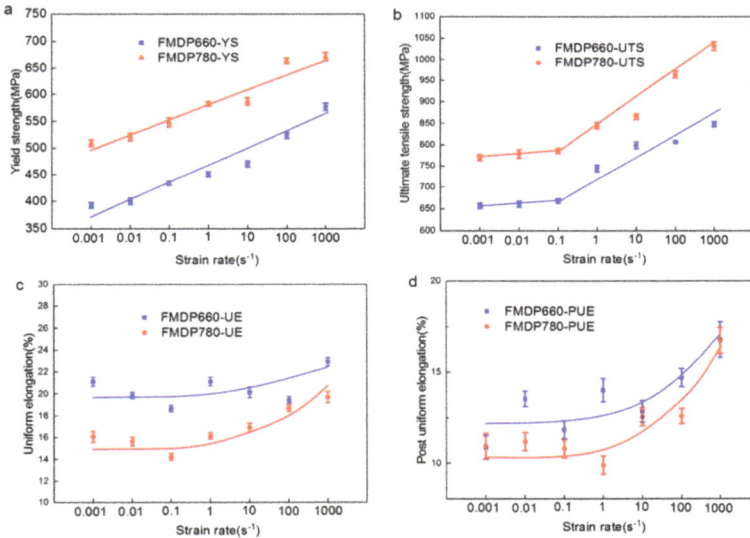

Figure 6. Values of (**a**) YS, (**b**) UTS, (**c**) UE and (**d**) PUE of the two steels as a function strain rate.

The energy absorption capacity of two steels was evaluated using the following equation [20]:

$$\Delta E = \int_0^\varepsilon \sigma \mathrm{d}\varepsilon \tag{1}$$

where ΔE is the absorbed energy during tensile testing, σ is the engineering stress and ε is the engineering strain. The calculated energy of the two steels at three strain levels is shown in Figure 7. At the strain of 10% (Figure 7a), the absorbed energy of both steels shows a monotonic increase with higher strain rates, and the FMDP780 steel possesses a much higher energy absorption than the FMDP780 steel over the whole strain rate range. This corresponds to the trend of the YS for the two steels. The behavior is different for the energy absorption determined at the strains corresponding to UE and TE (Figure 7b,c), where the values keep relatively stable at low strain rates (0.001–0.1 s^{-1}) and increase sharply at the higher strain rate range (0.1–1000 s^{-1}). At these two strain levels, the FMDP780 steel shows a lower energy absorption value at lower strain rates (up to 1 s^{-1}) compared with the FMDP660 steel. However, the energy absorption of the FMDP780 steel demonstrates a higher increasing slope at the high strain rate stage, which makes it surpass the value of the FMDP660 steel when the strain rate reaches 100 s^{-1}.

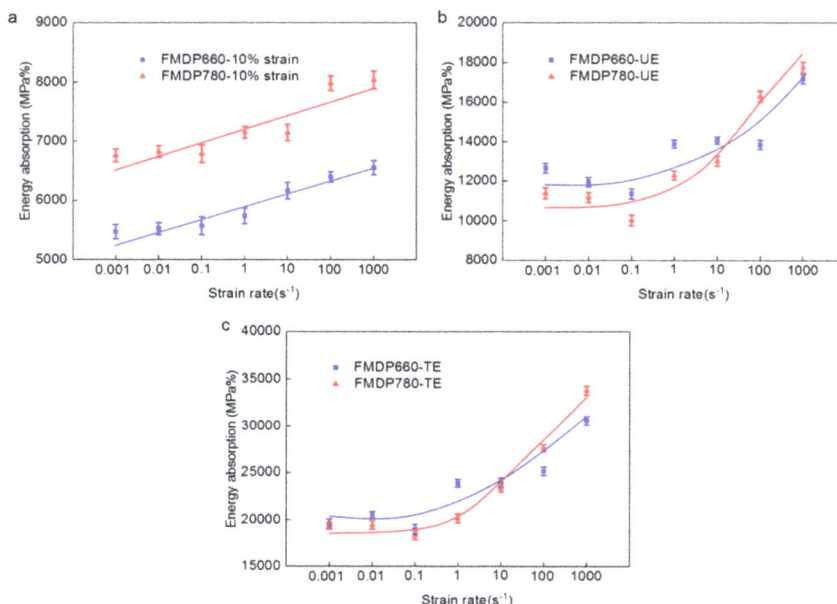

Figure 7. Absorbed energy of the two steels at the strain level of (**a**) 10% strain; (**b**) UE and (**c**) TE as a function of strain rate.

3.3. Strain Hardening Behavior

Figure 8 shows the strain hardening rate ($\frac{d\sigma}{d\varepsilon}$) along with the true stress–strain curves of the two steels at different strain rates. For all the strain rate levels, the strain hardening rate of the two steels show a one-stage decreasing trend until the occurrence of the plastic instability (i.e., necking). The work hardening rate of the two steels is not sensitive to the applied strain rate at low strain rate levels (0.001–0.1 s^{-1}), but then it starts to increase with higher strain rates above 0.1 s^{-1}. Such increase is more pronounced for the FMDP780 which contains a higher fraction of martensite. The different work hardening rate due to the variation of the strain rate and microstructure is the reason for the aforementioned behavior of UTS and UE.

In order to explain the change of the work hardening rate with applied strain rate, the adiabatic heating during the tensile testing should be firstly considered. For the low strain rate range, the adiabatic heating is normally ignored [10,14]. The adiabatic temperature rise (ΔT) at the strain rate from 1 to 1000 s^{-1} was calculated as follows:

$$\Delta T = \frac{\eta}{\rho} \int_0^\varepsilon \frac{\sigma}{c_v} d\varepsilon \qquad (2)$$

where ε is true strain, σ is true stress, η is the factor of the plastic work turning into thermal (taken as 1 [21]), ρ is the density of the FMDP steel (taken as 7.85 g/cm^3 [22]), c_v is the specific heat capacity at constant volume (0.48 J/(g·K) [21]). The calculated results of the two steels are shown in Figure 9. It is obvious that the adiabatic heating of the two steels during tensile testing increases with the increasing strain rates, and the change of adiabatic heating in FMFDP780 steel is more remarkable. It is well documented that higher temperature would promote dislocation recovery, thus resulting in a reduced work hardening rate. However, the calculated maximum adiabatic temperature rise is only around 50 °C for the two steels tested at the highest strain rate, i.e., 1000 s^{-1}, which is unlikely to make a large influence on dislocation storage [23]. The increasing work hardening rate of both steels with higher strain rates and the higher strain rate sensitivity of the work hardening rate of the FMDP780 steel also validate this point.

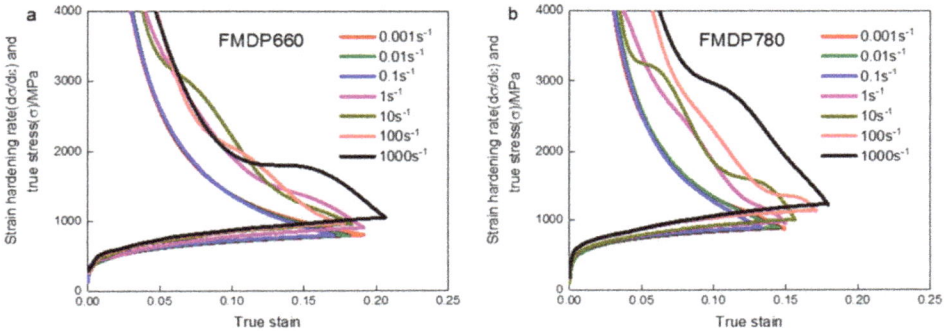

Figure 8. True stress–strain curve and strain hardening rate of the (**a**) FMDP660 and (**b**) FMDP780 steel at different strain rates from 0.001 to 1000 s^{-1}.

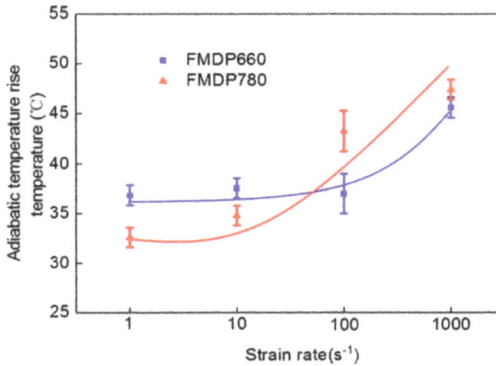

Figure 9. Adiabatic temperature rises of the two steels at the high strain rate stage (1–1000 s^{-1}).

Figure 10 shows the dislocation configurations in the vicinity of the fracture surface of the two steels under various strain rates. It can be concluded that the dislocation density in ferrite of the two steels does not change significantly at low strain rates (0.001–0.1 s^{-1}), as shown in Figure 10a–d. However, the dislocation density increases obviously in the high strain rate stage (1–1000 s^{-1}) and multiple slip systems are also activated in some ferrite grains (Figure 10c,e,g and Figure 10d,f,h). Furthermore, at the high strain rate stage, the entwining degree of dislocation increases with higher strain rates [24,25]. This can explain why the work hardening rate of both steels at low strain rates is maintained unchanged and increases at high strain rates. In addition, it seems that the FMDP780 possesses a higher dislocation density compared to the FMDP660 at the same strain rate, which is likely to be the reason for the higher work hardening rate of the FMDP780 sample. The higher martensite fraction in the FMDP780 sample essentially results in a higher local strain in the ferrite phase during deformation [26], which leads to a larger amount of statistically stored dislocations (SSDs). The higher fraction of ferrite–martensite interfaces also gives rise to a higher number of geometrically necessary dislocation (GNDs).

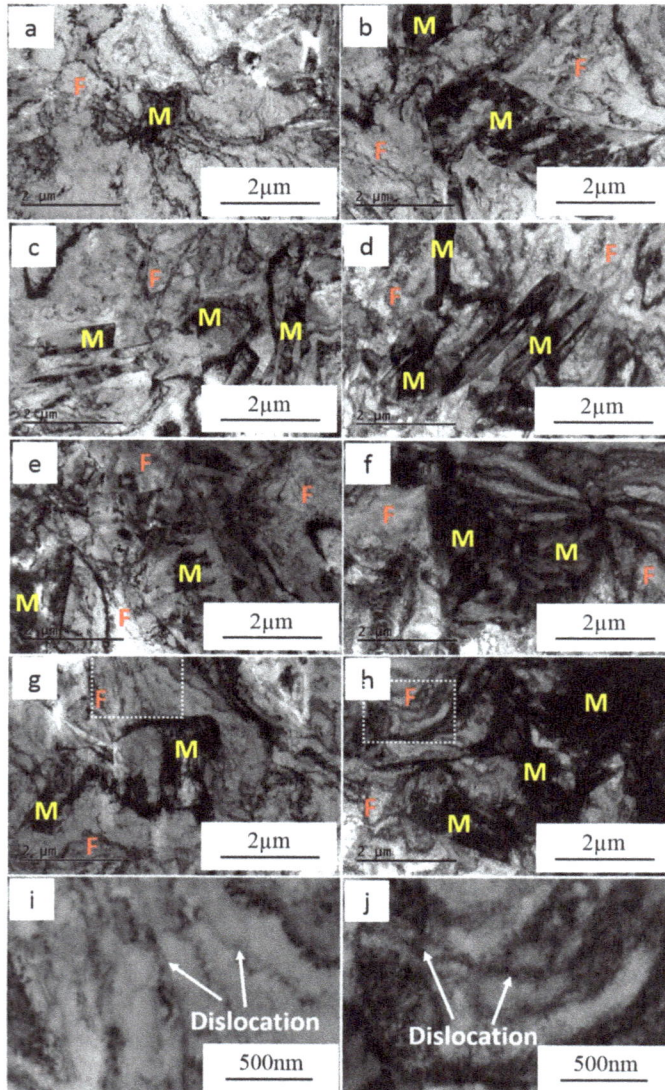

Figure 10. Dislocation configurations near the fracture tip of the (**a,c,e,g**) FMDP660 and (**b,d,f,h**) FMDP780 steel at the stain rates of (**a,b**) 0.001 s^{-1}; (**c,d**) 0.1 s^{-1}; (**e,f**) 100 s^{-1} and (**g,h**) 1000 s^{-1}. In addition, (**i,j**) are partially enlarged for (**g,h**), respectively.

3.4. Fracture Behavior

The fracture surface of the two steels is shown in Figure 11, which exhibits the typical micro-fractography of the fibrous zones at the strain rate of 0.001 and 1000 s^{-1}, respectively. It should be noted that we have examined the samples tested at all the strain rates, here we only present the representative samples to explain the fracture mechanism. The fracture of the two steels is characterized by the well-defined dimple-typed fracture, regardless of the strain rate. However, it can be observed that the dimple becomes deeper and larger for the sample tested at higher strain rate, which corresponds to a higher PUE value as shown in Figure 6d. This might be due to the increased adiabatic temperature rise at high strain rates. The adiabatic heating can reduce the bonding strength among atoms, decompose the cell structure and decrease the energy of the grain boundary as well as the particles diffusion, which will result in softening [7,14,19,27,28]. At the low strain rate stage, the adiabatic temperature rise is negligible; as such the change of PUE is limited. For the high strain rate stage, the increasing adiabatic temperature rise with higher strain rates results in a higher PUE value. The more sharply increased PUE value of the FMDP780 sample compared to the FMDP660 steel can also be explained by the higher increment of the adiabatic temperature rise [28]. On the other hand, the increase of the local strain hardening with strain rate at the high strain rate stage might also lead to an increased PUE [2].

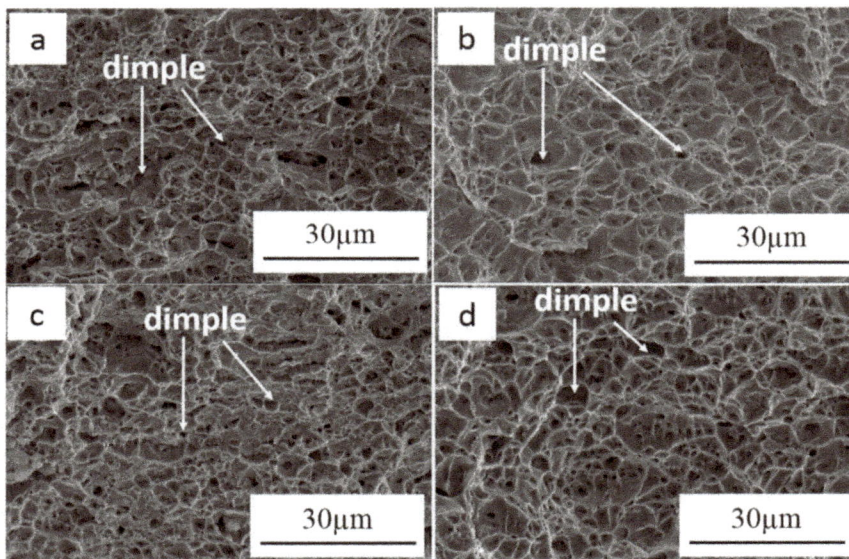

Figure 11. Fracture surface of the (**a,b**) FMDP660 and (**c,d**) FMDP780 steel at strain rates of (**a,c**) 0.001 s^{-1} and (**b,d**) 1000 s^{-1}.

Figure 12 shows the typical microstructure of the side surface in the vicinity of the fracture surface of the FMDP780 sample at the quasi-static and two different strain rates, revealing the damage formation sites and the distributions. The microstructure of the FMDP660 sample is not present here due to the similar damage behavior. It is found that the microvoids are mainly formed on the interfaces between ferrite and martensite, for all the strain rate conditions. Previous research on DP steels has shown that the microvoids can be nucleated both inside the martensite and on the ferrite–martensite interfaces [29–31], whereas in this study only the latter nucleation sites were observed. Obviously, the grain exhibits a more obvious elongating trend at 1000 s^{-1} than the quasi-static.

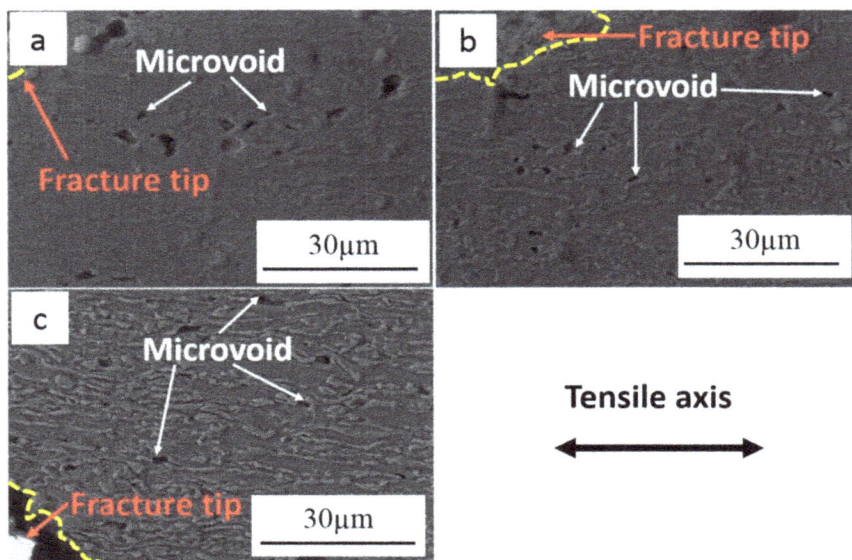

Figure 12. SEM micrographs of the damage of the FMDP780 steel after tensile test at the strain rates of (**a**) quasi-static; (**b**) 10 s^{-1} and (**c**) 1000 s^{-1}, showing the formation of microvoids at ferrite–martensite interfaces.

The formation of microvoids during tensile testing is mainly due to the heterogeneous deformation of martensite and ferrite under high strain rate conditions. It is more difficult to deform martensite than ferrite and the load is predominantly taken by the martensite phase during the deformation process. When the local stress exceeds the critical boundary binding stress, the microvoids would nucleate on the interfaces, followed by their growth and coalescence [31]. Ghadbeigi et al. [32], who use the Digital Image Correlation (DIC) and an in-situ tensile testing and found that the position of the nucleation is related to the strength of the interfaces and the stress distribution among different phases in the DP steel. The fracture mechanism of the investigated alloys tests at high strain rates is depicted by a schematic diagram shown in Figure 13.

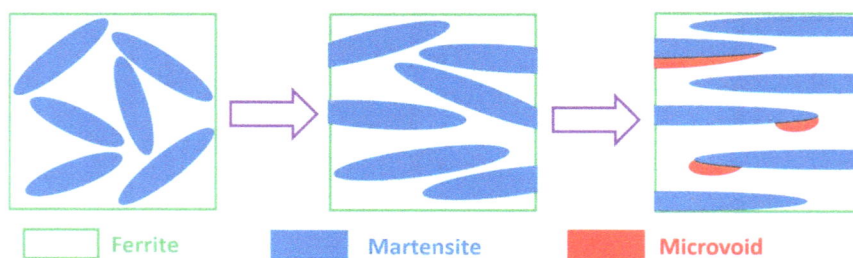

Figure 13. Schematic diagram of microvoid formation in ferrite–martensite dual-phase steel.

4. Conclusions

In the present work, the mechanical behavior of two ferrite–martensite dual phase steels (FMDP660 and FMDP780) under tensile testing at different strain rate (0.001–1000 s^{-1}) was investigated. The effect of the strain rate and microstructure on tensile properties and energy absorption, strain hardening behavior and associated dislocation motion, the adiabatic heating and the fracture mechanism was discussed. The main conclusions are as follows:

(1) The YS in both steels monotonically increased with strain rates. However, the values of UTS, UE, PUE and Z of both steels were maintained stable at the low strain rate range (0.001–0.1 s^{-1}), followed by a significant increase with strain rates at the high strain rate range (0.1–1000 s^{-1}). The FMDP780 steel with a higher fraction of martensite possessed a stronger strain rate sensitivity of tensile strength and elongation values at the high strain rate stage, compared with the FMDP660 sample. For all the strain rates, the FMDP780 sample possessed higher strength and lower elongation values compared with FMDP660.

(2) The energy absorption in the two steels at the strain of 10% increased with increasing strain rates. The energy absorption of the two steels at the stage of UE and TE showed a first stable and then increasing trend with increasing strain rates; such increase of the energy absorption was found to be more pronounced for the FMDP780 steel.

(3) The different UTS and UE values due to different strain rates and microstructures were associated with the change of work hardening rate. At the low strain rate stage, the work hardening rate of both steels was relatively low and unchanged with strain rates. It started to increase when the strain rate was above 0.1 s^{-1}, and the increase was more pronounced in the FMDP780 sample.

(4) The dislocation density of ferrite in the two steels did not change significantly at low strain rates, whereas it increased obviously at the high strain rate stage. Further, it showed that the FMDP780 steel presented a higher dislocation density compared with the FMDP660 at the same strain rate.

(5) The fracture surface of the steels was characterized by dimpled-type fracture associated with microvoid formation at the ferrite–martensite interfaces, regardless of the strain rates. The change of the PUE value of the two steels with strain rate was attributed to the effect of adiabatic heating during the tensile testing. Besides, the grain exhibits a more obvious elongating trend at 1000 s^{-1} than the quasi-static.

Acknowledgments: This research was supported by the National Natural Science Foundation of china (Grant No. 51574028), the Science and Technology Plan Project of Beijing (D151100003515002). The financial support of the Beijing Laboratory of Metallic Materials and processing for Modern Transportation is also gratefully acknowledged.

Author Contributions: Jiangtao Liang, Zhengzhi Zhao, and Di Tang conceived and designed the experiments; Jiangtao Liang and Hongwu performed the experiments; Jiangtao Liang, Zhengzhi Zhao, Binhan Sun, Baoqi Guo, and Juhua Liang analyzed the data; Jiangtao Liang and Chong Peng contributed reagents/materials/analysis tools; Jiangtao Liang wrote the paper; Zhengzhi Zhao and Binhan Sun revised the language in this paper.

Conflicts of Interest: The authors declare no conflict of interest.

References

1. Verleysen, P.; Peirs, J.; Van Slycken, J.; Faes, K.; Duchene, L. Effect of strain rate on the forming behaviour of sheet metals. *J. Mater. Process Technol.* **2011**, *211*, 1457–1464. [CrossRef]
2. Kim, S.B.; Huh, H.; Bok, H.H.; Moon, M.B. Forming limit diagram of auto-body steel sheets for high-speed sheet metal forming. *J. Mater. Process Technol.* **2011**, *211*, 851–862. [CrossRef]
3. Sun, X.; Soulami, A.; Choi, K.S.; Guzman, O.; Chen, W. Effects of sample geometry and loading rate on tensile ductility of TRIP800 steel. *Mater. Sci. Eng. A* **2012**, *541*, 1–7. [CrossRef]
4. Boyce, B.L.; Dilmore, M.F. The dynamic tensile behavior of tough, ultrahigh-strength steels at strain-rates from 0.0002 s^{-1} to 200 s^{-1}. *Int. J. Impact Eng.* **2009**, *36*, 263–271. [CrossRef]
5. Kim, J.H.; Kim, D.; Han, H.N.; Barlat, F.; Lee, M.G. Strain rate dependent tensile behavior of advanced high strength steels: Experiment and constitutive modeling. *Mater. Sci. Eng. A* **2013**, *559*, 222–231. [CrossRef]
6. Curtze, S.; Kuokkala, V.T.; Hokka, M.; Peura, P. Deformation behavior of TRIP and DP steels in tension at different temperatures over a wide range of strain rates. *Mater. Sci. Eng. A* **2009**, *507*, 124–131. [CrossRef]
7. Singh, N.K.; Cadoni, E.; Singha, M.K.; Gupta, N.K. Dynamic tensile behavior of multi phase high yield strength steel. *Mater. Des.* **2011**, *32*, 5091–5098. [CrossRef]

8. Qin, J.; Chen, R.; Wen, X.; Lin, Y.; Liang, M.; Lu, F. Mechanical behaviour of dual-phase high-strength steel under high strain rate tensile loading. *Mater. Sci. Eng. A* **2013**, *586*, 62–70. [CrossRef]

9. Liu, Y.; Dong, D.; Wang, L.; Chu, X.; Wang, P.; Jin, M. Strain rate dependent deformation and failure behavior of laser welded DP780 steel joint under dynamic tensile loading. *Mater. Sci. Eng. A* **2015**, *627*, 296–305. [CrossRef]

10. Dong, D.; Liu, Y.; Yang, Y.; Li, J.; Ma, M.; Jiang, T. Microstructure and dynamic tensile behavior of DP600 dual phase steel joint by laser welding. *Mater. Sci. Eng. A* **2014**, *594*, 17–25. [CrossRef]

11. Huh, H.; Kim, S.B.; Song, J.H.; Lim, J.H. Dynamic tensile characteristics of TRIP-type and DP-type steel sheets for an auto-body. *Int. J. Mech. Sci.* **2008**, *50*, 918–931. [CrossRef]

12. Yu, H.; Guo, Y.; Lai, X. Rate-dependent behavior and constitutive model of DP600 steel at strain rate from 10^{-4} to 10^3 s^{-1}. *Mater. Des.* **2009**, *30*, 2501–2505. [CrossRef]

13. Song, R.B.; Dai, Q.F. Dynamic Deformation Behavior of Dual Phase Ferritic-Martensitic Steel at Strain Rates From 10^{-4} to 2000 s^{-1}. *Iron Steel Res. Int.* **2013**, *20*, 48–53. [CrossRef]

14. Cadoni, E.; Singh, N.K.; Forni, D.; Singha, M.K.; Gupta, N.K. Strain rate effects on the mechanical behavior of two Dual Phase steels in tension. *Eur. Phys. J. Special Topic* **2016**, *225*, 409–421. [CrossRef]

15. Wang, W.R.; Li, M.; He, C.W.; Wei, X.C.; Wang, D.Z.; Du, H.B. Experimental study on high strain rate behavior of high strength 600–1000 MPa dual phase steels and 1200 MPa fully martensitic steels. *Mater. Des.* **2013**, *47*, 510–521. [CrossRef]

16. Zhang, J.C.; Di, H.S.; Deng, Y.G.; Li, S.C.; Misra, R.D.K. Microstructure and mechanical property relationship in an ultrahigh strength 980 MPa grade high-Al low-Si dual phase steel. *Mater. Sci. Eng. A* **2015**, *645*, 232–240. [CrossRef]

17. Calcagnotto, M.; Ponge, D.; Raabe, D. Effect of grain refinement to 1 μm on strength and toughness of dual-phase steels. *Mater. Sci. Eng. A* **2010**, *527*, 7832–7840. [CrossRef]

18. Zhang, J.C.; Di, H.S.; Deng, Y.G.; Misra, R.D.K. Effect of martensite morphology and volume fraction on strain hardening and fracture behavior of martensite–ferrite dual phase steel. *Mater. Sci. Eng. A* **2015**, *627*, 230–240. [CrossRef]

19. Wei, X.; Fu, R.; Li, L. Tensile deformation behavior of cold-rolled TRIP-aided steels over large range of strain rates. *Mater. Sci. Eng. A* **2007**, *465*, 260–266. [CrossRef]

20. Feng, F.; Huang, S.; Meng, Z.; Hu, J.; Lei, Y.; Zhou, M.; Wu, D.; Yang, Z. Experimental study on tensile property of AZ31B magnesium alloy at different high strain rates and temperatures. *Mater. Des.* **2014**, *57*, 10–20. [CrossRef]

21. Xu, Z.; Huang, F. Plastic behavior and constitutive modeling of armor steel over wide temperature and strain rate ranges. *Acta Mech. Solida Sin.* **2012**, *25*, 598–608. [CrossRef]

22. Rana, R.; Liu, C.; Ray, R.K. Low-density low-carbon Fe–Al ferritic steels. *Scr. Mater.* **2013**, *68*, 354–359. [CrossRef]

23. Sugimoto, K.I.; Kobayashi, M.; Hashimoto, S.I. Ductility and strain-induced transformation in a high-strength transformation-induced plasticity-aided dual-phase steel. *Metall. Trans. A* **1992**, *23*, 3085–3091. [CrossRef]

24. Huh, H.; Kim, S.B.; Song, J.H.; Yoon, J.H.; Lim, J.H.; Park, S.H. Investigation of elongation at fracture in a high speed sheet metal forming process. *Steel Res. Int.* **2009**, *80*, 316–322.

25. Dong, D.; Liu, Y.; Wang, L.; Su, L. Effect of strain rate on dynamic deformation behavior of DP780 steel. *Acta Metall. Sin.* **2013**, *49*, 159–166. [CrossRef]

26. Bouaziz, O.; Buessler, P. Iso-work increment assumption for heterogeneous material behavior modelling. *Adv. Eng. Mater.* **2004**, *6*, 79–83. [CrossRef]

27. Sung, J.H.; Kim, J.H.; Wagoner, R.H. A plastic constitutive equation incorporating strain, strain-rate, and temperature. *Int. J. Plast.* **2010**, *26*, 1746–1771. [CrossRef]

28. He, Z.; He, Y.; Ling, Y.; Wu, Q.; Gao, Y.; Li, L. Effect of strain rate on deformation behavior of TRIP steels. *J. Mater. Process Technol.* **2012**, *212*, 2141–2147. [CrossRef]

29. Mazaheri, Y.; Kermanpur, A.; Najafizadeh, A.; Saeidi, N. Effects of initial microstructure and thermomechanical processing parameters on microstructures and mechanical properties of ultrafine grained dual phase steels. *Mater. Sci. Eng. A* **2014**, *612*, 54–62. [CrossRef]

30. Ghatei Kalashami, A.; Kermanpur, A.; Ghassemali, E.; Najafizadeh, A.; Mazaheri, Y. Correlation of microstructure and strain hardening behavior in the ultrafine-grained Nb-bearing dual phase steels. *Mater. Sci. Eng. A* **2016**, *678*, 215–226. [CrossRef]

31. Mazaheri, Y.; Kermanpur, A.; Najafizadeh, A. A novel route for development of ultrahigh strength dual phase steels. *Mater. Sci. Eng. A* **2014**, *619*, 1–11. [CrossRef]
32. Ghadbeigi, H.; Pinna, C.; Celotto, S.; Yates, J.R. Local plastic strain evolution in a high strength dual-phase steel. *Mater. Sci. Eng. A* **2010**, *527*, 5026–5032. [CrossRef]

metals

MDPI

Article

The Prediction of the Mechanical Properties for Dual-Phase High Strength Steel Grades Based on Microstructure Characteristics

Emil Evin [1,*]**, Ján Kepič** [2]**, Katarína Buriková** [2] **and Miroslav Tomáš** [1]

[1]   Institute of Technology and Material Engineering, Faculty of Mechanical Engineering,
     Technical University of Kosice, Mäsiarska 74, 040 01 Košice, Slovakia; miroslav.tomas@tuke.sk
[2]   Institute of Materials Research, Slovak Academy of Sciences, Watsonova 47, 040 01 Košice, Slovakia;
     jkepic@saske.sk (J.K); kburikova@gmail.com (K.B.)
*   Correspondence: emil.evin@tuke.sk; Tel.: +421-55-602-3547

Received: 28 February 2018; Accepted: 3 April 2018; Published: 5 April 2018

Abstract: The decrease of emissions from vehicle operation is connected mainly to the reduction of the car's body weight. The high strength and good formability of the dual phase steel grades predetermine these to be used in the structural parts of the car's body safety zones. The plastic properties of dual phase steel grades are determined by the ferrite matrix while the strength properties are improved by the volume and distribution of martensite. The aim of this paper is to describe the relationship between the mechanical properties and the parameters of structure and substructure. The heat treatment of low carbon steel X60, low alloyed steel S460MC, and dual phase steel DP600 allowed for them to reach states with a wide range of volume fractions of secondary phases and grain size. The mechanical properties were identified by a tensile test, volume fraction of secondary phases, and grain size were measured by image analysis. It was found that by increasing the annealing temperature, the volume fraction of the secondary phase increased, and the ferrite grains were refined. Regression analysis was used to find out the equations for predicting mechanical properties based on the volume fraction of the secondary phase and grain size, following the annealing temperature. The hardening mechanism of the dual phase steel grades for the states they reached was described by the relationship between the strain-hardening exponent and the density of dislocations. This allows for the designing of dual phase steel grades that are "tailored" to the needs of the automotive industry customers.

Keywords: dual phase steel; annealing; volume fraction of secondary phase; grain size; strain-hardening exponent; yield strength; ultimate tensile strength; properties prediction

---

1. Introduction

For reasons of environmental protection, an increased emphasis has been placed on the reduction of exhaust emissions from car use in recent years. The reduction in vehicle weight is considered to be one of the decisive factors for improving fuel consumption and hence, reducing emissions [1,2]. The considerable potential for vehicle weight reduction is hidden in the body, which accounts for about 25% of the total mass of cars. In the segment of middle and lower vehicle classes, the base material is steel. In the higher-end segment, the concepts of an aluminum-based light alloy body or a combined body of steel, aluminum, and composite materials are applied. When aluminum alloys or composite materials are applied, the weight reduction is achieved, even at the expense of higher costs.

The intention of the automotive industry is to produce vehicles not only with reduced weight but also with a high level of safety characteristics such as strength, stiffness, and deformation work [3–5]. In comparison to other materials, the advantage of steel grades is the variability in performance properties (strength, stiffness, energy absorption ability, corrosion resistance of galvanized sheets, and so forth), technological properties (formability and weldability by application of various technologies), their recyclability at the end of the car's lifetime, and the lower production costs. To meet the oftentimes contradictory demands of the automotive industry on the utility properties, the steel industry is constantly developing new concepts of high-strength steel grades (DP—dual phase steel, CP—complex phase steel, TRIP—transformation induced plasticity steel, TWIP—twining induced plasticity steel, and so forth). It appears that thanks to a wide variety of combinations of strength, plasticity properties (yield strength $R_{p0.2}$ = 280–700 MPa, ultimate tensile strength R_m = 600–1000 MPa, ductility A = 12–34%, strain-hardening exponent n = 0.09–0.21, and normal anisotropy ratio r = 0.9–1), and cost from all the known high-strength steel sheets used in the construction of motor vehicles, dual phase steel grades take the largest share [6,7].

Dual phase steel grades (DP) consist of a fine-grained ferrite matrix with dispersed islands of martensite or lower bainite and often, with a certain share of residual austenite. The soft ferritic structure is the carrier of the plastic properties and the hard particles of the martensitic phase are the carriers of the strength properties. The share of martensite in dual phase steel grades ranges from 10 to 30%. With a greater share of martensite in the ferrite matrix, the clustering of martensitic islands may occur, which results in the deterioration of their strength-plastic properties combination [8–12].

The dual phase ferritic-martensitic structure can be obtained from any low-carbon steel by a controlled rolling or intercritical annealing method, provided that the transformation of austenite to perlite is avoided [13–15]. Perlite formation is suppressed by the Cr and Mo elements which, at the same time, support the formation of martensite. Further enhancement of the over-hardenability can be achieved by the addition of Mn, Si, and P. The silicon inhibits the perlite and carbides formation, Nb ensures ferritic grain refinement and increases the temperature of intercritical annealing Tnr [16].

2. Mechanics of Plastic Flow during Deformation

The mechanical properties of dual phase ferritic-martensitic steel depend on the chemical composition, the volume fraction of martensite, the volume fraction of ferrite, the carbon content in martensite, the grain size of martensite, and their strength [17,18]. To describe the behavior of dual phase ferritic-martensitic steel under plastic deformation, various constitutive equations were proposed [9,15,19–22]. Increasing the intercritical temperature increases the amount of austenite generated and this is transformed to martensite during rapid cooling. Thus, the strength and hardness of the material increases as well. The carbon content in martensite is larger for dual phase steels with low volume fractions of martensite. Otherwise, the carbon content in martensite decreases when the volume fraction of martensite increases. The carbon content in martensite controls the phase hardness and influences the final properties of the material. By controlling the metallurgical processes, it is possible to reach ferritic-martensitic structures with volume fractions of martensite from 35 to 50% with a wide combination of strength and plastic properties [17,18].

The effective use of dual phase steels in the automotive industry requires a better understanding of how they behave in crashes, as well as how they behave when processed by stamping to the structural parts of the safety zones. Nowadays, numerical simulations of crash tests and metal forming processes, based on the Finite Element Method, are widely used to predict the deformation behavior of materials. Thus, to describe the material behavior under deformation, the following constitutive equations are used [23]:

$$\text{Hollomon} \qquad \sigma_s = K\varphi_i^n \qquad (1)$$

$$\text{Swift} \qquad \sigma_s = K(\varphi_0 + \varphi_i)^n \qquad (2)$$

where σ_s is the true stress, K is the material constant, n is the strain-hardening exponent that expresses the intensity of the strain-hardening and the ability of the material to deform uniformly, ϕ_i is true strain $\varphi_i = \ln(1 + dL_i/L_0)$, and $]\phi_0$ is the pre-strain [24].

These models can be used to prepare the production of DP steel grades with precisely defined "tailor-made" properties for the components of the vehicle's deformation zones at the front and the side impact. When selecting material for the car-body safety zones, the main criterion is resistance to deformation (that is, deformation work) that is consumed at the crash. This can be determined by the tensile test record σ_s–ϕ (Figure 1):

$$dW_{Pls} = V_0 \int_{\varphi=0.002}^{\varphi_{UE}} \sigma_s d\varphi \tag{3}$$

where V_0 is the specimen volume on the initial length L_0, ϕ_{UE} is the uniform true strain (true strain at tensile strength), and $]\phi_{0.002} = 0.002$ is the true strain at yield strength.

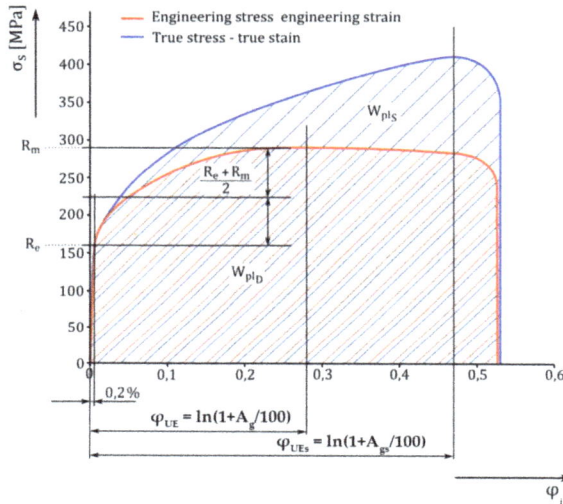

Figure 1. The record of the tensile test.

The parts of the body deformation zones are elastically and plastically deformed during impact and during their production. However, crash tests are only concerned with plastic deformation. After we insert Equation (1) into Equation (3) and make adjustments, we get the following:

$$dW_{Pls} = V_0K \int_{\varphi=0.002}^{\varphi_{UE}} \varphi_i^n d\varphi \tag{4}$$

After the integration and adjustment Equation (4) we get the following:

$$W_{Pls} = V_0K \frac{(\varphi_0 + \varphi_{UE})^{n+1} - (\varphi_0 + 0.002)^{n+1}}{n+1} \tag{5}$$

The values of the material constant K and the strain-hardening exponent n can be determined from the tensile test record by regression analysis. However, to gain a better understanding of the mechanics of the deformation process, the strain-hardening exponent *n*, and the material constant K can be determined from the mutual bonds of the mechanical properties of metallic materials. If Equation (1) (or, by analogy, Equation (3)) is subjected to a logarithmic operation, we get the following linear dependence:

$$\ln(\sigma_S) = \ln(K) + n\ln(\varphi_i) \tag{6}$$

and we express the contribution to strain-hardening at tensile strength R_m (R_m refers to the ultimate tensile strength) with respect to the yield strength, depending on the uniform deformation, in the interval from $\phi = 0.002$ to ϕ_{UE} [25] as follows:

$$n = \frac{d \ln \sigma_s}{d \ln \varphi} \tag{7}$$

which yields the exponent of the strain-hardening n:

$$n = \left[\frac{\ln\left(R_m e^{(\varphi_{UE})}\right) - \ln\left(R_e e^{(0.002)}\right)}{\ln(\varphi_{UE}) - \ln(0.002)} \right] \tag{8}$$

The n value is not constant throughout the uniform deformation, so it is necessary to expect a certain uncertainty in the calculation of the deformation work and the actual strength, especially in the case of minor strains. The exact determination of the strain-hardening exponent requires the division of an even deformation region into several intervals and the expression of the strain-hardening exponent in terms of deformation:

$$n = n_0 - p\varphi_i \tag{9}$$

where n_0 is the strain-hardening exponent found in the first interval (for example, ϕ_i is the true strain between 0.002 and 0.02), p is the constant determined by the approximation of the dependence of the strain-hardening exponent on the deformation at individual intervals.

From the Equation (6), it follows that the material constant K will be

$$\ln(K) = \ln(\sigma_S) - n \ln(\varphi_i) \tag{10}$$

Upon adjustment, we get the following:

$$K = \frac{R_m e^{(\varphi_{UE})}}{(\varphi_{UE})^n} \tag{11}$$

The above-mentioned mechanical properties of materials (the yield strength R_e, the tensile strength R_m, the material constant K, the strain-hardening exponent n, the maximum value of uniform deformation ϕ_{UE}, and so forth) are given by their internal structure—the structure of the material, which, in turn, depends on the chemical composition of steel and on its production technology. The production of "tailored" or "customized" steel grades, with exactly defined properties, requires knowledge of not only the above-mentioned relationships that determine the mechanical properties but also knowledge of the relationships between the structural parameters and the mechanical properties of the metallic materials.

Due to the fact that the structural parts of the safety zones are deformed at higher strain rates when a car crashes, the influence of the strain rate needs to be included in constitutive equations. In Reference [26] the authors included the influence of strain rate and temperature into these equations. Authors from References [27,28] included the strain rate influence into the constitutive equations when predicting the deformation work. It has been found that the influence of the strain rate was low at quasistatic strain rates [28], but a notable effect was found at higher strain rates and that it is connected to the evolution of the dislocation density.

In the literature, the structural nature of the material properties of ferritic-martensitic steel grades is given a great deal of attention [29–31]. Based on the dislocation theory, founded on the motion of dislocations and their interaction with various obstacles (grain boundaries, precipitates, interstitial atoms, fractions of different phases, as well as other dislocations), the actual stress necessary for plastic deformation flow can be expressed in terms of the individual contributors to hardening:

$$\sigma_s(\varphi) = \sigma_0 + \Delta\sigma_g + \Delta\sigma_{IN} + \Delta\sigma_S + \Delta\sigma_P + \Delta\sigma_{PR} + \Delta\sigma_{FP} + \sigma_D(\varphi) \tag{12}$$

where σ_0 is Peierls stress necessary to overcome the lattice friction stress, the resistance of alloying elements dissolved in solid solution, the precipitation matrix resistance, and the lattice defects [32]; $\Delta\sigma_g$ is the hardening effect depending on the size of the ferritic grain; $\Delta\sigma_S$ is the effect of substitute hardening; $\Delta\sigma_{IN}$ is the effect of interstitic hardening; $\Delta\sigma_P$ is the effect of precipitation hardening; $\Delta\sigma_{PR}$ is the effect of perlite hardening; $\Delta\sigma_{SG}$ is the effect subgrain hardening (also possible to be expressed as $\Delta\sigma_{FMaB}$—the hardening through bainitic or martensitic fractions or plates); $\Delta\sigma_D$ is the dislocation density hardening effect, and so forth [33,34].

For dual phase ferritic-martensitic (DP) steel grades, Equation (1) can be adjusted as follows:

$$\sigma(\varphi) = \sigma_0 + \Delta\sigma_g + \Delta\sigma_{MaB} + \sigma_D(\varphi) = \sigma_0 + \Delta\sigma_g + \Delta\sigma_{MaB} + \alpha Gb\sqrt{\rho(\varphi)} \tag{13}$$

The Peierls stress σ_0 [32,33]

$$\sigma_0 = 77 + 750(\%P) + 60(\%Si) + 80(\%Cu) + 45(\%Ni) + 60(\%Cr) + 80(\%Mn) \\ + 11(\%Mo) + 5000(\%N) \tag{14}$$

The hardening effect of ferritic grain size:

$$\Delta\sigma_g = \frac{k_y}{\sqrt{d_\alpha}} \tag{15}$$

The hardening effect of the martensitic or bainitic fractions:

$$\Delta\sigma_{MaB} = k_{MaB}V_m \tag{16}$$

The hardening effect of the dislocation density:

$$\sigma_D(\varphi) = \alpha Gb\sqrt{\rho(\varphi)} = \rho_D^{0.5}7.34 \times 10^{-6} \tag{17}$$

where d_α is the mean grain size of ferrite, k_y is the strengthening coefficient, α is a material constant, G is the shear modulus (80,000 MPa), b is Burger's vector, and ρ_D is the dislocation density.

The aims of the experimental research were to prepare materials with different volume fractions of martensite up to 50% from commercial steels, to describe the relationship between the mechanical properties and the temperature of the intercritical annealing, and to describe the relationship between properties that are sensitive to changes of the sub-structural parameter when cold deformed.

3. Materials and Methods

The deformation behavior of the dual phase steel types (Equations (13)–(17)) depends mainly on the chemical composition, the volume of martensite, the morphology and distribution in the ferrite matrix, as well as the ferrite grain size d_α. The aim of the experimental research was to prepare materials (states) with a martensite volume of up to 50% from commercially produced low carbon steel types of 3–3.3 mm thickness: A (X60), B (S460MC), and C (DP600), whose chemical composition and values of carbon equivalent C_E calculated from Equation (18) [35] are listed in Table 1.

$$C_E = C + 0.75 + 0.25\tanh[20(C - 0.12)]\left\{ \frac{Si}{24} + \frac{Mn}{6} + \frac{Cu}{15} + \frac{Ni}{20} + \frac{(Cr + Mo + Nb + V)}{5} + 5B \right\} \tag{18}$$

The microstructures of the initial materials A, B, and C used are shown in Figure 2. The low carbon steel microstructure (A) is ferritic-pearlite (Figure 2a). The low-carbon micro-alloyed steel (B) microstructure is ferritic-pearlite with a low perlite content (Figure 2b). The microstructure of steel C is a ferritic-martensitic one with a martensitic volume of 24% (Figure 2c). As can be seen from Table 1, the carbon content and the average size of the ferritic grain d_α are approximately equal for as-received steels A, B, and C. The dispersion of the mean ferrite grain size under the surface and in the middle of the sheet thickness of the as-received A, B, and C materials was ±10%.

Table 1. The chemical composition of the as-received steels (wt %).

Material	C	Mn	Si	P	S	Al	N	Cr	V	Nb	C_E	d_α [μm]
X60 (A)	0.082	1.44	0.29	0.012	0.005	0.038	0.0063	0.017	0.048	0.046	0.244	5.7
S460MC (B)	0.068	1.22	0.02	0.015	0.0037	0.05	0.0063	0.019	0.041	0.049	0.194	5.3
DP600 (C)	0.068	1.18	0.03	0.037	0.0054	0.037	0.0075	0.542	0.004	0.002	0.239	5.2

Prior to the heat treatment, the proper starting and final temperatures of the transformation of ferrite to austenite A_{C1}, A_{C3}, A_{r1}, and A_{r3} were set according to Andrews [36] (Table 2). The non-recrystallization temperature T_{nr} and the critical cooling time between 800 and 500 °C for the beginning of the perlite precipitation were calculated according to the equations listed in Reference [37].

Table 2. The calculated temperatures of phase transformations (°C).

Material	A_{C1}	A_{C3}	A_{r1}	A_{r3}	T_{nr}	B_s	B_f	M_s	M_f
A	716	845	508	770	995	677	557	425	259
B	712	860	537	791	1101	699	557	425	259
C	720	869	544	786	896	669	549	473	266

The as-received materials samples were prepared by single-step annealing in a flowing cantalum furnace REH-B-10-60 (Linn High Therm GmbH, Bad Frankenhausen, Germany) with a protective argon atmosphere. The samples made out of material A (marked as DPA) were annealed at temperatures 740, 790, and 840 °C. Then, considering the results reached, the samples made out of materials B and C (marked as DPB and DPC) were annealed at temperatures of 750 and 820 °C (which lie between the temperatures A_{C1}–A_{C3}) with the same steady-state 10 min for each temperature, followed by cooling in water with a cooling rate of 30 °C/s [38].

Samples for metallographic analysis were hot mounted in dentacrylate, wet grinded (sandpaper 220–1200), and polished by diamond grit in suspension. Then, the samples were etched in 2% Nital.

The grain size d_α was identified by the linear method according to the Slovak standard STN 42 0462 on the microscope Olympus GX71. The volume fraction of secondary phases (V_{FSP}) was measured by the grid method (square foil 15 × 15 cm with grid 1 × 1 cm) and by the image analysis method using the image analyzer Image J at a magnification 1000× [38].

Mechanical properties of as-received materials A, B, and C, and the samples after annealing DPA, DPB, and DPC were measured by static tensile tests according to STN EN ISO 6892-1 at room temperature on a testing machine TIRAtest 2300. These are shown in Table 3. The transversal feed was 1 mm·min^{-1} and the corresponding quasistatic strain rate was 0.003 s^{-1}. Five specimens for each material and annealing state were tested. The specimen's shape is shown in Figure 3 and the dimensions were as follows: L_0 = 20 and 35 mm, L_C = 50 mm, a = 14 mm, h = 15 mm, b = 8 ÷ 10 mm [38].

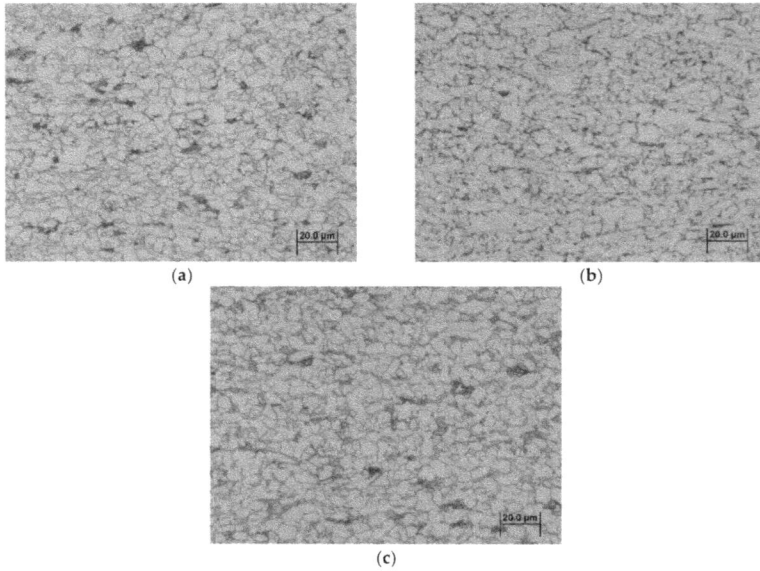

Figure 2. The microstructure of the as-received steels: (**a**) A—X60; (**b**) B—S460MC; (**c**) C—DP600.

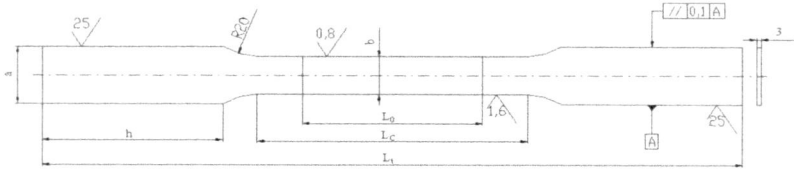

Figure 3. A specimen for the tensile test (unit: mm).

Table 3. The mechanical properties of as-received steels and annealed states.

Material	Annealing Temp. (°C)	$V_{FSP} \pm 4$ (%)	$d_\alpha \pm 0.3$ (µm)	$R_{P0.2}$ (MPa)	R_m (MPa)	A_g (%)	A (%)	HV5 $\pm$ 3
	As-received A	-	5.7	540	605	15.4	29.0	171
DPA$_{740}$ [1]	740	23.4	7.7	406	737	15.1	25.6	197
DPA$_{790}$	790	50.1	6.4	439	755	11.5	21.2	214
DPA$_{840}$	840	68.2	4.7	486	750	13.2	25.4	210
	As-received B	-	5.3	469	541	13.7	29.6	162
DPB$_{750}$	750	23.1	7.3	389	712	15.3	25.6	187
DPB$_{820}$	820	55.6	5.2	471	717	12.6	18.5	201
	As-received C	24.2	5.2	405	639	15.5	28.0	220
DPC$_{750}$	750	23.1	7.3	396	712	15.4	25.6	189
DPC$_{820}$	820	64.4	5.2	479	717	9.6	18.5	201

[1] Reference material.

4. Results and Discussion

The range of annealing conditions within the temperatures of A_{C1}–A_{C3} applied to commercially available low carbon steel X60 (A), low-alloyed steel S460MC (B), and dual phase steel DP 600 (C) were allowed to reach a wide range of microstructure states, with a martensite volume between 20.4% and

68.2% and a ferritic grain size between 4.7 and 7.7 μm, as seen in Figures 4–6, respectively. Hereinafter, these phases are designated as DPA_{740}, DPA_{790}, DPA_{820}, DPB_{750}, DPB_{820}, DPC_{750}, and DPC_{820}.

Figure 4. The microstructure of material A after annealing: (**a**) DPA_{740}; (**b**) DPA_{840}.

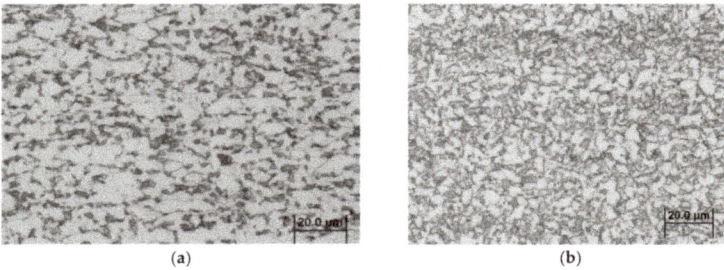

Figure 5. The microstructure of material B after annealing: (**a**) DPB_{750}; (**b**) DPB_{820}.

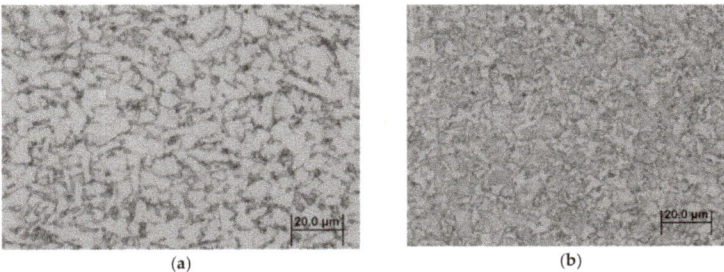

Figure 6. The microstructure of material C after annealing: (**a**) DPC_{750}; (**b**) DPC_{820}.

In determining the volume fraction of ferrite and martensite, the ferrite fraction was evaluated as the dominant phase, while the sum of all other phases (martensite, residual austenite, and bainite) represents the fraction of the secondary phase particles (FSP). This means that the fraction of the purely martensitic phase is slightly overestimated. However, the fraction of bainite, cementite, and residual austenite in the analyzed states of DPA, DPB, and DPC was ±3% within the distribution of the volume fraction of martensite. In the samples of the DPA_{740}, DPA_{790}, and DPA_{840} states obtained by the heat treatment from the initial material A, the volume fraction of the martensite ranged between 23.4% and 68.2% and the ferritic grain size ranged from 4.7 to 7.7 μm; in the samples of the DPB_{750} and DPB_{820} states obtained by the heat treatment from the initial material B, the volume fraction of the martensite ranged between 22.4% and 58.6%, the ferritic grain size ranged from 7.7 to 3.1 μm; and in the samples of the DPC_{750}, DPC_{820} states obtained by the heat treatment from the initial material C, the volume fraction of the martensite ranged between 23.8% and 64.4% and the ferritic grain size ranged from 5.2 to 7.3 μm. Thus, the assumption that the states obtained from the A and C materials with higher values of carbon equivalent would result in greater percentages of the secondary phase fractions—shown in Table 2 and Figure 7a—depending on the annealing temperature in the range between 740 and 820 °C, has been confirmed.

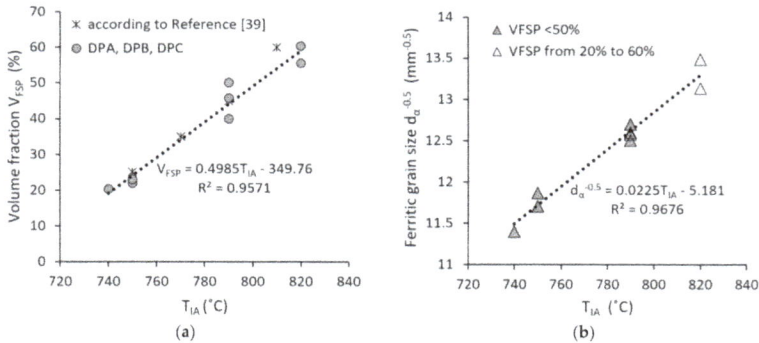

Figure 7. The dependencies of the structure parameters: (**a**) the volume fraction of the secondary phases; (**b**) the ferritic grain size and the annealing temperature.

The dependence of the volume fractions of the secondary phase (Figure 7a) can be described by the regression model:

$$V_{FSP} = 0.499\, T_{IA} - 348 \qquad [\%] \tag{21}$$

Independent of the observed structural differences, increased annealing temperature T_{IA} resulted in refinement of the mean size of the ferritic grain in examined states—Figure 7b. This tendency of grain size refinement in the annealing temperature interval between 740 and 820 °C has been described by the regression model:

$$d_{\alpha,IA}^{-0.5} = 0.0225\, T_{IA} - 5.18 \qquad [mm^{-0.5}] \tag{22}$$

The interaction between the fractions of the secondary phase and the size of the ferrite grains as it is shown in Figure 8 is described by the regression model:

$$d_{\alpha}^{-0.5} = 0.043\, V_{FSP} + 10.72 \qquad [mm^{-0.5}] \tag{23}$$

According to References [38,39], the increase in the secondary phase fractions of the dual phase ferritic-martensitic steel grades is mostly due to the number of grains of the secondary phase fractions rather than due to their volume in the structure. Regardless of the annealing temperature, it is possible to further increase the martensite fraction in the volume and the refinement of ferrite grain by increasing the rate of cooling [40].

The decisive criterion for the choice of steel sheets for the structural parts of the deformation zone of the car-body is the deformation work, which expresses the absorption capacity during a crash. The deformation work can be determined with greater uncertainty from the conventional stress-strain diagram as seen in Figure 1.

$$W_{PLd} = V_0 \frac{R_e + R_m}{2} \frac{A_g}{100} \tag{24}$$

or with less uncertainty (more precisely) from the true stress–true strain diagram by application of Equation (3).

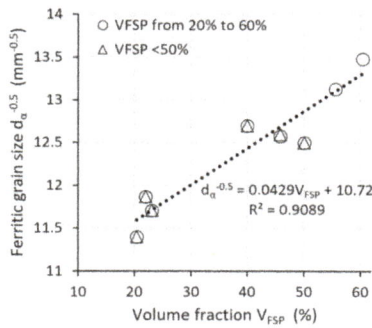

Figure 8. The dependence of the ferritic grain size on the volume fractions of the secondary phase.

For the obtained states, the attention was focused on the analysis of the relationships between the mechanical properties (the yield strength Re, the tensile strength Rm, the uniform elongation A_g, the total elongation A, the material constant K, and the strain-hardening exponent n) and the parameters of the dual phase ferritic-martensitic steels structure. It follows from the measured results (Table 2 and Figure 9) that the increase in the volume fraction of the secondary phase resulted in the increased yield strength values by about 80 MPa and the increased tensile strength by about 78 MPa in the DPA states. For the DPB states, the yield strength increased by 92 MPa and the tensile strength increased by 81 MPa; for the DPC, the yield strength increased by 85 MPa and the tensile strength increased by 96 MPa. Higher strength properties (yield strength, tensile strength) of the obtained states are mainly related to the volume fraction of the secondary phase. The results obtained indicate a linear dependence of the tensile strength on the volume fraction of the secondary phase, described by the following regression equation:

$$R_e = 1.876 V_{FSP} + 352 \qquad [MPa] \qquad (25)$$

However, the dependency of the tensile strength on the volume fraction of the secondary phase was found to be a linear, ranging between 20% and 50%, described by the regression equation as follows:

$$R_m = 3.74 V_{FSP} + 640 \qquad [MPa] \qquad (26)$$

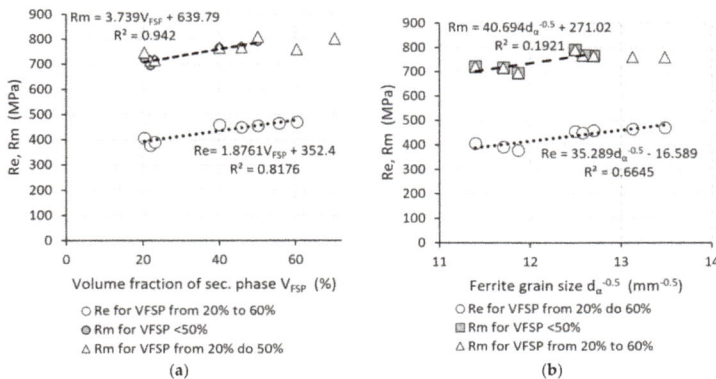

Figure 9. The dependence of the yield strength and tensile strength on (**a**) the volume fractions of the secondary phase; (**b**) the size of the ferritic grains.

No increase in tensile strength has been observed for the volume fraction of the secondary phase greater than 50%. Rather, a decrease in tensile strength values has been noted. We assume that on the one hand, martensite contributes to an increase in tensile strength due to the increased volume of the harder phase (martensite), on the other hand, the carbon content of martensite decreases with the increasing volume of martensite. As is well known, the strength of martensite is mainly determined by its carbon content. Another reason for the reduced tensile strength values may be due to the size of martensite islands and martensite distributions in the ferrite matrix [11,32,39]. At lower annealing temperatures T_{IA} 740 or 750 °C, martensite was dispersed along the borders of ferritic grains, however, at the annealing temperature of 820 °C, the size of martensitic fractions was greater in comparison to the states obtained under the annealing temperature of 740 or 750 °C (Figures 4–6).

It follows, from Figure 9b, that the lower the size of the ferritic grain $d_\alpha^{-0.5}$, the larger the yield strength and tensile strength values that have been observed. These curves have been described by the regression models in Equations (27) and (28).

$$R_e = 35.29 d_\alpha^{-0.5} - 16.6 \qquad \text{[MPa]} \qquad (27)$$

and similarly, in the 20% to 50% interval for the tensile strength

$$R_m = 40.69 d_\alpha^{-0.5} + 271 \qquad \text{[MPa]} \qquad (28)$$

However, it should be noted that the tensile strength dependency on the size of the ferritic grain $d_\alpha^{-0.5}$ is more of a tendency because the residual dispersion value $R^2 = 0.192$ was low.

The influence of strengthening contributors in terms of the individual structural parameters on the strength properties of dual phase ferritic-martensitic materials cannot be assessed separately. For this reason, attention was focused on expressing the summary influence of the structural parameters on the yield strength according to Equation (13) by the unit sum S_j of the individual parameters of the structure. The unit sum S_j of the parameters of the structure was determined as the ratio of yield strengths expressed by the Equations (25) and (27) with respect to the yield strength value of the reference material DPA_{750} (R_e = 406 MPa) as follows:

$$S_{iR_e} = \sigma_{0i} + \frac{1}{iR_{e,ref}} \sum_{i=1}^{3} (k_{y\alpha i} d_\alpha^{-0.5} + k_{VFSPi} V_{FSP}) \qquad (29)$$

where i is the number of parameters of the structure (i = 3), $R_{e,ref}$ is the yield strength value of the reference material, and $k_{y\alpha i}$ is a constant expressing the influence of the ferritic grain size.

The dependence of the yield strength on the unit sum of the parameters of the structure S_{iRe} is given in Figure 10 and expressed through the following regression model:

$$R_e = 433.4 \, S_{iR_e} - 16.9 \qquad \text{[MPa]} \qquad (30)$$

Then, after inserting Equation (29) into Equation (30), while taking into account the relationships given by Equations (25) and (27) and after the subsequent adjustment, we arrive at the following result:

$$R_e = 18.8 d_\alpha^{-0.5} + 1 V_{FSP} + 162 \qquad \text{[MPa]} \qquad (31)$$

If we insert Equation (22) into Equation (31) as d_α and we insert Equation (21) as V_{FSP}, we obtain the relation from which we can predict the yield strength dependent on the annealing temperature:

$$R_e = 0.922 T_{IA} - 186 \qquad \text{[MPa]} \qquad (32)$$

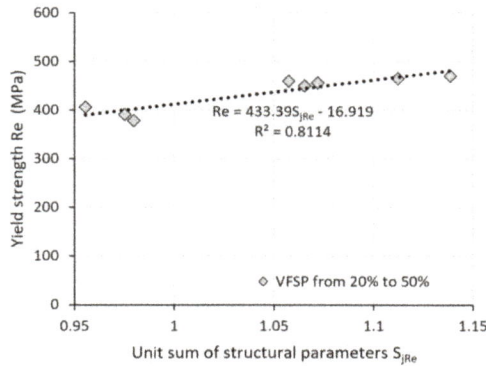

Figure 10. The dependence of the yield strength on the unit sum of the parameters of the structure.

To determine the deformation work, it is necessary to know the value of the uniform elongation A_g. Figure 11 shows that the increasing volume fraction of the secondary phase causes the values of the total elongation and the uniform elongation to drop. A similar trend was found in Reference [32]. The uniform deformation values A_g ranged from 8.8 to 16.3 and total deformation A ranged from 17.1 to 25.3. At the annealing temperature of 840 °C, the lower A_g and A values were recorded compared to the states obtained at 750 °C. We assume that this tendency may be related in particular to the morphology and the distribution of secondary phase fractions.

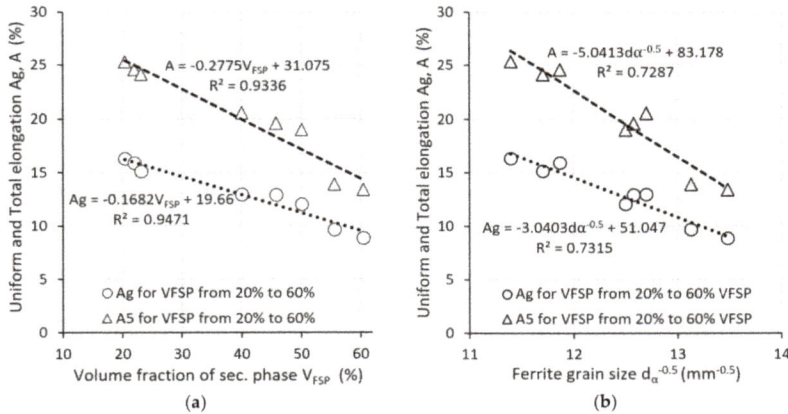

Figure 11. The dependence of the uniform and total elongation on (**a**) the volume fraction of the secondary phase; (**b**) the ferritic grain size.

It should be noted that in most metallic materials, the dependence force on the elongation, or the conventional strength on the deformation, is flat in the area of the maximum uniform elongation A_g. If we determine the elongation ΔL_i from the conventional diagram (Figure 1) at the moment when

$$\frac{\sigma_i - \sigma_{i-1}}{\varepsilon_i - \varepsilon_{i-1}} \leq 0 \tag{33}$$

The A_g value will be determined by the following relation:

$$A_g = \frac{\Delta L_i}{L_0} 100 \qquad [\%] \qquad (34)$$

Then, the value of the uniform deformation does not allow for the precise deformation work to be determined by Equation (3). The deformation work in the interval from $A = 0.2\%$ to the maximum uniform deformation A_g, does not express the overall deformation work of the material, as the material resistance to the deformation increases even with a greater deformation than A_g (Figure 1). For this reason, we recommend using a reduced elongation value A_{gs}, to be determined from the tolerance range as $1/3$ of the difference of the total elongation A and the uniform elongation A_g relative to the standard quadratic deviation of the measured values of A and A_g (STDEVA, A_g), according to the six-sigma method:

$$\Delta A_{g1/3} = \frac{\overline{A - Ag}}{2} + \frac{(A - A_g)}{3 \text{STDEV}} \qquad [\%] \qquad (35)$$

Then the reduced value of the uniform elongation will be:

$$A_{gs} = \overline{A_g} + \Delta A_{g1/3} \qquad [\%] \qquad (36)$$

and the true strain (or real deformation) will be:

$$\varphi_{\text{UES}} = \ln\left(1 + \frac{A_{gs}}{100}\right) \qquad (37)$$

Trend analyses of the dependence of the immediate stress value on deformation in the interval from $A = 0.002$ to the maximum uniform deformation of A_{gs} (Figure 1) allow the designers in the automotive industry to understand the differences of mechanical behavior in conventional and advanced high-strength steel grades in crashes and it allows designers to optimize the choice of the materials for individual "tailored" parts of deformation zones. When compared to the initial material A, the (gradient) curve directions (Figure 12) under the DPA, DPB, and DPC states show that the obtained states of DPA, DPB and DPC exhibit a more favorable course of material resistance to deformation and thus, the course of the deceleration at a crash in comparison to the curve of the steel A.

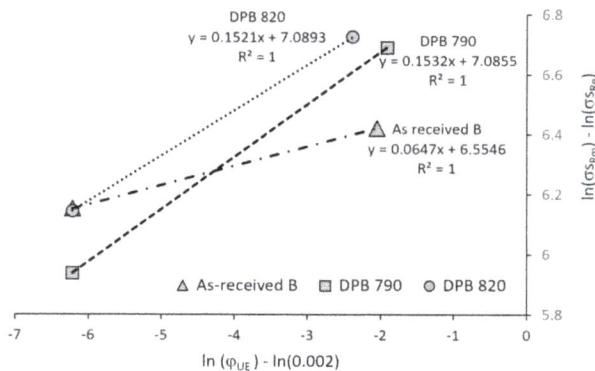

Figure 12. The dependence of the true stress on the deformation in logarithmic coordinates.

The direction or the slope of the curves express the degree of steel deformability and also the intensity of the deformation resistance upon deformation. Figure 12 shows that the states obtained at lower annealing temperatures, whose strain-hardening exponent values are higher than in the states obtained at higher temperatures, exhibit the greatest deformation resistance.

Figures 13 and 14 show that the material constant K and the strain-hardening exponent n depend on the thermomechanical history of steel, with the strain-hardening exponent n being more sensitive to changes in the parameters of the structure than the material constant K. In terms of physics, the strain-hardening exponent n determines the ability of the steel to distribute the stress along the tensile specimen. For low-carbon steel grades, used in the production of complex car-body shapes, the required value of *n* is >0.22. The higher the *n*-value, the more uniform the deformation distribution, the greater the steel's resistance to deformation, and the better its formability [39–41].

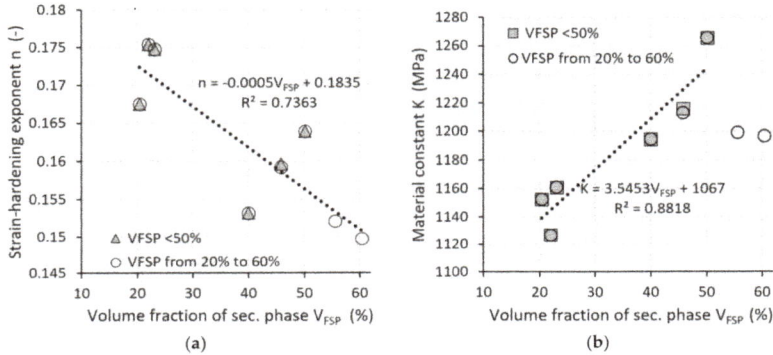

Figure 13. The dependence of the material parameters on the volume fractions of secondary phases: (a) the strain-hardening exponent n; (b) the material constant K.

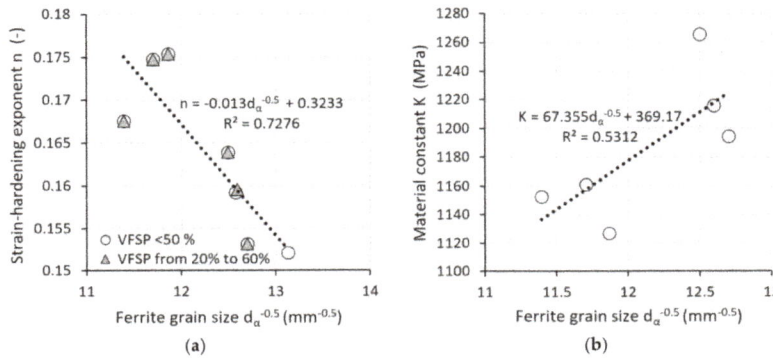

Figure 14. The dependence of the material parameters on the ferritic grain size: (a) the strain-hardening exponent n; (b) the material constant K.

Equation (13) allows for the estimation of the real strength of the material using empirical models based only on the structural parameters of ferritic-martensitic steel grades. This model describes the deformation behavior dependent on the volume fraction of the secondary phase, the ferrite grain size, and the lattice friction stress required for the dislocation motion. Another important parameter that affects the deformation behavior of metallic materials is the density of dislocations. The density of the dislocations is different for each material. The combined effects of the structural parameters of model (13) differ in the density of dislocations. We can express the density of dislocations from the actual difference in tensile strength:

$$\sigma(\varphi_{UE}) = R_m e^{(\varphi_{UE})} = \sigma_0 + \Delta\sigma_g + \Delta\sigma_{MaB} + \rho_{D,\varphi EU}^{0.5} 7.34 \times 10^{-6} \tag{38}$$

and the actual strength at the yield strength:

$$\sigma(\varphi_{0.002}) = R_e e^{(0.002)} = \sigma_0 + \Delta\sigma_g + \Delta\sigma_{MaB} + \rho_{D,\varphi 0.002}^{0.5} 7.34 \times 10^{-6} \tag{39}$$

After adjustment, we obtain the contribution coming from the density of dislocations upon deformation at the tensile strength limit:

$$\Delta\rho_{D\varphi EU-\varphi 0.002} = \left(\rho_{D,\varphi EU}^{0.5} - \rho_{D,\varphi 0.002}^{0.5}\right) = \frac{R_m e^{(\varphi_{UE})} - R_e e^{(0.002)}}{7.34 \times 10^{-6}} \tag{40}$$

The high values of the dislocation densities in the dual phase steel grades listed in Table 4 are not surprising since the transformation of austenite into martensite is the cause of great stress in ferrites. Near the martensite fractions, the dislocation density may be even higher (Figure 15). For example, Reference [34] states that, depending on the deformation in DP 500, the dislocation density values can range from 1.5×10^{14} m^{-2} up to 1.7×10^{15} m^{-2}.

Table 4. The calculated values of material constant K, the strain-hardening exponent *n*, and the dislocation density.

Material	Annealing Temp. (°C)	Material Constant K (-)	Strain-Hardening Exponent *n* (-)	Dislocation Density $\Delta\rho_{D\varphi EU-\varphi 0.002}$ (m^{-2})
	As-received A	702	0.06	4.48×10^{14}
DPA$_{740}$ [1]	740	1153	0.168	3.48×10^{15}
DPA$_{790}$	790	1266	0.164	3.42×10^{15}
DPA$_{840}$	840	1201	0.147	2.98×10^{15}
	As-received B	702	0.065	3.91×10^{14}
DPB$_{750}$	750	1127	0.175	3.38×10^{15}
DPB$_{820}$	820	1255	0.152	2.52×10^{15}
	As-received C	969	0.140	2.15×10^{15}
DPC$_{750}$	750	1161	0.175	3.47×10^{15}
DPC$_{820}$	820	1197	0.150	2.35×10^{15}

[1] Reference material.

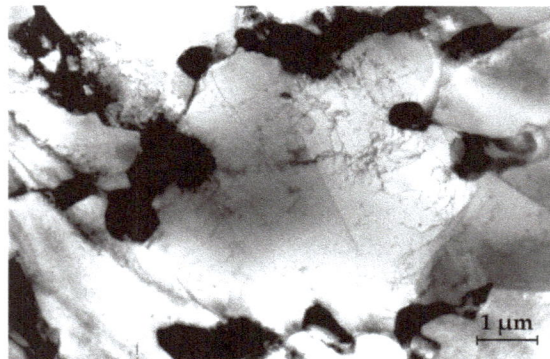

Figure 15. The dislocation density in the ferrite grain near the martensitic fraction.

5. Conclusions

The experimental work in this paper was focused on reaching dual phase steel types with different volume fractions of martensite, within a range of 20–70%. This was done by changing the intercritical annealing temperatures. The reached states were analyzed using optical microscopy and transmission electron microscopy and the mechanical properties were measured by the tensile test. The size of the martensitic islands depended on the volume fraction of martensite. It was found that increasing the volume fraction of martensite increases the strength and lowers the ductility. The mechanical properties are strongly influenced by the morphology of the disperse martensitic phase. The results for the prediction material constant K and the strain-hardening exponent n, reached from constitutive relations, were acceptable in comparison with the results reported in Reference [34].

The volume fraction of the secondary phase grew with the increased temperature of the intercritical annealing and the refinement of the ferritic grain appeared. The dependencies of both the volume fraction of the secondary phase and the ferritic grain size on the intercritical annealing temperature have been described by means of regression equations. Thus, the paper proposes equations predicting the yield strength, the uniform elongation and the true stress (the true strain curve depending on both the volume fraction of the secondary phase V_{FPS} and the mean size of the ferrite grain d_α). A complex relationship between the yield strength and the microstructure parameters makes it possible to describe the deformation behavior of dual phase steels during deformation in physical terms.

Based on the relationships obtained, material engineers and designers will be able to design the dual phase steel grades with a wide range of strength-plastic properties, that is, "tailor-made" to the requirements of the automotive industry. The nature of the deformation behavior of the steel resides in the stress increment, expressed by the strain-hardening exponent. The results obtained show that the strain-hardening exponent n depends on the structural parameters (volume fraction of the secondary phase and grain size) and the state of the substructure (dislocation density). The proposed model might be verified and used by the engineers during the selection of material for car-body structural parts of the safety zones, especially to gain compatibility in the crash situations of different classes of cars.

Acknowledgments: The authors are grateful for the support given to the experimental works by the Slovak Research and Development Agency, under project APVV-0273-12 "Supporting innovations of auto body components from the steel sheet blanks oriented to the safety, the ecology and the car weight reduction", as well as the grant agency for the support of the project VEGA 2/0113/16 "Influence of laser welding parameters on structure and properties of welded joints of advanced steels for the automotive industry".

Author Contributions: Emil Evin and Ján Kepič conceived and designed the experiments; Katarína Buriková and Miroslav Tomáš performed the experiments; Emil Evin, Ján Kepič, and Miroslav Tomáš analyzed the data; Emil Evin and Miroslav Tomáš wrote the paper.

Conflicts of Interest: The authors declare no conflict of interest.

References

1. Jung, J.; Jun, S.; Lee, H.-S.; Kim, B.-M.; Lee, M.-G.; Kim, J.H. Anisotropic Hardening Behaviour and Springback of Advanced High-Strength Steels. *Metals* **2017**, *7*, 480. [CrossRef]
2. Ramazani, A.; Mukherjee, K.; Abdurakhmanov, A.; Abbasi, M.; Prahl, U. Characterization of Microstructure and Mechanical Properties of Resistance Spot Welded DP600 Steel. *Metals* **2015**, *5*, 1704–1716. [CrossRef]
3. Švec, P.; Schrek, A.; Dománková, M. Microstructural characteristics of fibre laser welded joint of dual phase steel with complex phase steel. *Kovove Mater.* **2018**, *56*, 29–40. [CrossRef]
4. Cornette, D.; Hourman, T.; Hudin, O.; Laurent, J.; Reynaert, A. High strength steels for automotive safety parts. *SAE Tech. Pap.* **2001**. [CrossRef]
5. Mihaliková, M.; Német, M. The Effect of Strain Rate on the Mechanical Properties of Automotive Steel Sheets. *Acta Polytech.* **2013**, *53*, 384–387.

6.   Amigo, F.J.; Camacho, A.M. Reduction of Induced Central Damage in Cold Extrusion of Dual-Phase Steel DP800 Using Double-Pass Dies. *Metals* **2017**, *7*, 335. [CrossRef]

7.   Wagoner, R.H.; Smith, G.R. *Report: Advanced High Strength Steel Workshop*; Columbus, OH, USA, 22–23 October 2006. Available online: http://li.mit.edu/Stuff/RHW/Upload/AHSSDRAFTReport10-29-06.pdf (accessed on 5 April 2018).

8.   Zhao, J.Z.; Mesplont, C.; De Cooman, C. Calculation of the phase transformation kinetics from a dilatation curve. *J. Mater. Proc. Technol.* **2002**, *129*, 345. [CrossRef]

9.   Džupon, M.; Parilák, Ľ.; Kollárová, M.; Sinaiová, I. Dual Phase Ferrite-Martensitic Steel Micro-Alloyed with V-Nb. *Metalurgija* **2007**, *46*, 15–20.

10. Sodjit, S.; Uthaisangsuk, V. A micromechanical flow curve model for dual phase steels. *J. Met. Mater. Miner.* **2012**, *22*, 87–97.

11. Evin, E.; Tomáš, M.; Kmec, J.; Németh, S.; Katalinč, B.; Weselý, E. The Deformation Properties of High Strength Steel Sheets for Auto-Body Components. In Proceedings of the Procedia Engineering-24th DAAAM International Symposium on Intelligent Manufacturing and Automation 2013, Zadar, Croatia, 23–26 October 2013; Volume 69, pp. 758–767.

12. Bleck, W.; Papaefthymiou, S.; Frehn, A. Microstructure and Tensile Properties in Dual Phase and Trip Steels. *Steel Res. Int.* **2004**, *75*, 704–710. [CrossRef]

13. Tavares, S.S. M.; Pedroza, P.D.; Teodósio, J.R.; Gurova, T. Mechanical properties of a quenched and tempered dual phase steel. *Scr. Mater.* **1999**, *40*, 887–892. [CrossRef]

14. Lis, J.; Lis, A.K.; Kolan, C. Processing and properties of C–Mn steel with dual-phase microstructure. *J. Mater. Proc. Technol.* **2005**, *162–163*, 350–354. [CrossRef]

15. Speich, G.R.; Demarest, V.A.; Miller, R.L. Formation of austenite during intercritical annealing of dual phase steels. *Metall. Trans. A* **1981**, *12*, 1419–1428. [CrossRef]

16. Waterschoot, T.; de Cooman, B.C.; vanderschueren, D. Influence of run-out table cooling patterns on transformation and mechanical properties of high strength dual phase and ferrite–bainite steels. *Ironmak. Steelmak.* **2013**, *28*, 185–190. [CrossRef]

17. Mohaved, P.; Kolahgar, S.; Marashia, S.P.H. The effect of intercritical heat treatment temperature on the tensile properties and work hardening behavior of ferrite-martensite dual phase steels sheets. *Mater. Sci. Eng. A* **2009**, *518*, 1–6. [CrossRef]

18. Maffei, B.; Salvatore, W.; Valentini, R. Dual-phase steels rebars for high-ductile r.c. structures, part I: Microstructural and mechanical characterization of steel rebars. *Eng. Struct.* **2007**, *29*, 3325–3332. [CrossRef]

19. Gerbase, J.; Embury, J.D.; Hobbs, R.M. The mechanical behavior of some dual-phase steels—With emphasis in the initial work hardening rate. *Struct. Prop. Dual-Phase Steels* **1979**, 118–144.

20. Crawley, A.; Shahata, M.T.; Pussegoda, N. Processing, properties and modeling of experimental batch-annealed dual-phase steels. *Fundam. Dual-Phase Steels* **1981**, 181–197.

21. Sherman, A.M.; Davies, R.G.; Donlon, W.T. Electron microscopic study of deformed dual-phase steels. *Met. Soc. AIME* **1981**, 85–94.

22. Kumara, A.; Singh, S.B.; Rayb, K.K. Influence of bainite/martensite content on the tensile properties of low carbón dual-phase steels. *Mater. Sci. Eng. A* **2008**, *474*, 270–282. [CrossRef]

23. Wang, W.; He, C.; Zhao, Z.; Wei, X. The limit drawing ratio and formability prediction of advanced high strength dual-phase steels. *Mater. Des.* **2011**, *32*, 3320–3327. [CrossRef]

24. Colla, V.; De Sanctis, M.; Dimatteo, A.; Lovicu, G.; Solina, A.; Valentini, R. Strain Hardening Behavior of Dual-Phase Steels. *Metall. Mater. Trans.* **2009**, *40*, 2557–2567. [CrossRef]

25. Davis, J.R.; Semiatin, S.L.; American Society for Metals. *ASM Handbook Volume 14: Forming and Forging*, 9th ed.; ASM International: Geauga County, OH, USA, 1988; ISBN 0-87170-007-7.

26. Belingardi, G.; Chiandussi, G.; Ibba, A. Identification of strain-rate sensitivity parameters of steel sheet by genetic algorithm optimisation. *WIT Trans. Built Environ.* **2006**, *85*, 201–210.

27. Larour, P.; Rusinek, A.; Klepaczko, J.R.; Bleck, W. Effects of Strain Rate and Identification of Material Constants for Three Automotive Steels. *Steel Res. Int.* **2007**, *78*, 348–358. [CrossRef]

28. Evin, E.; Tomáš, M.; Výrostek, M. Quasistatic strain rates' effect to the properties of advanced steels for automotive industry. *Acta Metall. Slovaca* **2016**, *22*, 14–23. [CrossRef]

29. Nanda, T.; Kumar, B.R.; Singh, V. A simplified micromechanical modeling approach to predict the tensile flow curve behavior of dual-phase steels. *J. Mater. Eng. Perform.* **2017**, *26*, 5180–5187. [CrossRef]

30. Rodriguez, R.M.; Gutierrez, I. Unified formulation to predict the tensile curves of steels with different microstructures. *Mater. Sci. Forum* **2003**, *426–432*, 4525–4530. [CrossRef]
31. Uthaisangsuk, V.; Prahl, U.; Bleck, W. Micromechanical modelling of damage behaviour of multiphase steels. *Comput. Mater. Sci.* **2008**, *43*, 27–35. [CrossRef]
32. De la Concepción, V.L.; Lorusso, H.N.; Svoboda, H.G. Effect of carbon content on microstructure and mechanical properties of dual phase steels. *Procedia Mater. Sci.* **2015**, *8*, 1047–1056. [CrossRef]
33. Gutierrez, I. AME modelling the mechanical behaviour of steels with mixed microstructures. *Metalurgija* **2005**, *11*, 201–214.
34. Bergström, Y.; Granbom, Y.; Sterkenburg, D. A Dislocation-Based Theory for the Deformation Hardening Behavior of DP Steels: Impact of Martensite Content and Ferrite Grain Size. *J. Metall.* **2010**. [CrossRef]
35. Yurioka, N.; Kasuya, T. A Chart Method to Determine Necessary Preheat Temperature in Steel Welding. *Q. J. Jpn. Weld. Soc.* **1995**, *13*, 347–357. [CrossRef]
36. Andrews, K.W. Empirical Formulae for the Calculation of Some Transformation Temperatures. *J. Iron Steel Inst.* **1965**, *203*, 721–727.
37. Boratto, F.; Barbosa, R.; Yue, S.; Jonas, J.J. Effect of Chemical Composition on Critical Temperatures of Microalloyed Steels. In Proceedings of the Iron and Steel Institute of Japan, THERMEC '88, Tokyo, Japan, 6–10 June 1988; pp. 383–390.
38. Buriková, K. Structural Nature of Multi-Phase Steels. Ph.D. Thesis, Institute of Materials Research-Slovak Academy of Sciences, Košice, Slovakia, 2009.
39. Evin, E. *Formability of Dual Phase Steels*; Habilitation Work; Technical University of Košice-Faculty of Mechanical Engineering: Košice, Slovakia, 1996.
40. Hrivňák, A.; Evin, E. *Formability of Steels*, 1st ed.; Elfa: Košice, Slovakia, 2004; ISBN 80-89066-93-3.
41. Tomáš, M.; Hudák, J. Material formability of steel sheets for automotive industry. In Proceedings of the International Scientific Conference of Progressive Technologies and Materials, Bezmiechowa, Poland, 6–8 July 2009; Politechnika Rzeszowska: Rzeszów, Poland, 2009; pp. 299–302.

![metals logo] *metals*

MDPI

Article

Microstructural, Mechanical, Texture and Residual Stress Characterizations of X52 Pipeline Steel

Olivier Lavigne [1], Andrei Kotousov [1,*] and Vladimir Luzin [2]

[1] School of Mechanical Engineering, The University of Adelaide, Adelaide, SA 5005, Australia; olivier.lavigne@adelaide.edu.au

[2] Australian Nuclear Science and Technology Organization, Lucas Heights, NSW 2234, Australia; Vladimir.Luzin@ansto.gov.au

* Correspondence: andrei.kotousov@adelaide.edu.au; Tel.: +61-8-8313-5439

Received: 13 June 2017; Accepted: 2 August 2017; Published: 9 August 2017

Abstract: In this paper, the microstructural and mechanical properties of a high-strength low-alloy (HSLA) API 5L X52 steel, which is widely utilized in the construction of gas pipelines, were characterized with optical microscopy, electron backscatter diffraction, and standard mechanical tests. The outcomes of these characterizations were used to evaluate the strengthening contributions of the solid solution, grain size, dislocations, and precipitates to the overall strength of the steel. In addition, texture and residual stresses were determined with neutron diffraction. The residual stresses were found to be low in comparison with the expected stresses due to the operating pressure. However, these stresses could contribute to the initiation and propagation of stress corrosion cracking at the outer surface of the pipe. Neutron diffraction results also suggested that the outer surface of the pipe had a texture that is expected to have a low resistance to high pH stress corrosion cracking. Both conclusions were found to be consistent with field observations.

Keywords: HSLA steel; API X52; gas pipeline; microstructure; neutron diffraction; strength; stress corrosion cracking

1. Introduction

The strong environmental incentives for the use of natural gas as a source of clean energy has boosted the mass production of high-strength low-alloy (HSLA) steels over the past decades. These steels are now widely utilized in the construction of long-distance, high-pressure gas pipelines and other pressure equipment worldwide. A great deal of research has also been directed to obtaining an excellent combination of mechanical properties in HSLA through alloy design as well as optimizing thermo-mechanical controlled processing (TMCP) parameters and corresponding microstructures. TMCP promotes the formation of an acicular ferrite (AF)-based microstructure, which is the preferred microstructure for pipe steels. Steels with AF microstructure normally possess higher strength and toughness, as well as superior stress corrosion and fatigue resistance than steels with ferrite and pearlite (P) microstructure [1,2].

API X52 steel is used widely as pipeline material in Australia and other countries. For example, it is the most-used material in the existing European gas pipeline network. The importance of the characterization of the material properties and susceptibility to various failure modes is progressively increasing with the prospective plans of the EU to utilize the existing pipeline network for the transportation of natural gas and hydrogen mixtures [3,4]. Therefore, the investigation of the microstructure and mechanical properties of pipeline steels is an important aspect required for the understanding of failure mechanisms including plastic collapse, fracture, and stress corrosion cracking (SCC) [5,6]. All of these failure modes represent a significant threat to the integrity of pipelines and pressure equipment.

In the present study, microstructural and mechanical characterizations of HSLA API 5L X52 steel were conducted to assess its mechanical properties. Special attention was given to a theoretical model for the prediction of yield strength (as an indication of the susceptibility of pipes to plastic collapse failure) from the steel composition, microstructure, and measured hardness data.

The API 5L X52 pipes are normally formed by spiral welding technique, which inevitably leads to the generation of residual stresses. Although the residual stresses do not affect the plastic collapse conditions, their presence could significantly influence the susceptibility of the pipe to stress corrosion or the initiation of fatigue cracks. With the help of neutron diffraction, the texture and the field of residual stresses were determined to assess the susceptibility of the API 5L X52 to these failure mechanisms. The outcomes of this study can contribute to the understanding of the stress corrosion resistance, strength, and remaining life of the existing pipeline network made of X52 steel, which is expected to be in service for the foreseeable future.

2. Materials and Methods

2.1. Material

API 5L X52M PSL2 steel was obtained from a pipe section with a wall thickness of 12.7 mm and an outer diameter of 508 mm. Table 1 shows the chemical composition of the steel.

Table 1. Measured chemical composition (wt %) for the X52 pipeline steel.

Element	C	P	Mn	S	Cu	Ni	Cr	Mo	Al	Ti	V	Si	Nb	Fe
wt %	0.075	0.014	0.95	0.003	0.01	0.01	0.015	Tr.	0.021	0.014	0.0015	0.226	0.013	Bal.

2.2. Microstructural Analysis

2.2.1. Optical Analysis

A semi-automatic Tegramin polishing machine (Struers, Ballerup, Denmark) was used for polishing the sample surface (cross-section of the pipe) down to 1 μm diamond paste. The samples were then etched with 2% Nital solution, and the revealed microstructure was inspected with an Axio Imager optical microscope (Zeiss, Oberkochen, Germany).

2.2.2. Electron Back-Scatter Diffraction (EBSD) Measurements

For EBSD measurements, final polishing was achieved using a porous neoprene disc with a colloidal silica suspension (0.04 μm). The EBSD scans were collected using a FEI Helios Nanolab 600-SEM (Thermo Fisher Scientific, Waltham, MA, USA) equipped with an EBSD detector (EDAX Hikari™, AMETEK, Berwyn, IL, USA). The acceleration voltage and the electron beam current of the SEM were 20 kV and 2.7 nA, respectively. The step size was 1 μm with a hexagonal scan grid (scans were 400 × 400 μm²). EDAX OIM™ Data Analysis 5.2 software (AMETEK, Berwyn, IL, USA) was used for the data collection and analyses.

2.2.3. Hardness Measurements

The bulk hardness of the steel was determined using a Vickers indenter with a load of 10,000 gf. The micro-Vickers hardness measurements were conducted using a LM700AT (Leco, St. Joseph, MO, USA) at different loads from 10 gf to 1000 gf [7] in order to determine the dislocation density in the steel, by following the method exposed in [7,8].

2.2.4. Residual Stress Measurements

The residual stresses profile was measured with neutron diffraction (ND) in the pipe through wall thickness. The measurements were performed on a strain scanning diffractometer (KOWARI,

ACNS, ANSTO, Lucas Heights, Australia) [9]. A monochromatic beam with λ = 1.67 Å from Si{400} monochromator reflection was used in this analysis. This choice of wavelength resulted in a scattering angle of 90° of the sample Fe(211) reflection. A nominal gauge volume of $0.3 \times 0.3 \times 8 \text{ mm}^3$ was used to measure two principal stress directions (normal and hoop) and to reconstruct the hoop stress component (under an assumption of zero normal stress). Measurement steps were 0.4 mm through thickness and started at 0.3 mm from the outer surface. The sample was a plate of $100 \times 100 \text{ mm}^2$ dimension cut from the pipe section.

2.2.5. Texture Measurements

The KOWARI diffractometer was also used to perform texture measurements in the pipe through wall thickness (at seven locations). Samples were prepared by cutting a slice at certain depths measured from the outer surface of the pipes of about 0.5 mm thick and 6 mm wide. The slice was then cut into small coupons and coupons were glued together to form a cube of about 6 mm side. The orientation of each coupon was preserved when cutting and gluing them together. A monochromatic beam with λ = 1.676 Å was used to measure three pole figures (110), (200), and (211) on a grid close to $5 \times 5°$, and orientation distribution functions (ODF) at $\phi_2 = 45°$ were plotted from these three pole figures [5].

3. Results

3.1. Microstructural Analysis

3.1.1. Optical Analysis

Figure 1 shows that the microstructure of the steel consisted of ferrite and ~4.7% of pearlite (determined with ImageJ analysis software (v1.48, open source)).

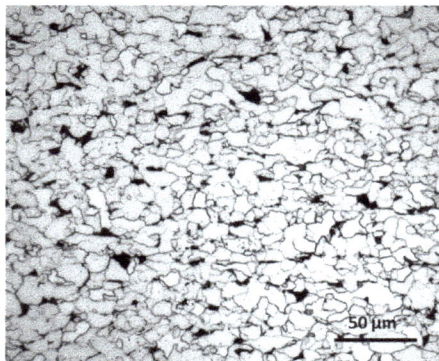

Figure 1. Microstructure of X52, typically consisting of ferrite (in white) and pearlite (in black).

3.1.2. EBSD Measurements

The EBSD measurements were performed at 0.8 mm from the pipe outer surface, and are presented in Figure 2. The inverse pole figure (IPF) showing the orientation of the grains that constitute the steel is presented in Figure 2a. The average grain size diameter from this scan was evaluated as 9.35 μm with a standard deviation of 3.19 μm, and the average grain size of ferrite determined by the linear intercept method with random test lines drawn on the scan was measured as 5.53 μm. The grain boundaries character is shown in Figure 2b. The fractions of low angle grain boundaries (LABs, between 2° and 15°) and high angle boundaries (HABs, between 15° and 180°) were measured as, respectively, 0.08 and 0.92. The average grain boundary misorientation angle was measured as 39.1°.

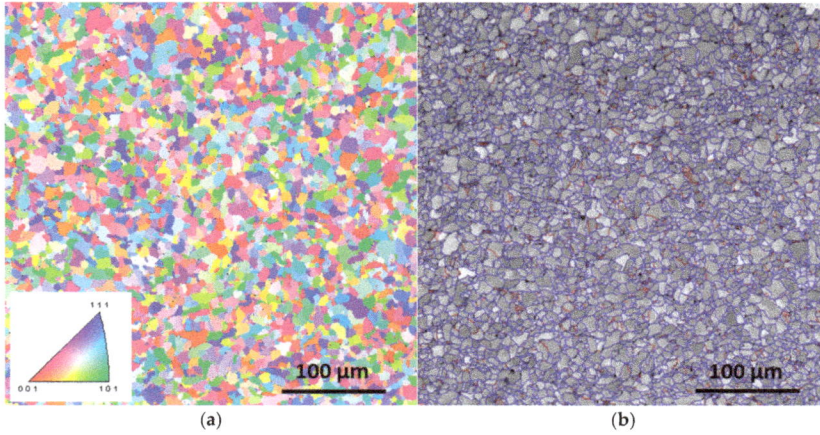

Figure 2. Electron back-scatter diffraction (EBSD) scan: (**a**) Qualitative view of the texture (001 inverse pole figure (IPF) map); (**b**) grain boundary orientations.

3.2. Mechanical Properties

3.2.1. Tensile Test Results

The mechanical properties of the pipe steel provided by the manufacturer are presented in Table 2. The specified minimum yield strength (SMYS) for the API 5L X52M PSL2 is 52 ksi or 359 MPa.

Table 2. Tensile tests results (provided by the manufacturer).

Orientation	Yield Strength (MPa)	Tensile Strength (MPa)	Elongation (%)
Longitudinal	400	500	32
Transverse	395	490	42

3.2.2. Hardness Measurements and Determination of Dislocation Density

The dislocation density can be estimated from indentation measurements for materials displaying indentation size effect (ISE) [7,8]. The hardness value is related to the indentation depth by the equation [10]:

$$\left(\frac{H}{H_0}\right) = 1 + h^*\left(\frac{1}{h}\right),\tag{1}$$

where H_0 is the hardness in the limitation of infinite depth (bulk hardness, determined in this work as 153 HV10), h^* is a characteristic length, and H is the hardness value corresponding to indentation depth h. By fitting Equation (1) to the experimental hardness values (see the regression curve in Figure 3), a value of h^* ~2086 nm could be obtained. h^* and the dislocation density statistically stored in the lattice, ρ, are related by the equation [7,8,10]:

$$\rho = \frac{3}{2}\frac{1}{f^3}\frac{\tan^2\theta}{bh^*},\tag{2}$$

where θ is the angle between the surface of the material and the surface of the indenter (22°) and f is a correction factor (=1.9) for the size of the plastic zone [7,8]. By introducing these values in Equation (2), ρ was approximately estimated as 6.9×10^9 cm^{-2}.

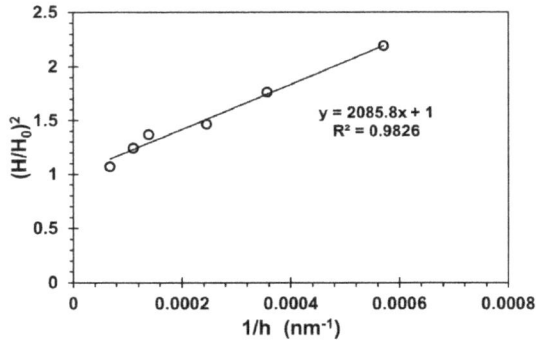

Figure 3. Indentation size effect and fitting of Equation (1) for the X52 steel.

3.3. Neutron Diffraction Measurements

3.3.1. Residual Stress Measurements

Figure 4a shows the profile of the residual stresses in the tangential (or hoop) direction measured by ND for the X52 steel, mainly resulting from the steel strip production. Very weak stress distribution in the through-wall-thickness direction of the pipe was recorded, in the range of ±20 MPa with uncertainties ±10 MPa on the stress values. Although statistical oscillations are pronounced with these error bars, the overall stress distribution has features most likely originating from the production process: the V-shape in the middle is typical for hot deep rolling, while sharp drop to compression close to surfaces is associated with colder temperature surface treatment (rolling). An additional stress component (linear bending stress distribution) existed in the as-built pipe that was removed by sample cutting. In the uncut pipes, the tensile stress was reconstructed through accurate measurement of the pipe opening after making the hoop cut. It was estimated to be 41 MPa higher on the outer surface and 42 MPa lower on the inner surface. Figure 4b shows the total residual stress obtained by the combination of the steel strip production and bending pipe process. The residual stresses near the outer surface of the pipe were thus tensile, and had a maximum value of approximately 5.3% of the YS or 5.8% of the SMYS.

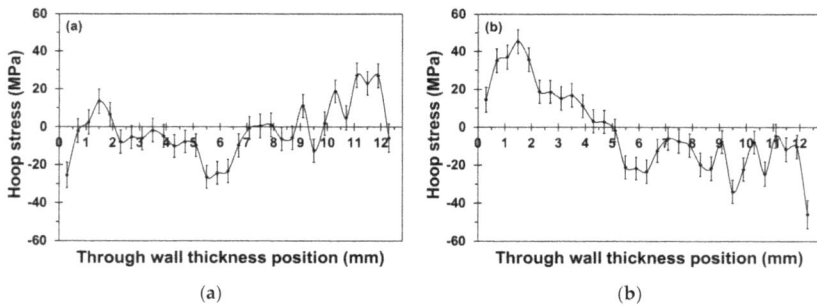

Figure 4. Residual stresses profiles in the hoop direction for the X52 steel: (**a**) Measured for the sample cut out of the pipe ring; (**b**) Plotted including the bending pipe linear distribution stress.

3.3.2. Texture Measurements

The orientation distribution functions obtained from the three pole figures (110), (200), and (211) allows the quantitative description of the texture of the crystalline phase. In body-centred cubic steels,

ODFs at $\phi_2 = 45°$ section displays the major texture components [5,11]. These ODFs are presented in Figure 5 at seven locations in the pipe through-wall (Figure 5a–g) along with the color scale bar showing the orientations intensity and the ideal locations of the main texture components on the ODF section (schematically shown in Figure 5h). Although ODF profiles were smoothed by the ODF harmonic reconstruction method, an excellent grain statistic was provided in the neutron texture experiment (10 to 100 million grains considered for each measurement). It can be seen that the texture of the steel presents a layer-by-layer characteristic in the through-wall direction, directly linked to the rolling process of the slab. A relatively weak texture was found at the outer surface of the pipe while a typical texture of a rolled strip was found in the central line of the pipe (i.e., along the α- and γ-fibers).

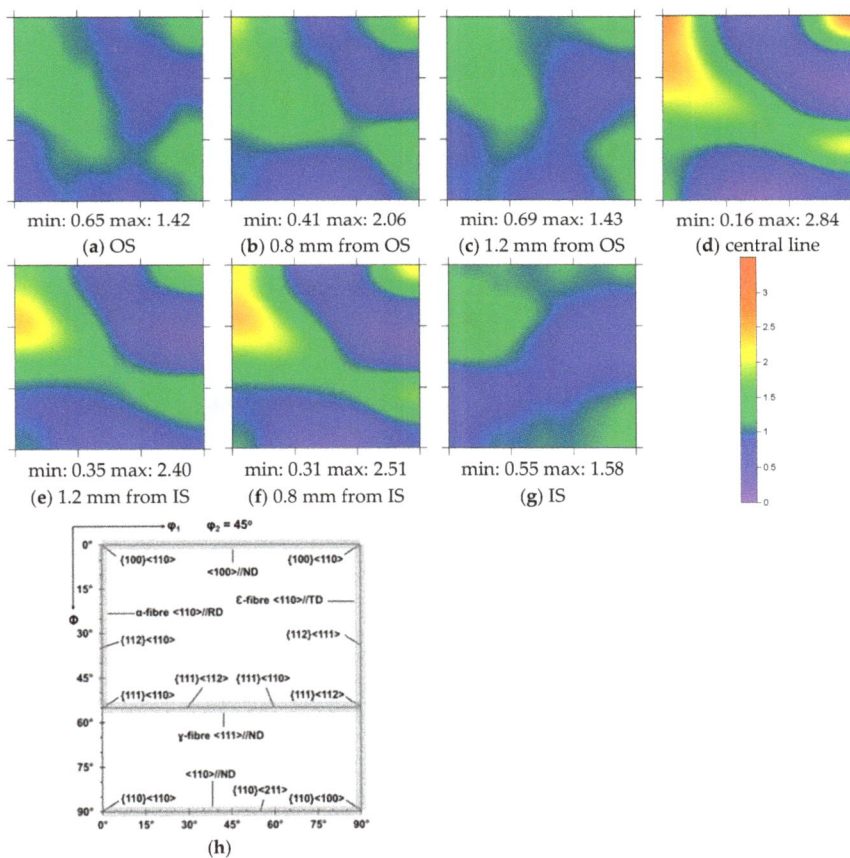

Figure 5. ODFs at $\phi_2 = 45°$ section: (**a**) At the outer surface (OS); (**b**) At 0.8 mm from OS; (**c**) At 1.2 mm from OS; (**d**) At the pipe central line; (**e**) At 1.2 mm from inner surface (IS); (**f**) At 0.8 mm from IS; (**g**) At the inner surface; (**h**) Schematic representation of the major texture components (where ND, TD, and RD are, respectively, the normal, transverse, and rolling directions).

4. Discussion

4.1. Strengthening Mechanisms

In accordance with the structure-based strength calculation model [12,13], the yield strength, σ_Y, of ferritic steels can be factorized into a number of intrinsic components:

$$\sigma_Y = \sigma_0 + \sigma_{SS} + \sigma_{GS} + \sigma_{dis} + \sigma_{ppt} \text{ (MPa)}, \tag{3}$$

where σ_0 is the intrinsic lattice friction stress (=54 MPa for pure iron single crystal [12,14]), σ_{SS}, σ_{GS}, σ_{dis}, and σ_{ppt} are contributions from, respectively, solid solution, grain size, dislocation, and precipitation strengthening.

4.1.1. Solid Solution Strengthening

The solid solution strengthening can be obtained by considering the effect of only Si and Mn atoms, using the equation proposed by Pickering [14]:

$$\sigma_{SS} = 83.2 \times (\text{mass\% Si}) + 32.3 \times (\text{mass\% Mn}) \text{ (MPa)} \tag{4}$$

Using the values presented in Table 1, σ_{SS} was thus determined as approximately 49 MPa. The effect of carbon was not considered, since its solubility in ferrite is negligible [15]. Carbon is expected to be found essentially in the pearlite phase (Figure 1) and as NbC precipitates [16]. However, the fraction of pearlite measured here (Section 3.1.1) is not expected to have any effect on the yield strength [13,17], and the effect of the NbC precipitates on the yield strength is exposed hereafter.

4.1.2. Grain Size Strengthening

The grain size strengthening in ferrite can be expressed from the Hall-Petch equation as [13,16]:

$$\sigma_{GS} = 15.1 d^{-1/2} \text{ (MPa)}, \tag{5}$$

where d is the average grain size of ferrite (in mm) determined by the linear intercept method with random test lines in the EBSD analysis ($\approx 5.53 \times 10^{-3}$ mm, see Section 3.1.2). Consequently, the value of σ_{GS} was estimated as approximately 203 MPa.

4.1.3. Dislocation Strengthening

The strengthening contribution due to dislocations can be calculated using the following relation [13,18]:

$$\sigma_{dis} = 0.38 G b \rho^{-1/2} \text{ (MPa)}, \tag{6}$$

where G is the shear modulus of the ferrite (81.6 GPa [12]), b is the Burgers vector of the dislocations (0.248 nm [12]), and $\rho = 6.9 \times 10^9$ cm^{-2} (see Section 3.2.2). σ_{dis} was therefore determined as approximately 64 MPa.

4.1.4. Precipitation Strengthening

The degree of strengthening due to the precipitates is dependent on both their fraction and size in the microstructure, and can be expressed by the following simplified relation [13]:

$$\sigma_{ppt} = B \text{ (\%solute) (MPa)}, \tag{7}$$

where the values of B are presented in Table 3 [13].

Table 3. Coefficient (*B*) of precipitation strengthening, from [13].

Solute and Precipitate	B_{max} (MPa/wt %)	B_{ave} (MPa/wt %)	Solute Concentration (wt %)
V as V_4C_3	1000	500	0–0.15
V as VN	3000	1500	0–0.06
Nb as Nb(CN)	3000	1500	0–0.05
Ti as TiC	3000	1500	0.03–0.18

The concentration of Ti and V being negligible in this X52 steel (Table 1), σ_{ppt} was determined as approximately 19 MPa by considering the average value of *B* for Nb as NbC precipitates.

Adding up all the intrinsic contributions in the analytical expression of the yield strength (Equation (3)) results in an overall calculated strength of approximately 389 MPa, which correlates well with the experimental value from the tensile tests (395 MPa).

Similarly, the hardness depends on the average ferrite grain size. This dependence has been shown to follow a Hall-Petch relation for a ferrite/cementite steel [19], and has been determined by regression analysis as:

$$HV = 56.55 + 214.19d^{-1/2}, \tag{8}$$

where *d* is expressed in µm. With *d* = 5.53 µm, *HV* is analytically determined as 148, which is also in good agreement with the measured bulk hardness value of 153 HV10.

It can be noted that the yield and the tensile strengths are empirically related to the hardness by the following equations (derived for an X65 steel) [20]:

$$\sigma_Y = 2HV + 105, \tag{9a}$$

$$TS = 1.3HV + 344. \tag{9b}$$

With *HV* = 153, σ_Y and *TS* are estimated as 411 and 543 MPa, respectively, which slightly overestimates the measured values (Table 2) of 4% and 11%, respectively.

4.2. Microstructure and Stresses Implications in Stress Corrosion Cracking Susceptibility

HSLA X65 steels were found to have an increased resistance to high pH SCC when a high intensity of texture along the <110>//ND was measured [5,11]. At the outer surface of the considered X52 steel, the measured texture is weak and does not present these orientations, suggesting that this steel is normally susceptible to high pH SCC. Moreover, a high fraction of HABs is also measured, which can promote high pH SCC to initiate and propagate [5]. The microstructure of this steel is thus typically susceptible to high pH SCC [5,11,21,22].

Pipelines are operated to a maximum pressure level that is a percentage of the specified minimum yield strength. For Class 1 locations, for example (offshore, or location unit that has 10 or fewer buildings intended for human occupancy), the maximum allowable operating pressure (MAOP) is 80% of SMYS [22]. However, the areas of pipelines presenting greatest risk of high pH SCC have been defined for MAOP >60% of SMYS [23]. For high pH SCC to occur, the stress or the stress intensity factor must be above a threshold stress, which is reduced by small amplitude and low frequency stress fluctuations superimposed on the mean stress [22]. It has been shown that the threshold stress for an X52 steel decreased to 40% of the SMYS when stress fluctuations of 15% of the mean stress were applied every 12 days [24]. Overall, this study determined the threshold stress values to comprise between 40% and 110% of the SMYS, depending on the amplitude and frequency of the stress fluctuations. Pitting and rougher surfaces can also contribute to the decrease of the threshold stress values [25].

It can therefore be understood that the pipeline considered in this study presenting a susceptible microstructure, operating as level of stress as low as 35% of the SMYS, and presenting additional levels of residual stress at the outer surface (~6% of SMYS) is theoretically susceptible to the development of high pH SCC. This is consistent with observed on-site high pH SCC failure of pipeline made of X52 steel [21].

5. Conclusions

This paper presents the outcomes of micro- and macro-characterizations of API X52 steel, which is one of the most common structural materials of the existing pipeline network in Europe and globally. As outlined in the previous section, the structure-based strength calculation model provides a simple way to effectively evaluate the yield strength of X52 steels through the evaluation of individual strengthening contributions. It is demonstrated that this model can adequately predict the yield strength of steels with ferritic/pearlitic microstructure. However, the theoretical calculations would probably require some adjustment in order to apply to higher strength steels such as X70, the microstructure of which usually incorporates smaller grain sizes and different grain shapes corresponding to acicular ferrite/bainite micro-constituents.

The present characterizations also indicated the potential contribution of the microstructure, relatively weak texture, and residual stresses due to manufacturing to high pH stress corrosion cracking susceptibility. This suggestion is consistent with field observations of failed X52 pipes, and more generally with failed pipes made of higher strength API steels such as X65 [5]. Therefore, the evaluation of the remaining life of the existing X52 pipeline network and its utilization for alternative gas mixture transportation or storage should consider the possible damage accumulation due to stress corrosion cracking as a result of the past operation.

Acknowledgments: This work was funded by the Energy Pipelines CRC, supported through the Australian Government's Cooperative Research Centres Program. The funding and in-kind support from the APIA RSC is gratefully acknowledged. The authors acknowledge the facilities, and the scientific and technical assistance of the Australian Microscopy & Microanalysis Research Facility at the University of Adelaide as well as Michael Law (ANSTO) for the sample preparation for the texture measurements. The authors would also like to thank Geoff Callar (APA Group) and Erwin Gamboa (The University of Adelaide) for discussion.

Author Contributions: Olivier Lavigne performed the OM, EBSD, and indentations experiments; Vladimir Luzin performed the ND experiments; Olivier Lavigne, Vladimir Luzin and Andrei Kotousov analyzed the data; Olivier Lavigne and Andrei Kotousov wrote the paper.

Conflicts of Interest: The authors declare no conflict of interest.

References

1. Costin, W.L.; Lavigne, O.; Kotousov, A. A study on the relationship between microstructure and mechanical properties of acicular ferrite and upper bainite. *Mater. Sci. Eng. A* **2016**, *663*, 193–203. [CrossRef]
2. Costin, W.L.; Lavigne, O.; Kotousov, A.; Ghomashchi, R.; Linton, V. Investigation of hydrogen assisted cracking in acicular ferrite using site-specific micro-fracture tests. *Mater. Sci. Eng. A* **2016**, *651*, 859–868. [CrossRef]
3. Fernandes, T.R.C.; Chen, F.; Da Graça Carvalho, M. "HySociety" in support of European hydrogen projects and EC policy. *Int. J. Hydrog. Energy* **2005**, *30*, 239–245. [CrossRef]
4. Mulder, G.; Hetland, J.; Lenaers, G. Towards a sustainable hydrogen economy: Hydrogen pathways and infrastructure. *Int. J. Hydrogen Energy* **2007**, *32*, 1324–1331. [CrossRef]
5. Lavigne, O.; Gamboa, E.; Costin, W.; Law, M.; Luzin, V.; Linton, V. Microstructural and mechanical factors influencing high pH stress corrosion cracking susceptibility of low carbon line pipe steel. *Eng. Fail. Anal.* **2014**, *45*, 283–291. [CrossRef]
6. Carretero Olalla, V.; Bliznuk, V.; Sanchez, N.; Thibaux, P.; Kestens, L.A.I.; Petrov, R.H. Analysis of the strengthening mechanisms in pipeline steels as a function of the hot rolling parameters. *Mater. Sci. Eng. A* **2014**, *604*, 46–56. [CrossRef]
7. Faraji, G.; Mashhadi, M.M.; Bushroa, A.R.; Babaei, A. TEM analysis and determination of dislocation densities in nanostructured copper tube produced via parallel tubular channel angular pressing process. *Mater. Sci. Eng. A* **2013**, *563*, 193–198. [CrossRef]

8. Graça, S.; Colaço, R.; Carvalho, P.A.; Vilar, R. Determination of dislocation density from hardness measurements in metals. *Mater. Lett.* **2008**, *62*, 3812–3814. [CrossRef]
9. Alipooramirabad, H.; Paradowska, A.M.; Ghomashchi, R.; Kotousov, A.; Hoye, N. Prediction of welding stresses in WIC test and its application in pipelines. *Mater. Sci. Technol.* **2016**, *32*, 1462–1470. [CrossRef]
10. Nix, W.D.; Gao, H. Indentation size effects in crystalline materials: A law for strain gradient plasticity. *J. Mech. Phys. Solids* **1998**, *46*, 411–425. [CrossRef]
11. Arafin, M.A.; Szpunar, J.A. A new understanding of intergranular stress corrosion cracking resistance of pipeline steel through grain boundary character and crystallographic texture studies. *Corros. Sci.* **2009**, *51*, 119–128. [CrossRef]
12. Kamikawa, N.; Sato, K.; Miyamoto, G.; Murayama, M.; Sekido, N.; Tsuzaki, K.; Furuhara, T. Stress-strain behavior of ferrite and bainite with nano-precipitation in low carbon steels. *Acta Mater.* **2015**, *83*, 383–396. [CrossRef]
13. Liu, G. Designing with Carbon-, Low-, and Medium-alloy Steels. In *Handbook of Mechanical Alloy Design*; Totten, G.E., Xie, L., Funatani, K., Eds.; Taylor & Francis Inc., Marcel Dekker Inc.: New York, NY, USA, 2004; pp. 73–89.
14. Pickering, F.B. *Physical Metallurgy and the Design of Steels*; Applied Science Publishers: London, UK, 1978; p. 63.
15. Bhadeshia, H.K.D.H. *Models for the Elementary Mechanical Properties of Steel Welds*; Institute of Materials: London, UK, 1997; pp. 229–284.
16. Altuna, M.A.; Iza-Mendia, A.; Gutierrez, I. Precipitation strengthening produced by the formation in ferrite of Nb carbides. In Proceedings of the 3rd International Conference on Thermomechanical Processing of Steels, Padova, Italia, 10–12 September 2008.
17. Gladshtein, L.I.; Larionova, N.P.; Belyaev, B.F. Effect of ferrite-pearlite microstructure on structural steel properties. *Metallurgist* **2012**, *56*, 579–590. [CrossRef]
18. Yang, J.R.; Bhadeshia, H.K.D.H. The dislocation density of acicular ferrite in steel welds. *Weld. Res. Suppl.* **1990**, *69*, 305s–307s.
19. Zhao, M.C.; Hanamura, T.; Qiu, H.; Nagai, K.; Yang, K. Grain growth and Hall-Petch relation in dual-sized ferrite/cementite steel with nano-sized cementite particles in a heterogeneous and dense distribution. *Scr. Mater.* **2006**, *54*, 1193–1197. [CrossRef]
20. Hashemi, S.H. Strength–hardness statistical correlation in API X65 steel. *Mater. Sci. Eng. A* **2011**, *528*, 1648–1655. [CrossRef]
21. Saleem, B.; Ahmed, F.; Rafiq, M.A.; Ajmal, M.; Ali, L. Stress corrosion failure of an X52 grade gas pipeline. *Eng. Fail. Anal.* **2014**, *46*, 157–165. [CrossRef]
22. Zheng, W.; Elboujdaini, M.; Revie, R.W. Stress corrosion cracking in pipelines. In *Stress Corrosion Cracking*; Raja, V.S., Shoji, T., Eds.; Woodhead Publishing Ltd.: Cambridge, UK, 2011; pp. 749–771.
23. Standard Practice. *Stress Corrosion Cracking (SCC) Direct Assessment Methodology*; NACE International: Houston, TX, USA, 2008; ISBN 1-57590-191-9.
24. Fessler, R.R. Combination of conditions causes stress-corrosion cracking. *Oil Gas J.* **1976**, *74*, 81–83.
25. Wells, D.B. SCC threshold stress in line pipe steels. In Proceedings of the 8th Symposium on Line Pipe Research, Houston, TX, USA, 26–29 September 1993.

metals

MDPI

Article

Precipitation and Grain Size Effects on the Tensile Strain-Hardening Exponents of an API X80 Steel Pipe after High-Frequency Hot-Induction Bending

Rafael A. Silva [1,*], André L. Pinto [2], Alexei Kuznetsov [3] and Ivani S. Bott [1]

[1] Departamento de Engenharia Química e de Materiais—PUC-Rio—DEQM, Pontifícia Universidade Católica do Rio de Janeiro, 222541-900 Rio de Janeiro-RJ, Brazil; bott@puc-rio.br
[2] Centro Brasileiro de Pesquisas Físicas (CBPF), 22290-180 Rio de Janeiro-RJ, Brazil; pinto@cbpf.br
[3] Instituto Nacional de Metrologia, Qualidade e Tecnologia—INMETRO/RJ, 25250-020 Rio de Janeiro-RJ, Brazil; okuznetsov@inmetro.gov.br
* Correspondence: rafael.engmet@gmail.com; Tel.: +55-16-98117-9981

Received: 30 November 2017; Accepted: 24 February 2018; Published: 9 March 2018

Abstract: This study discusses the use of the Morrison model to estimate the strain-hardening exponent (n) in the presence of precipitation hardening for an API X80 steel pipe. As the grain size becomes larger, high values of n are expected according to the Morrison equation. However, the grain size alone is not sufficient to explain the changes of the strain-hardening exponent (n) after hot-induction bending. The vanadium in the ferritic solid solution has an important influence on the decrease of the precipitation hardening, and consequently on the increase of the values of n, despite the refinement of the grain size and high dislocation densities. Therefore, the effects of grain boundaries on the capability to uniformly distribute deformations within the plastic regime become negligible, which limits the application of the Morrison model to estimate the values of n.

Keywords: API X80 steel; strain-hardening exponent; high-frequency hot-induction bending; thermal treatments

---

1. Introduction

The American Petroleum Institute (API) establishes standards for steels used in oil-and-gas pipelines, such as the API X80 grade steel. In pipeline construction, some steel pipes of the API class can be subjected to hot-induction bending, which is the usual process to obtain smaller radii of curvature and larger bending angles of up to 90°. The hot-induction bending changes the microstructure and, therefore, the mechanical properties of the hot-bent section of the pipe.

The Hollomon equation (Equation (1)) can in many cases be used to fit the data of the true stress-strain curves in the non-linear section between the yield strength (YS) and ultimate tensile strength (UTS). In this equation, K is the strength coefficient, n is the strain-hardening exponent, and σ and ε are the true stress and true strain, respectively. By definition, n is the slope of the straight line in Equation (2), where the pre-exponential, K, can be found by extrapolating to $\varepsilon = 1.0$ [1].

$$\sigma = K\varepsilon^n \qquad (1)$$

$$\ln \sigma = \ln K + n \ln \varepsilon \qquad (2)$$

The K parameter indicates the strength level of the material, and the exponent n indicates the capability to uniformly distribute the deformation. In other words, n evaluates the strain-hardening capability of the material. The values of n can be influenced by processing, test temperatures, as well as the strain rate that was used during the tensile tests. Typical values of n tend to be in the range of 0.15–0.18 for high-strength low-alloy (HSLA) steels and in the range of 0.20–0.23 for low-carbon steels. High-strength materials have lower values of *n* than low-strength materials [1] because the hardening mechanisms interact with the mobile dislocations, and, therefore, influence the values of *n*. If the value of *n* is low, the strain-hardening rate initially is high, but this rate decreases rapidly with further loading.

The effect of grain size (*d* in mm) on the values of *n* can be evaluated based on the Morrison model [2], as shown in Equation (3). The refinement of the grain size plays an important role in the mechanical properties of the steels—acting as obstacles for the movements of mobile dislocations and affecting the work-hardening mechanisms of steels.

$$n = 5/\left(10 + d^{-1/2}\right) \tag{3}$$

The formation of dislocation loops around precipitates is responsible for the high initial strain-hardening rate of age-hardened alloys. These loops will repel the subsequent dislocations. Therefore, the stresses that are required to keep the new mobile dislocations in movement increase as the distances between the precipitates decrease. The increase in strength can be affected by both a decrease in the size of the precipitates as well as an increase in the volume fraction.

In this study, we discuss the limitations of the Morrison model for estimating the values of *n* in the presence of precipitation hardening for an API X80 steel microalloyed with titanium, niobium, and vanadium. The motivation for this study comes from the effects of microstructural changes on the mechanical behavior of this HSLA steel under high-frequency hot-induction bending—which leads to a high level of grain refinement, but with a loss of mechanical properties, especially a loss of the YS. High contents of titanium and niobium can result in the formation of coarse precipitates during the solidification process or during the thermo-mechanical controlled processing (TMCP). Titanium and niobium precipitates are of very difficult solubilization during hot-induction bending. This process has a short austenitizing period (nearly 2 min) within the temperature range of 1060–950 °C, followed by external quenching. The temperature gradient is distributed through the wall thickness of the pipe. Within this scenario, the effect of hot-induction bending on the plasticity of the studied steel pipe and the impact of the solubilization of the vanadium precipitates on the mechanical properties are features of interest for the development of this study.

2. Experimental

2.1. Material

The chemical composition and Carbon Equivalent (Pcm) of the API X80 steel pipe are shown in Table 1. The as-received material (straight stretch of the pipe) was produced from a plate treated by TMCP without accelerated cooling and submitted to the UOE forming process (U for U-ing cold forming from the plate, O for O-ing cold forming from the U shape and, E for cold expansion to meet the geometric tolerances). The nominal wall thickness and diameter of the pipe under study are 19 mm and 508 mm, respectively.

Table 1. Chemical composition (wt %) of the API X80 steel pipe.

C	Mn	Si	P	S	Ni	Cr	Mo	Ti	Nb	V	N	Pcm
0.05	1.74	0.21	0.018	0.002	0.011	0.147	0.177	0.014	0.069	0.022	0.005	0.17

2.2. Heat Treatments

Table 2 describes the different routes applied to obtain the samples: TMCP + UOE processes (sample A), hot-induction bending (samples B and C), hot-induction bending followed by tempering (samples D and E), and normalizing heat treatment at 900 °C for 30 min followed by still-air cooling (sample F).

When the hot-bent section does not reach the expected YS, a tempering—that is often used for stress relief of the hot-bent pipe—can also be applied to harden the hot-bent section by precipitation of microalloying elements (at appropriate temperatures). The samples D and E represent this stage of the hot-bending process. All of the processing routes, especially the normalizing (sample F), were used to obtain the samples that allowed the relationship between the experimental values of n and the average microhardness values of the ferritic grains to be evaluated.

The hot-induction bending was performed with the following parameters: 105 kW, 1050 °C, 2500 Hz, and bending speed of 0.6 mm/s. The hot-bent section was externally cooled with water jets at 0.3 kg/cm^2. The regions of the hot-bent section tested were the *extrados* (external section of the curve arc) and the *intrados* (internal section of the curve arc).

The tempering of the hot-bent section consisted of heating cycles from 200 °C to the soaking plateau temperatures (600 and 650 °C), with a heating rate of 100 °C/h. In these tempering heat treatments, the soaking times were 1 h and the final cooling was performed in still air.

The Morrison model and a modified model were evaluated to estimate the values of n for these samples obtained by the different processing conditions described in Table 2.

Table 2. Identification of the samples obtained using different processing conditions.

Samples	Description	Condition
A	Straight stretch	As-received
B	Intrados	Hot-bent
C	Extrados	Hot-bent
D	Extrados tempered at 600 °C	Hot-bent + as-tempered
E	Extrados tempered at 650 °C	Hot-bent + as-tempered
F	Straight stretch normalized at 900 °C	As-normalized

2.3. Characterization

During the processing of the hot bending, the Joule effect, due to the induced currents, promotes temperature gradients, and therefore a microstructural gradient is formed within the pipe wall (in the hot-bent section) after the cooling stage (superficial quenching). The thickness of the layer affected by the induced currents only depends on the frequency that is applied, in this case, 2500 Hz, while the thickness of the quenched layer depends on the bending temperature, the hardenability of the steel, as well as the post-bending quenching features. Thus, all of the metallographic specimens, as well as the tensile test specimens, were sampled in the middle of the pipe wall. The transverse cross-sectional samples (20 × 20 × 19 mm^3) were cut from different sections of the hot-bent pipe for all of the metallographic analyses. After etching in 2% Nital solution, the samples were examined by optical microscopy (OM) in a Zeiss Axioskop microscope (Carl Zeiss Microscopy GmbH, Göttingen, Germany), and also by scanning electron microscopy (SEM) in a JSM-6510LV JEOL microscope (JEOL Ltd., Tokyo, Japan) operating at 20 kV. Grain sizes were measured in 15 randomly selected fields, using the linear intercept method standardized by the American Society for Testing and Materials (ASTM; West Conshohocken, PA, USA) in Standard E112-96 [3].

Thin-foil samples for transmission electron microscopy (TEM) were prepared from a very thin piece cut from the bulk. These pieces were initially ground to a thickness of 120 μm, and 3 mm diameter disks were punched out of these pieces. Further thinning was obtained by electro polishing in a Tenupol-5 apparatus (Struers ApS, Ballerup, Denmark) at approximately 20 V and 15 °C, using a solution of 95% acetic acid and 5% perchloric acid. The bright-field images were obtained with EM 2010 and EM 2100F microscopes (JEOL Ltd., Tokyo, Japan) operating at 200 kV. TEM analyses of the steel in hot-bent and tempered-and-bent conditions were performed for the *intrados* and for the *intrados* tempered at 650 °C, respectively. The objective was to show the interactions between the local precipitation and the dislocations.

The dislocation densities were obtained from the X-ray diffraction (XRD) patterns after the crystallite size effects on the XRD peaks were separated (and removed) from the microstrains (effects). The dislocation density for the steel in the as-received condition (sample A) had already been characterized by TEM in a previous work [4]. The conditions of the XRD measurements were: CoK_α radiation (λ = 1.7889 Å), Bragg angles (2θ) ranged from 30° to 135°, angular step of 0.02° and a count time of 5 s per step. The XRD measurements were performed with a D8-Focus diffractometer (Bruker AXS, Karlsruhe, Germany). The standard reference material (SRM) applied for the correction of instrumental broadening was a sample of Corundum SRM 1976 from NIST (National Institute of Standards and Technology, Gaithersburg, MD, USA). Before XRD analyses, the plates of $10 \times 10 \times 3$ mm^3 were submitted to traditional metallographic preparation and then electropolished at 15 V to remove the 50 μm superficial layers, using a solution (at 15 °C) of 90% acetic acid and 10% perchloric acid. Due to the anisotropic conditions of the samples, the modified Williamson-Hall methodology was used to obtain the dislocation densities [5,6].

The amounts of the microalloying elements in the ferritic solid solution were determined by ICP-OES techniques using the acidic dissolution method for the preparation of the samples. The tensile specimens were dissolved using a hydrochloric acid solution to extract insoluble carbonitrides of titanium, niobium, vanadium, and molybdenum from the steel [7–9]. The supernatants—obtained by centrifugation of the acidic solutions containing the dissolved samples—were analyzed via ICP-OES, and any traces of these elements that were in the ferritic solid solution were determined [10–12]. The ICP-OES analyses were performed in an Optima model 7300DV device (PerkinElmer Inc., Waltham, MA, USA) and the vanadium emission line at 292.402 nm was specifically used to determine any vanadium traces in the samples. Given that the chemical composition of the steel is known (Table 1) and is based on the principle of mass conservation (for these elements), the volume fraction of the nano-precipitates ($\leq$10 nm) could be estimated using the Equation (4) [9], where: f_v is the volume fraction of the nano-precipitates (total volume of nano-precipitates in 1 μm^3 of steel), ρ_{Fe} and ρ_{ppt} are the densities of iron (7.87 g/cm^3) and the precipitate, respectively, and $wt\%_{ppt}$ is the mass percentage of the nano-precipitates in the steel.

$$f_v = \left(1 \ \mu m^3 \times \rho_{Fe} \times wt\%_{ppt}\right)/\rho_{ppt} \tag{4}$$

2.4. Mechanical Tests

Tensile tests were carried out at room temperature using a universal testing machine EMIC model DL20000 (EMIC, São José dos Pinhais, Brazil). The test speed was 1 mm/min and the strain rate was 0.04 min^{-1}. The test speed remained constant over the strain interval when the values of n were determined, allowing accurate measurements of the loads and of the displacements during the tensile tests [13].

The preparations of the specimens and the tensile tests were performed in accordance with API 5 L standard [14] and ASTM A370 [15]. Three standard round tensile specimens from the reduced section (transverse sub-size specimens from API 5L) were used to characterize the tensile mechanical properties. The values of n were obtained from the stress-strain curves, according to the ASTM E646-07 standard [13]. In the plastic region of the stress-strain curves, seven points uniformly distributed between the YS and the UTS were used to calculate the values of n.

Finally, the Vickers microhardness values (HV) were taken from 100 g of the ferrite using a HMV-2/HMV-2T/SHIMADZU microdurometer (SHIMADZU Corporation, Kyoto, Japan), applying a load of 10 g for 15 s. The sizes of the individual grains of the ferrite were measured simultaneously with the microhardness measurements. Using samples prepared and etched by traditional metallography techniques and applying the same methodology as the measurement used to obtain the diagonals of the microindenters, the grains of the ferrite could be visualized and their sizes measured.

Hence, before taking the microhardness measurements, the individual size of each grain was measured. This procedure was repeated 100 times, where 100 g of the ferrite were randomly chosen and then measured. The resolution of the measurements was 0.01 μm, using a 40× objective lens. The average microhardness values of the samples were correlated with the average values of the experimental n to obtain the modified equations (see next section) that describe the behavior of n associated with the different processing routes that are evaluated in this work.

3. Results and Discussion

Initially, the effect of grain size refinement on n was evaluated based on the Morrison equation [2], and lower values of n for the samples with smaller grain sizes were expected [16–18]. However, despite the TMCP, the sample in the as-received condition revealed to have an average grain size larger than the steel in the as-bent condition, as can be seen in the micrographs obtained by OM and SEM in Figures 1 and 2. The microstructure of the steel shows a ferritic matrix with dispersed particles of the martensite-austenite (M-A) constituents and other microphases.

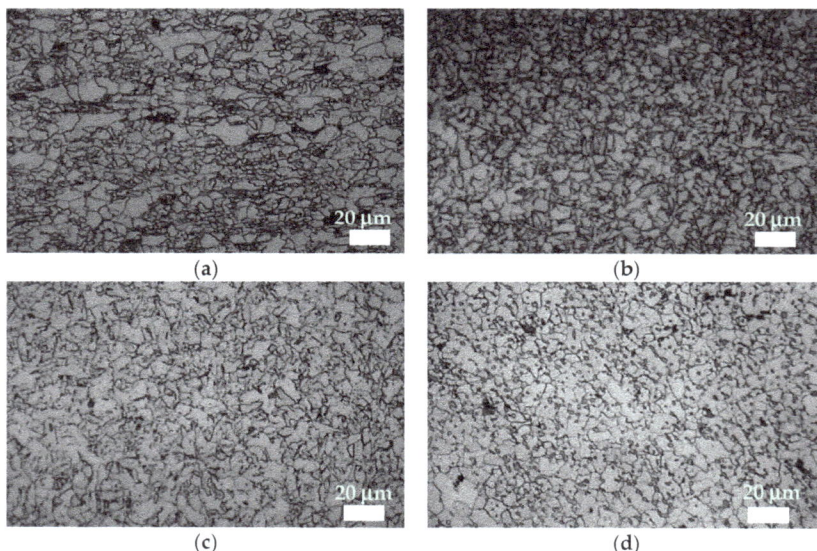

Figure 1. Optical micrographs from (**a**) straight stretch, (**b**) *extrados*, (**c**) *extrados* tempered at 600 °C and, (**d**) straight stretch normalized at 900 °C.

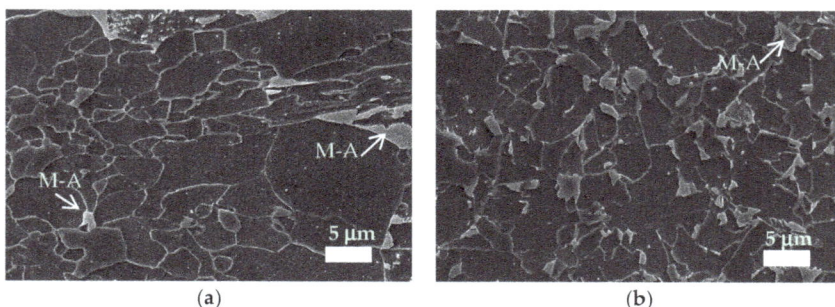

Figure 2. *Cont.*

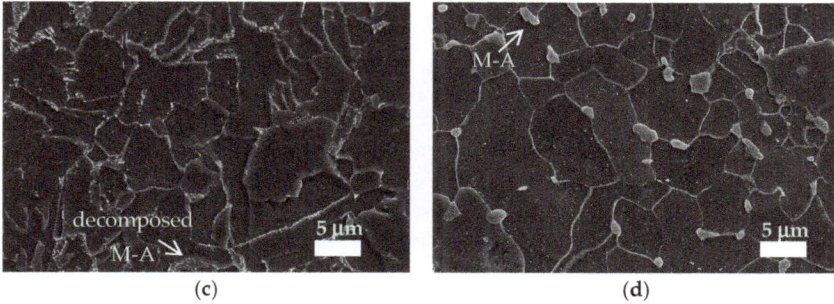

Figure 2. SEM micrographs from (**a**) straight stretch, (**b**) *extrados*, (**c**) *extrados* tempered at 600 °C, and (**d**) straight stretch normalized at 900 °C.

Figure 3 shows the engineering stress-strain curves of the samples that were evaluated in this study. The percentages of the microalloying elements from the chemical composition that precipitated are shown in Table 3. Table 4 summarizes the main properties of the samples and shows that the effect of the grain size alone was not sufficient to explain the changes in the values of *n* after the hot-induction bending.

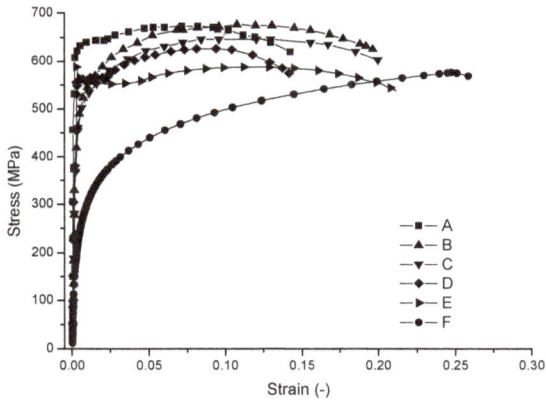

Figure 3. Engineering stress-strain curves.

Table 3. Effects of the hot-induction bending and tempering on the percentage of the alloying elements that precipitated.

Samples	V (%)	Cr (%)	Cu (%)	Mo (%)	Ti (%)	Nb (%)
A	41.7	35.0	66.1	83	100	99.9
B	25.9	28.1	66.8	82.7	100	99.9
C	12.7	19.2	65.2	80.4	99.9	99.8
D	29.5	22.3	66.3	81	100	99.8
E	30.5	21.9	67.5	81	100	99.9

A refined dispersion of precipitates can act as an obstacle for dislocations, and when this occurs the additional effect of the grain boundaries becomes negligible. The strain hardening of precipitation-hardened materials is initially very high and rapidly decreases, leading to low values of *n* [1]. However, the grain size, the precipitation, and the dislocation density—all being primary characteristics responsible for different strengthening mechanisms—are dependent on the processing conditions. These hardening mechanisms have a significant influence on the microhardness of the phases. The values of *n* were obtained from the stress-strain curves that are shown in Figure 3. Equation (5) was obtained from the correlation between the experimental values of *n* and the average microhardness of the ferritic grains. The microhardness values that were used to obtain this fitted curve (Figure 4) are related to the processing conditions. Equations (6)–(8) show that it is possible to replace the constants of the Morrison model with the microhardness dependent variables. These modified equations were able to satisfy the experimental results in this work. Therefore, the effects of the processing and also of all the hardening mechanisms were considered additionally to the effect of the refinement of the grain size (*d* in mm), despite the redundancy introduced by the grain size effects on the hardness of the ferrite phase.

$$n \; = \; -0.118 \, + \, 1.49 \text{Exp}(-0.0081 \, \times \, \text{HV} \, (10 \, \text{g})) \tag{5}$$

$$n \; = \; a\text{f}(\text{HV} \, (10 \, \text{g})) / \left\{ b\text{f}(\text{HV} \, (10 \, \text{g})) + d^{-1/2} \right\} \tag{6}$$

$$a \; = \; \{b\text{f}(\text{HV} \, (10 \, \text{g}))\}/2 \tag{7}$$

$$b \; = \; -1.316 \, + \, 214.46 \text{Exp}(-0.0156 \, \times \, \text{HV} \, (10 \, \text{g})) \tag{8}$$

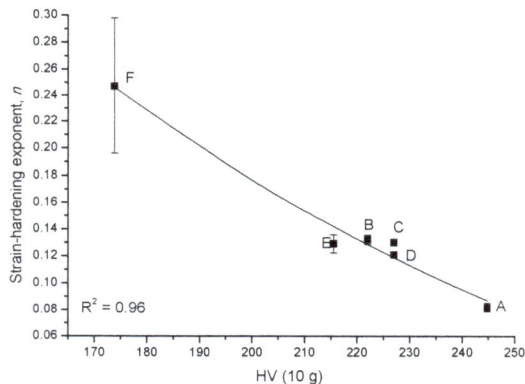

Figure 4. Relationship between experimental values of *n* and the average microhardness HV (10 g) of 100 ferritic grains. Where: A = straight stretch, B = *Intrados*, C = *Extrados*, D = *Extrados* tempered at 600 °C, E = *Extrados* tempered at 650 °C and F = straight stretch normalized at 900 °C.

Table 3 shows the influences of hot-induction bending and tempering on the percentage of the alloying elements that are precipitated. For example, for the steel in the as-received condition, 41.7% of the vanadium precipitated, while 58.3% of the vanadium is in ferritic solid solution. The percentage of precipitated vanadium decreases due to the hot-induction bending (to 25.9% in the *intrados* and to 12.7% in the *extrados*), increasing again after the post tempering (to approximately 30% in the tempered *extrados*). Thus, the tempering changes the amount of the vanadium in the ferritic solid solution of this steel in the hot-bent condition due to the precipitation process.

Table 3 shows that most of the titanium, as well as the niobium, remained in the precipitates, possibly as carbonitrides like in the as-received steel where the coarser precipitates were not dissolved during hot-bending. Amounts of copper and molybdenum remain similar for the steel in all conditions studied. The solubilization behaviors of the microalloying elements in the austenite and their precipitates are extensively reported in the literature [8,19,20]. The amount of vanadium that precipitates varies due to the hot-induction bending and also due to the tempering. This variation is reflected in the improvement of the YS after tempering of the steel in the hot-bent condition, as shown in Table 4, where the higher volume fractions of $VC_{0.75}$ are related to the higher values of YS.

Table 4. Tensile mechanical properties (YS and UTS), average microhardness (HV), average grain size (d), amount of vanadium in ferritic solid solution (V%), volume fraction (f_v) of $VC_{0.75}$, dislocation density (ρ), strain-hardening exponent (n) from the experimental data, the Morrison model and the modified model (Equation (5)) for the API X80 steel submitted to different processing conditions.

Samples	YS (MPa)	UTS (MPa)	HV (10 g)	d (μm)	V% (wt %)	f_v $VC_{0.75}$ (%)
A	604 ± 20	679 ± 9.0	245 ± 22	5.35 ± 0.9	0.013	0.0150
B	477 ± 10	673 ± 1.5	222 ± 19	3.90 ± 0.4	0.016	0.0093
C	472 ± 16	658 ± 8.6	227 ± 17	3.70 ± 0.6	0.019	0.0046
D	558 ± 1.7	628 ± 1.7	227 ± 14	3.20 ± 0.6	0.016	0.0106
E	550 ± 19	594 ± 2.1	215 ± 13	5.80 ± 0.7	0.015	0.0110
F	239 ± 9.5	573 ± 4.0	174 ± 14	5.71 ± 0.8	-	-

Samples	ρ (m^{-2})	n (Experimental)	n (Morrison)	Error (%)	n (Equation (5))	Error (%)
A	1.81×10^{14}	0.082 ± 0.00301	0.211	158	0.087	6.6
B	4.26×10^{14}	0.132 ± 0.00351	0.192	46	0.129	2.3
C	4.12×10^{14}	0.130 ± 0.00252	0.189	45	0.119	8.3
D	2.94×10^{14}	0.121 ± 0.00206	0.181	49	0.119	1.5
E	1.38×10^{14}	0.129 ± 0.00667	0.216	68	0.142	10
F	-	0.247 ± 0.051	0.215	13	0.247	0.1

The cooling rate that is induced by the superficial quenching was high enough to delay the formation of the precipitates, including the chromium, vanadium, titanium, molybdenum, and niobium precipitates. Additionally, the precipitation of molybdenum delays the precipitation of chromium, vanadium, titanium and niobium [21]. Thus, after the hot-induction bending, the ferrite became enriched in vanadium and chromium and the precipitation of these elements was induced by post tempering, improving the YS at the hot-bent section, as mentioned above.

Figure 5 shows how the Morrison model and the modified equations compare with each other, and shows that the experimental data combines closely and accurately with the modified equations that use an average microhardness value of the material to estimate the values of n and not only an average grain size, as applied by the Morrison model.

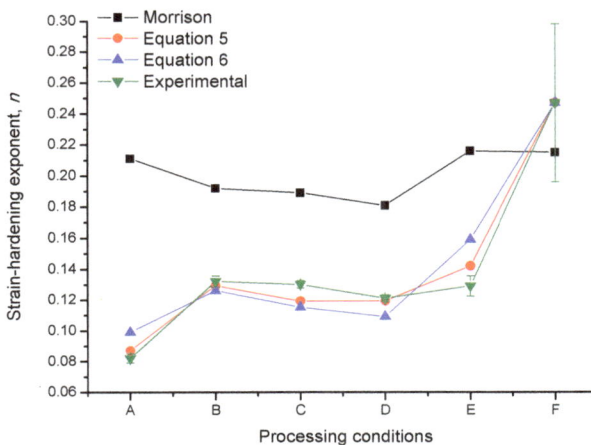

Figure 5. Strain-hardening exponent (n) from Morrison model, modified equations and experimental data. Where: A = straight stretch, B = *Intrados*, C = *Extrados*, D = *Extrados* tempered at 600 °C, E = *Extrados* tempered at 650 °C and F = straight stretch normalized at 900 °C.

The distribution of the *n* values for the ferrite grains as a function of their individual sizes is shown in Figure 6, where the values of *n* were obtained from Equation (5) using the individual values of the microhardness HV (10 g) of the ferritic grains. It is interesting to note that the values of *n* tend to increase with the growth of the grain size for all of the conditions under study. However, Figure 6 shows that within a range of the ferritic grain sizes from 10 to 15 μm, the steel in the as-received condition has ferritic grains with lower values of *n* than those from hot-bent section, and from the as-normalized steel. In other words, the values of *n* that were obtained for these samples (from the same steel) show the influence of the processing and therefore of the associated hardening mechanisms. Although this behavior can be expected, based on the current knowledge of physical metallurgy, experimental examples of this effect are not usually found in the literature. For example, small precipitates act as barriers for the movement of dislocations [1], leading to low values of *n* as those obtained for the steel in the as-received condition. When compared with the values of *n* characterized for the steel in hot-bent and as-normalized conditions, the steel in the as-received condition showed a lower capability to uniformly distribute the plastic deformation. This is because of the high volume fraction of small precipitates that hinder the movement of dislocations more rapidly. This precipitation causes an increase in the shear stresses, hindering the movement of dislocations that interact with the stress fields that surround the precipitates. Thus, for the steel studied here, the precipitation of the microalloying elements is a limiting factor for the Morrison model, as shown by the decreasing trend in the values of *n* with the intensification of the refined precipitation due to the tempering of *extrados* at 600 °C. For the hot-bent section, an increase in the volume fraction of $VC_{0.75}$ occurred at tempering temperatures in the range from 600 to 650 °C, however, at 650 °C, some grain growth was observed (Table 4). Thus, the influence of the tempering on the capability of the API X80 steel to uniformly distribute the plastic deformation depends on heat-treatment temperature.

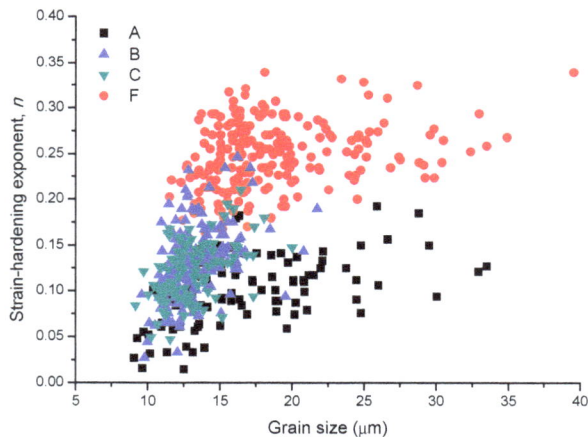

Figure 6. Distribution of the strain-hardening exponents (*n*) of the individual grains of ferrite, obtained from Equation (5), plotted as a function of the individual grain size. Where: A = straight stretch, B = *Intrados*, C = *Extrados* and F = straight stretch normalized at 900 °C.

The interaction between the M-A constituents and the ferritic matrix has some effect on the work-hardening behavior of steels. After the tempering of the *extrados* at 600 °C, a certain increase of the values of *n* could be expected due to the decomposition of the hard particles of the M-A constituents (Figure 2c). However, for the steel in the as-tempered condition (at 600 °C) the main effect on *n* was observed due to the precipitation of the microalloying elements. Thus, the precipitation of vanadium decreased the values of *n* for the steel studied here (Table 4), while the effects of the decomposition of the M-A constituents (due to tempering) were not strong enough to increase the values of n for the steel in the as-tempered condition.

Figure 7 shows the dislocation structures (irregular, random and cluster) for the steel under study in the hot-bent condition (*intrados*) and in the tempered-and-bent condition (*intrados* tempered at 650 °C). The interaction of dislocations with the local precipitates and the formation of the loops around the precipitates can be seen in Figure 7. Additionally, Figure 7b shows (inside the square) one precipitate—possibly a titanium-niobium carbonitride due to its morphological features—that was not dissolved by the hot-induction bending, which is in agreement with the results of Table 3. It is assumed that the titanium-niobium carbonitrides have similar contributions to the precipitation hardening in both the as-received and the hot-bent conditions. The solubilization of the vanadium carbides also proved to be significant as the volume fractions of $VC_{0.75}$ decreased for all of the samples in the hot-bent condition (Table 4).

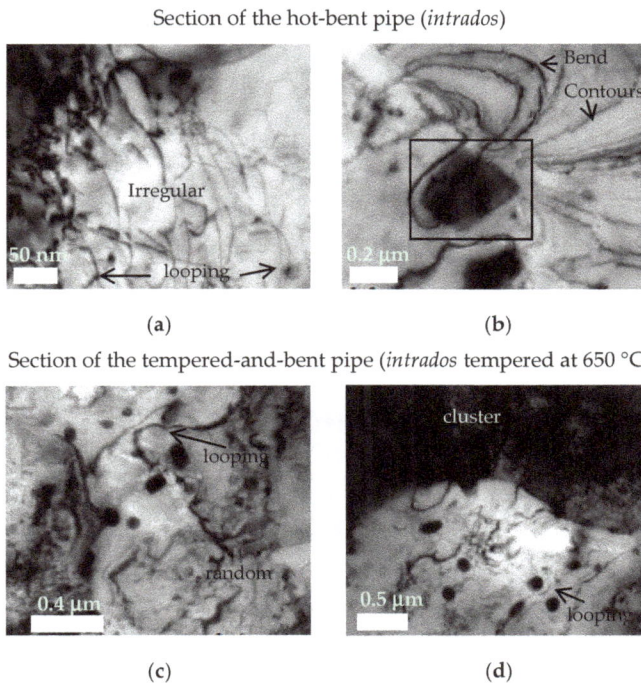

Section of the hot-bent pipe (*intrados*)

(a)   (b)

Section of the tempered-and-bent pipe (*intrados* tempered at 650 °C)

(c)   (d)

Figure 7. Dislocation structures (irregular, random and cluster) and the interaction of the dislocations with the local precipitation. (**a,b**) *intrados* and (**c,d**) *intrados* tempered at 650 °C.

The trend of precipitation is often represented by the solubility products of the precipitates. The effectiveness of the precipitation hardening depends on the solubility limits of these microalloying elements in the austenite. The limit of solubility controls the content of the microalloying element that can be dissolved in the austenite and then to be available to precipitate during the cooling stage. Due to the high solubility of $VC_{0.75}$ when compared with titanium and niobium precipitates (carbides, nitrides, and carbonitrides), the steels that are microalloyed with vanadium can be hardened after normalizing at temperatures close to 950 °C, where the titanium and niobium precipitates have very low solubilities [22]. The steel in the as-normalized condition discussed in this work is hardened mainly by the grain refinement because the normalizing temperature of 900 °C is not high enough to dissolve the coarser precipitates of the microalloying elements in the austenite. Despite the refined grain size in the normalized condition, the tensile properties of the normalized condition, especially YS, are very low when compared to the tensile properties of this steel processed under the other conditions. Also, both models agree closely with the experimental value of n of the studied steel in its as-normalized condition (Table 4 and Figure 5).

Precipitates are known to interact with the dislocations by blocking and/or generating new dislocations during deformation in the tensile test. The mobile dislocations predominantly cut through small coherent precipitates, creating particles of the sheared precipitates (Friedel cutting). When the precipitates are incoherent with the ferritic matrix the dislocations make loops around them (Orowan looping) [1,23–25].

The Orowan mechanism becomes the main mechanism of interaction between the precipitates and the dislocations, after the size of the precipitate reaches the critical value. In this context, the increase of the strain-hardening rates is associated with the increase of the stresses that are required to loop the dislocations around the precipitates [20].

The samples of the hot-bent section showed high experimental values of n, despite the higher dislocation densities and larger grain refinement than the as-received steel. These results were associated with the different levels of precipitation hardening, which resulted from the various processing routes. In materials with high initial dislocation densities, the sessile dislocations increase the work hardening, since this type of the dislocation can hinder the movement of mobile dislocations [1]. Sessile dislocations do not have their Burgers vector lying in the fault plane with which they are associated, and are incapable of gliding [26]. Consequently, the sessile dislocations can decrease the capability of the steel to uniformly distribute the plastic deformation (n). However, the results in Table 4 suggest that the values of n are more significantly reduced by precipitation hardening than by work hardening.

As mentioned above, the cutting mechanism of the precipitates by the action of the dislocations occurs for coherent and small particles (less than 10 nm). The presence of pairs of dislocations is a proof of this mechanism [27]. However, the authors in [28] reported that the small precipitates of vanadium can be incoherent and interact with the dislocations through the formation of Orowan loops. In the present case, Figure 7a shows dislocations pinned by nano-precipitates suggesting the formation of Orowan looping. Additionally, Figure 7c,d shows the formation of dislocation loops (Orowan mechanism) around coarse precipitates whose interfaces with the ferritic matrix are generally of the incoherent type. Some bend contours appear as broad, fuzzy dark lines in the bright-field images [29] in Figure 7b, where there is the presence of coarse precipitation that was not dissolved during hot-induction bending. This coarser precipitation does not contribute to the mechanical strength of this steel.

4. Conclusions and Final Considerations

1. The impact of precipitation hardening on the stable plastic behavior of the steel increased as the precipitation hardening became more intense. The effect of precipitation hardening on n become dominant, when compared with the effects of hardening mechanisms by dislocations and grain size.

2. The values of n could not be accurately estimated by the Morrison equation due to the different levels of precipitation shown—after different processing routes—by the pipeline microalloyed steel of this work.

3. The values of n calculated from the modified equations agree well with the experimental results, revealing that the application of the microhardness of the material can be an alternative to estimate the capability to uniformly distribute the plastic deformation due to good correlations involving microhardness, processing conditions, and the hardening mechanisms.

Acknowledgments: This work was supported by the Conselho Nacional de Desenvolvimento Científico e Tecnológico (CNPq), Coordenação de Aperfeiçoamento de Pessoal de Nível Superior (CAPES) and Fundação Carlos Chagas Filho de Amparo à Pesquisa do Estado do Rio de Janeiro (FAPERJ). Authors would like to thank LABNANO of CBPF for the use of electron microscopy facilities, and Fundação de Amparo à Computação Científica (FACC) for providing the article processing charge funds, and Ana Luiza Rocha for her help with JEOL 2010 microscope of PUC-Rio and TEM samples preparation.

Author Contributions: Ivani S. Bott conceived and supervised the research. Rafael A. Silva completed the DSC for the influence of induction hot bending parameters on the microstructure evolution of an API X80 Pipe steel. Alexei Kuznetsov contributed to the XRD results and André L. Pinto contributed to TEM results and analysis. All the authors contributed in writing this paper.

Conflicts of Interest: The authors declare no conflict of interest.

References

1. Hosford, W.F. *Mechanical Behavior of Materials*, 2nd ed.; Cambridge University Press: Cambridge, UK, 2010; p. 163.

2. Morrison, W.B. The effect of grain size on the stress-strain relationship in low carbon steel. *Trans. ASM* **1966**, *59*, 824–846.

3. ASTM E112-96. *Standard Test Methods for Determining Average Grain Size*; American Society for Testing and Materials: West Conshohocken, PA, USA, 2004.

4. Morales, E.V.; Silva, R.A.; Bott, I.S.; Paciornik, S. Strengthening mechanisms in a pipeline microalloyed steel with a complex microstructure. *Mater. Sci. Eng. A* **2013**, *585*, 253–260. [CrossRef]

5. Ungar, T.; Dragomir, I.; Révész, Á.; Borbély, A. The contrast factors of dislocations in cubic crystal: The dislocation model of strain anisotropy in practice. *J. Appl. Cryst.* **1999**, *32*, 992–1002. [CrossRef]

6. Ungár, T.; Gubicza, J.; Hanák, P.; Alexandrov, I. Densities and character of dislocations and size-distribution of subgrains in deformed metals by X-ray diffraction profile analysis. *Mater. Sci. Eng. A* **2001**, *319–321*, 274–278. [CrossRef]

7. Silva, R.A. Correlation between the Induction Hot Bending Parameters for API X80 Pipe and the Resulting Mechanical and Microstructural Properties. Master's Thesis, Pontifical Catholic University of Rio de Janeiro, Rio de Janeiro, Brazil, 21 October 2009; p. 207. (In Portuguese)

8. Park, J.S.; Ha, Y.S.; Lee, S.J.; Lee, Y.K. Dissolution and precipitation kinetics of Nb(C,N) in austenite of a low-carbon Nb-microalloyed steel. *Metall. Mater. Trans. A* **2009**, *40*, 560–568. [CrossRef]

9. Lu, J. Quantitative Microstructural Characterization of Microalloyed Steels. Ph.D. Thesis, University of Alberta, Edmonton, AB, Canada, November 2009; pp. 213–214.

10. Garcia-Sanchez, R.; Bettmer, J.; Ebdon, L. Development of a new method for the separation of vanadium species and chloride interference removal using modified silica capillaries-DIN-ICP-MS. *Microchem. J.* **2004**, *76*, 161–171. [CrossRef]

11. Karbasi, M.H.; Jahanparast, B.; Shamsipur, M.; Hassan, J. Simultaneous trace multielement determination by ICP-OES after solid phase extraction with modified octadecyl silica gel. *J. Hazard. Mater.* **2009**, *170*, 151–155. [CrossRef] [PubMed]

12. Aydin, I.; Aydin, F.; Hamamci, C. Vanadium fractions determination in asphaltite combustion waste using sequential extraction with ICP-OES. *Microchem. J.* **2013**, *108*, 64–67. [CrossRef]

13. ASTM E646-07. *Standard Test Method for Tensile Strain-Hardening Exponents (n-Values) of Metallic Sheet Materials*; American Society for Testing and Materials: West Conshohocken, PA, USA, 2007.

14. API 5L. *Specification for Line Pipe*; American Petroleum Institute: Washington, WA, USA, 2004.

15. ASTM A370. *Standard Test Methods and Definitions for Mechanical Testing of Steel Products*; American Society for Testing and Materials: West Conshohocken, PA, USA, 2003.

16. Spindola, M.O.; Ribeiro, E.A.S.; Gonzalez, B.M.; Santos, D.B. Modeling of work hardening behaviour of high Mn and low C polycrystalline austenitic steel with TWIP effect. *Rev. Mater.* **2010**, *15*, 145–152.

17. Lucas, J.P.; Gerberich, W.W. Low temperature and grain size effects on the cyclic strain hardening exponent of an HSLA steel. *Scr. Metall.* **1981**, *15*, 327–330. [CrossRef]

18. Antoine, P.; Vandeputte, S.; Vogt, J.B. Empirical model predicting the value of the strain-hardening exponent of a Ti-IF steel grade. *Mater. Sci. Eng. A* **2006**, *433*, 55–63. [CrossRef]

19. Gao, N.; Baker, T.N. Influence of AlN precipitation on thermodynamic parameters in C-Al-V-N microalloyed steel. *ISIJ Int.* **1997**, *37*, 596–604. [CrossRef]

20. Hong, S.G.; Jun, H.J.; Kang, K.B.; Park, C.B. Evolution of precipitates in the Nb-Ti-V microalloyed HSLA steels during reheating. *Scr. Mater.* **2003**, *48*, 1201–1206. [CrossRef]

21. Lee, W.-B.; Hong, S.-G.; Park, C.-G.; Park, S.-H. Carbide precipitation and high-temperature strength of hot-rolled high-strength, low-alloy steels containing Nb and Mo. *Metall. Mater. Trans. A* **2002**, *33*, 1689–1698. [CrossRef]

22. Pickering, F.B. *Physical Metallurgy and Design of Steels*; Applied Science Publishers: London, UK, 1978.

23. ASM Handbook. *Metallography and Microstructures*; ASM International: Materials Park, OH, USA, 1985; p. 116.

24. Padilha, A.F. *Materiais de Engenharia—Microestrutura e Propriedades*; Hemus: Curitiba, Brazil, 2000; pp. 257–258.

25. Lui, M.-W.; Le May, I. On the "Friedel Relation" in precipitation hardening. *Scr. Metall.* **1975**, *9*, 587–589. [CrossRef]

26. Smallman, R.E.; Bishop, R.J. *Modern Physical Metallurgy and Materials Engineering: Science, Process, Applications*, 6th ed.; Butterworth-Heinemann: Oxford, UK, 1999; pp. 102–103.

27. Morales, E.V.; Gallego, J.; Kestenbachz, H.-J. On coherent carbonitride precipitation in commercial microalloyed steels. *Philos. Mag. Lett.* **2003**, *83*, 79–87. [CrossRef]

28. Morales, E.V.; Galeano Alvarez, N.J.; Morales, A.M.; Bott, I.S. Precipitation kinetics and their effects on age hardening in an Fe–Mn–Si–Ti martensitic alloy. *Mater. Sci. Eng. A* **2012**, *534*, 176–185. [CrossRef]

29. Fultz, B.; Howe, J. *Transmission Electron Microscopy and Diffractometry of Materials*, 3rd ed.; Springer: Berlin/Heidelberg, Germany, 2008; p. 353.

metals

MDPI

Article

Local Buckling Behavior and Plastic Deformation Capacity of High-Strength Pipe at Strike-Slip Fault Crossing

Xiaoben Liu [1,2], Hong Zhang [1,*], Baodong Wang [1], Mengying Xia [1,2,*], Kai Wu [1], Qian Zheng [1] and Yinshan Han [1]

[1] College of Mechanical and Transportation Engineering, China University of Petroleum-Beijing, Beijing 102249, China; liuxiaoben1991@126.com (X.L.); wangbaodong58@163.com (B.W.); wk0609@126.com (K.W.); zhengqian2981@163.com (Q.Z.); cathaya_han@163.com (Y.H.)

[2] Department of Civil and Environmental Engineering, University of Alberta, Edmonton, AB T6G 2W2, Canada

* Correspondence: hzhang@cup.edu.cn (H.Z.); xiamengying322@163.com (M.X.); Tel.: +86-010-89733274 (H.Z.); +86-010-89731239 (M.X.)

Received: 17 November 2017; Accepted: 26 December 2017; Published: 31 December 2017

Abstract: As a typical hazard threat for buried pipelines, an active fault can induce large plastic deformation in a pipe, leading to rupture failure. The mechanical behavior of high-strength X80 pipeline subjected to strike-slip fault displacements was investigated in detail in the presented study with parametric analysis performed by the finite element model, which simulates pipe and soil constraints on pipe by shell and nonlinear spring elements respectively. Accuracy of the numerical model was validated by previous full-scale experimental results. Insight of local buckling response of high-strength pipe under compressive strike-slip fault was revealed. Effects of the pipe-fault intersection angle, pipe operation pressure, pipe wall thickness, soil parameters and pipe buried depth on critical section axial force in buckled area, critical fault displacement, critical compressive strain and post buckling response were elucidated comprehensively. In addition, feasibility of some common buckling failure criteria (i.e., the CSA Z662 model proposed by Canadian Standard association, the UOA model proposed by University of Alberta and the CRES-GB50470 model proposed by Center of Reliable Energy System) was discussed by comparing with numerical results. This study can be referenced for performance-based design and assessment of buried high-strength pipe in geo-hazard areas.

Keywords: high-strength X80 steel; buried pipeline; strike-slip fault; strain-based design; compressive strain capacity; local buckling; finite element method

1. Introduction

Buried steel pipelines serve as the main means of transportation of both raw and processed hydrocarbon fluids worldwide. High-strength line pipe steels are preferred by operators for their higher profit induced by the increased throughput of the products [1]. The large expanse of pipe routine makes crossing some geo-hazard areas inevitable, such as active seismic faults. Large ground-relative displacements along the fault trace will cause axial compression (or tension) and lateral bending in pipe, which may be large enough to initiate local buckling or tensile fracture failure [2]. As crucial lifelines, the integrity of steel pipelines at tectonic fault crossings has been paid close attention to by both academic and industrial spheres.

Newmark and Hall conducted the pioneer work for analytical strain analysis of pipe at fault crossing [3], in which the pipe was assumed to deform like a cable. Due to its high simplicity, the Newmark method is still widely used in industry for the primary design of a pipe subjected to fault movement. Thereafter, a series of analytical or semi-analytical models were developed for refined analysis of pipe strain or stress under fault displacement. Kennedy et al. [4] and Wang et al. [5] considered pipe bending with assumptions that pipes deform as curved arcs and elastic beams. In more recent decades, more complicated semi-analytical approaches were proposed by Karamitros et al. [6], Trifonov et al. [7], and Zhang et al. [8], who considered the elastoplastic characteristic of pipe steel's constitutive model and their effects on pipe's nonlinear stress distributions in sections of large deformed pipe segments near fault trace. Although these analytical methods can be more easily popularized in guidelines for engineering applications, they also have severe weaknesses. Even the latest analytical methods are still based on the beam assumption of pipe, which is incapable of describing the local deformation of pipes subjected to compression or combined with bending loads.

For pipes subjected to compression at strike-slip fault crossings, numerical and experimental models were mostly utilized for pipe performance analysis. Due to the restrictions of hydraulic loading facilities, experimental investigations were all focusing on low- to medium-strength steel pipes or high-density polyethylene (HDPE) pipes with similar nonlinear stress-strain response as steels. Ha et al. [9,10] and O'Rourke et al. [11,12] conducted centrifuge and full-scale tests of buried HDPE under compression strike-slip faults. Results show that these two methods derived similar pipe response with the similar parameters. Jalali et al. [13–15] performed a full-scale experiment on responses of both American Petroleum Institute (API) Grade-B steel and HDPE pipes under reverse fault displacements. Both global and local buckling behavior were captured in his investigation. Valuable conclusions on pipe soil interactions were also obtained in their experiments.

Based on the experimental results, a lot of calibrated numerical models were developed for further investigations on pipe performance subjected to compression strike-slip fault movements. Generally, numerical models can be classified by the simulation methods for pipe and soil constraint on pipe. Pipes can be simulated by pipe (elbow) elements or shell elements. Pipe (elbow) elements have the advantage in calculation efficiency, but they are incapable of demonstrating the pipe wall folding phenomenon in pipe, when local buckling occurs. Soil constraints on pipe can be modeled by discrete nonlinear soil spring elements or the surface-to-surface contact between the shell elements simulating pipe and the continuum solid elements simulating surrounded soil [16]. Adopting the numerical models, stress and strain behaviors of different kind buried pipes under various types of active faults have been studied. Using finite element (FE) models with pipe (elbow) elements and nonlinear spring elements, Xie et al. validated the numerical models by comparing them with centrifuge experiment results [17,18]. Joshi et al. conducted parametric analysis on beam buckling behavior of steel pipe at reverse fault crossing [19]. Uckan et al. proposed a simplified model to derive the response of curved pipeline at strike-slip fault crossing [20]. Liu et al. developed a prediction model on peak pipe compressive strain based on the numerical results and nonlinear regression method [21]. Melissianos et al. conducted performance assessment of buried pipe at fault crossing [22,23]. Using Finite Element (FE) models with shell elements and nonlinear spring elements, Karamitros et al. validated his analytical model [6]. Liu et al. proposed a semi-empirical equation for strain demand of pipeline at oblique-reverse fault crossing [24]. Xu et al. investigated the wrinkling phenomenon under reverse fault [25]. More recently, using FE models with shell elements and continuum solid soil elements, Kaya et al., simulated the failure behavior of a welded steel pipe at Kullar fault crossing [26]. Zhang et al. discussed the collapse behavior of pipeline buried in rock under strike-slip fault movement and reverse fault displacements [27,28]. Vazouras conducted series of numerical models to investigate the failure behavior of pipeline under strike-slip fault movement in detail [29–32]. Trifonv et al. also built a rigorous numerical model to analyze the influence of the trench on the pipeline performance at fault crossing [7].

However, although extensive research is available, relatively little literature exists on the topic of high-strength pipe buckling behavior specifically. Some attempts were made by Liu et al. to investigate the effects of yield strength and strain hardening parameter on the local failure behavior of high-strength pipes subjected to reverse faulting [33]. Kainat et al. investigated the influences of geometric imperfections on high strength pipe's buckling behaviors [34]. Neupane et al. captured the anisotropic characteristic of high-strength pipe steel in longitudinal and circumferential directions, and discussed its effects on buckling response of pipe induced by frost heave in northern areas [35,36]. However, comprehensive investigations on local buckling of high-strength pipes in various engineering cases are urgently needed. Thus, a systematic analysis was conducted in this paper on the local buckling behavior of high-strength of X80 steel pipeline subjected to strike-slip fault displacements. Mechanical response of X80 pipe at onset moment when local buckling occurs was studied. Influences of the pipe-fault intersection angle, pipe diameter, pipe wall thickness, pipe operation pressure, soil parameters, pipe buried depth on pipe's buckling behavior were all investigated in detail. Based on the finite element results, applicability of some well-recognized compressive strain criteria on local buckling failure recognition for high-strength pipe was evaluated in a wide range of common parameters.

2. State of the Art in Buckling Failure Criteria for Line Pipes

Compressive local buckling is a limit state for pipeline under compressive load. In strain-based design, the principle purpose is to keep the strain demand (i.e., the maximum longitudinal compressive strain in pipe) in pipe less than the compressive strain capacity of pipe (i.e., the longitudinal pipe strain at the onset moment of pipe buckling) [1]. Thus, to provide suitable references for pipeline designers and operators, a series of prediction models for compressive strain capacity have been established and adopted by different guidelines and standards. In this section, three well-recognized models for compressive strain capacity of plain pipe are presented briefly.

2.1. CSA Z662 Model (2015)

The CSA Z662 model [37] was initially based on Gresnigt's research in 1980s [38]. Its latest version was released in 2015. Some modifications have been made to Gresnigt's model to improve the model's accuracy for pipes with high internal pressure. This model is the one most widely used so far. A lot of codes and standards directly adopt this model to calculate the compressive strain capacity of steel pipes ε_c^{crit}, such as the guideline for design of buried steel pipeline [39], and the aseismic guidelines for pipes in India [40]. The model considers the effects of pipe diameter to wall thickness ratio D/t and operation pressure P on pipe's strain capacity directly. It should be noted that the CSA formula was derived by regression of lower-bound test data. Thus, material property parameters and pipe geometrical imperfections were also implicitly considered. The CSA formula is given by Equation (1).

$$\varepsilon_c^{crit} = \begin{cases} 0.5\frac{t}{D} - 0.0025 + 3000\left[\frac{PD}{2tE}\right]^2 & \frac{PD}{2t} \leq 0.4\sigma_y \\ 0.5\frac{t}{D} - 0.0025 + 3000\left[\frac{0.4\sigma_y}{E}\right]^2 & \frac{PD}{2t} > 0.4\sigma_y \end{cases} \tag{1}$$

where D is the pipe diameter, t is pipe wall thickness, $D/t \leq 120$, P is the pipe operation pressure, σ_y is the yield strength of pipe steel.

2.2. UOA Model (2006)

The UOA models [41] were developed by structure group at the University of Alberta in 2006 [41]. The equations were established throughout regression analysis on a large number of finite element models and full-scale experimental results. The considered steel grades in parametric analysis included X52, X65 and X80. Four groups of CSA Z662 equations were finally obtained for plain pipes with yield plateau type or round house type stress-strain curve and girth weld pipe with yield plateau type or round house type stress-strain curve. For in the numerical analysis of presented paper, the pipe was considered as a plain pipe with roundhouse-type stress-strain curve; the equation for this kind of pipes was listed here (Equation (2)).

$$\varepsilon_c^{crit} = 100 \left(\frac{2.9398}{D/t} \right)^{1.5921} \left(1 - 0.8679 \frac{PD}{2t\sigma_y} \right)^{-1} \left(\frac{E}{\sigma_y} \right)^{0.8542} \left[1.2719 - \left(\frac{h_g}{t} \right)^{0.1501} \right] \tag{2}$$

where h_g is the height of the geometry imperfection which is defined as the peak-to-valley height of the surface undulation, $50 \leq D/t \leq 90$.

2.3. CRES-GB 50470 Model (2017)

The CRES model [42] was established by the Center for Reliable Energy Systems in 2013 under the support of the US Department of Transportation [42]. Similar to the UOA model, it is developed based on a large amount of finite element models. This model has been recently recommended by the latest version of China's national standard for seismic design and assessment of oil and gas transmission pipelines (GB 50470) [43]. The influence factors considered in the CRES-GB 50470 model is the most comprehensive one in reported research so far, including ratio of pipe diameter to pipe wall thickness D/t, operation pressure, ratio of yield strength to tensile strength of pipe material σ_y/σ_u, geometry imperfection h_g, girth weld, Lüder's strain, and net-section stress σ_a. For plain pipes with roundhouse-type stress-strain curves, the CRES equation is as follows:

$$\varepsilon_c^{crit} = \left(2.7 - 2.0 \frac{\sigma_y}{\sigma_u} \right) \left(1.84 - 1.6 \left(\frac{h_g}{t} \right)^{0.2} \right) F_{NF} F_{DP} \tag{3}$$

$$F_{DP} = \begin{cases} 9.8 \left[0.5 \left(\frac{D}{t} \right)^{-1.6} + 1.9 \times 10^{-4} \right] & \frac{PD}{2t\sigma_y} < 1.8 \times 10^{-4} \left(\frac{D}{t} \right)^{1.6} \\ 9.8 \left(1.06 \frac{PD}{2t\sigma_y} + 0.5 \right) \left(\frac{D}{t} \right)^{-1.6} & \frac{PD}{2t\sigma_y} \geq 1.8 \times 10^{-4} \left(\frac{D}{t} \right)^{1.6} \end{cases} \tag{4}$$

$$F_{NF} = \begin{cases} 1.2 \left(\frac{\sigma_a}{\sigma_y} \right)^2 + 1 & \frac{\sigma_a}{\sigma_y} \geq 0 \\ 1, & \frac{\sigma_a}{\sigma_y} < 0 \end{cases} \tag{5}$$

where σ_a is the applied net-section stress in the longitudinal direction of pipe, which can be derived from the stress demand analysis. If σ_a is not available, $\sigma_a = 0$; $20 \leq D/t \leq 104$, $0.76 \leq \sigma_y/\sigma_u \leq 0.96$.

Applicability of these models for high-strength steel pipe will be further evaluated by the numerical results of the buried X80 pipeline under compression strike-slip fault in subsequent sections. It should be noticed that some formulas focusing on compressive strain capacity of offshore pipes with small diameter to wall thickness ratio were recommended by the API-1111-1999 [44], DNV OS F101 [45]. For their incapable application to common onshore high-strength pipelines with large diameter to wall thickness ratio, they are not discussed here.

3. Finite Element Model

3.1. Modelling Pipe-Soil Interaction with Discrete Nonlinear Soil Springs

For buried steel pipelines, constraints will be induced on them by the surrounding soil in the axial, horizontal and vertical directions, if they have relative motions. The soil-applied force on pipe or pile structures is commonly nonlinear, which can be described by the *p-y* curves derived by geotechnical researchers [12]. There is a lot of available literature on this topic, among which the design recommendations specified by the ASCE Guideline [46] and ASCE-ALA Guideline [39] is the most widely used one for design and assessment of buried pipelines. According to the ASCE-ALA Guideline, soil constraints on pipe can be modeled using nonlinear soil springs with elastic plastic constitutive properties in three perpendicular directions, i.e., the axial, lateral and vertical directions of pipe central axis as illustrated in Figure 1, where continuous lines represent the real characteristic and dashed lines represent the approximation implemented in the model. The parameters P_u, T_u, and $Q_u(Q_d)$ represent the peak soil resistances per unit length of pipes, while Δp, Δt, and Δq_u (Δq_d) represent the corresponding yield displacements in three directions (i.e., lateral, axial, and vertical), respectively. Values of all these parameters can be readily obtained according to the ASCE-ALA Guideline with soil property parameters derived from field investigation. For numerical modelling, most commercial finite element software has developed spring elements capable of coupling a force with relative displacement nonlinearly. In this paper, the 3D SPRING elements (SPRING2) in general FE package ABAQUS were utilized for numerical investigation [47].

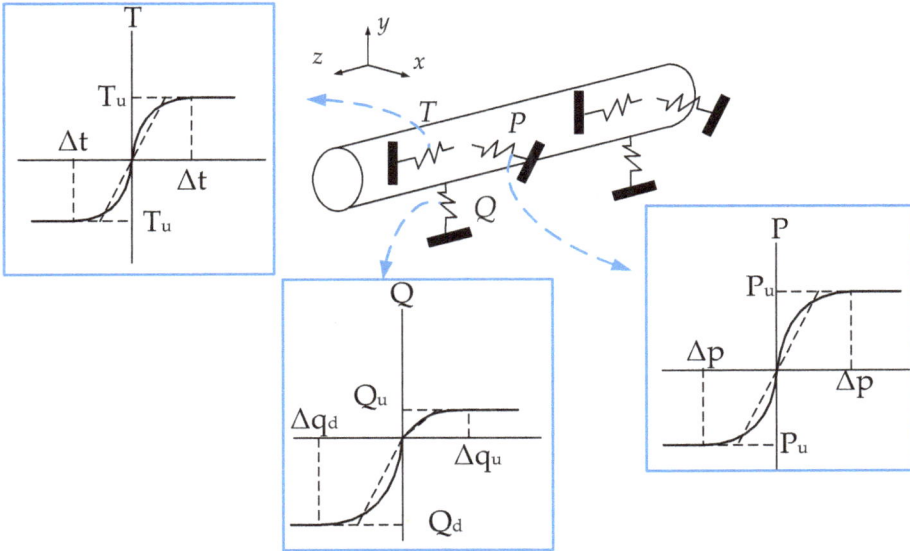

Figure 1. Schematic diagram of nonlinear soil springs on pipe, T for axial springs, P for lateral springs, and Q for vertical springs [39].

3.2. Modelling of Pipe

In this study, pipe segments were all modeled by four node shell elements with reduced integration (S4R) in general FE software ABAQUS (Dassault Systèmes, version 6.14, Johnston, RI, USA). The Second West-to-East Gas Pipeline in China was taken as the prototype for this study. The modeled pipe in numerical model was set to be 100 m in total, 50 m each at both sides of the fault trace, with pipe nodes at the two axial ends connected with nonlinear springs (SPRING2) to model the axial constraint of longitudinally adjacent pipes on the 100 m long pipe at fault trace (Figure 2). A constitutive model of the equivalent spring will be described in the subsequent section (Section 3.3). A refined mesh was achieved by sensitivity analysis with 54 elements discretized in circumferential direction. In addition, setting the longitudinal length of all the shell elements to be 0.04 m was proven to be accurate for simulating the wrinkling behavior of pipe induced by local buckling [33].

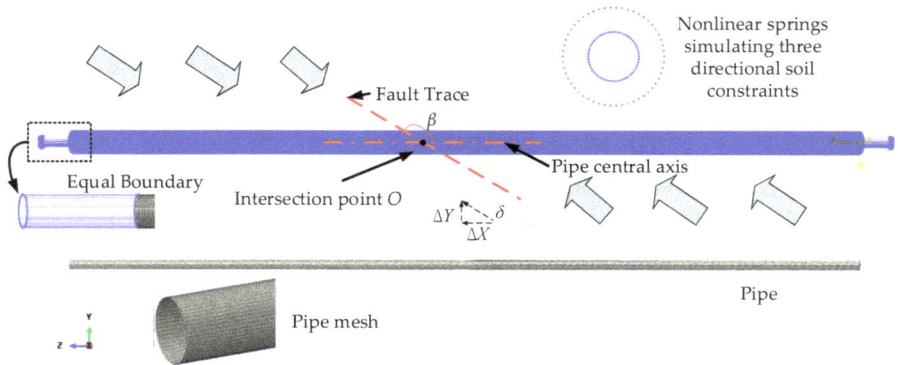

Figure 2. Established finite element model for pipeline crossing strike slip faults.

The material stress-strain model for X80 line pipeline considered in this study is plotted in Figure 3. A large-strain von Mises plasticity model with isotropic hardening is used for the steel in numerical investigation. The curve can also be expressed by the Ramberg-Osgood model (Equation (6)), with yield strength as 550 MPa, strain hardening exponent r as 17 and the yield offset parameter α as 0.94 [48].

$$\varepsilon_{true} = \frac{\sigma_{true}}{E} + \alpha \frac{\sigma_{true}}{E} \left(\frac{\sigma_{true}}{\sigma_y} \right)^{r-1}$$

(6)

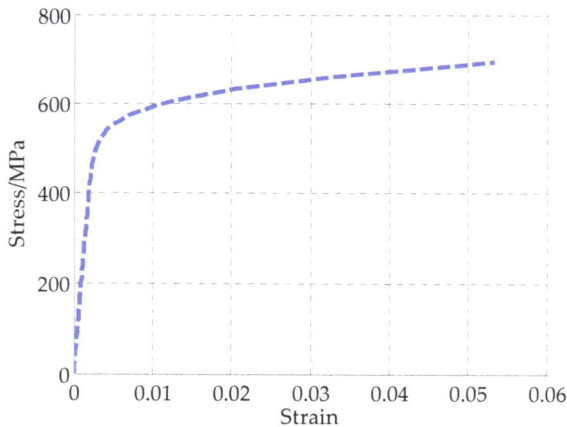

Figure 3. True stress strain curve of X80 line pipe steel.

The finite element model was calculated in two nonlinear static steps. In the first step, all soil nodes are fixed, and operation pressure was applied in the inner surface of the entire pipe. In the second step, all the soil nodes on the right of the fault trace were displaced with fault displacements both longitudinally and laterally in a horizontal plane, while all the soil nodes on the left of the fault trace were all kept motionless. As shown in Figure 2, the longitudinal displacement ΔX and the lateral displacement ΔY are $\delta \cos \beta$ and $\delta \sin \beta$, respectively. In order to insure the convergence of the iterative calculation of the FE model, nonlinear stabilization algorithm was utilized in the second step [49], the initial step size of the fault displacement load step is set to be 0.05, with the minimum allowable step size set to be 10^{-6}. If there is no pressure in the pipe, the first step should be removed. It also should be mentioned that the proposed model can be suitable for dynamic simulations by replacing the static analysis step to explicit dynamic step. With the performed dynamic analysis, the model can be utilized to predict the transients associated with buckling phenomena, thus possibly allowing integration with the techniques adopted for leak detection [50–53].

3.3. Constitutive Model of Equivalent Boundary

When buried pipelines are subjected to fault movement, relative displacements appear between the pipe and the surrounding soil. The equivalent boundary model was used to reduce the element numbers in numerical calculation by using few nonlinear springs simulating the axial constraints on pipe induced by the adjacent pipes, as show in Figure 4. Liu et al., developed the constitutive model of the equivalent boundary [54], which represents the relationship of the axial force F and the pipe extension at Point B ΔL (Equation (7)).

$$F(\Delta L) = \begin{cases} \sqrt{\frac{3EAT_u}{2}\Delta t^{-\frac{1}{6}}\Delta L^{\frac{2}{3}}} & 0 \le \Delta L \le \Delta t \\ \sqrt{2EA\left(\Delta L - \frac{1}{4}\Delta t\right)T_u} & \Delta t \le \Delta L \le \frac{\sigma_y^2 A}{2ET_u} + \frac{\Delta t}{4} \end{cases} \tag{7}$$

where E is the elastic modulus, A is the pipe cross section area, T_u is the slip friction force on unit length of pipe, i.e., the peak axial soil resistant force of soil spring, Δt is the in the axial direction, σ_y is the yield strength of pipe material.

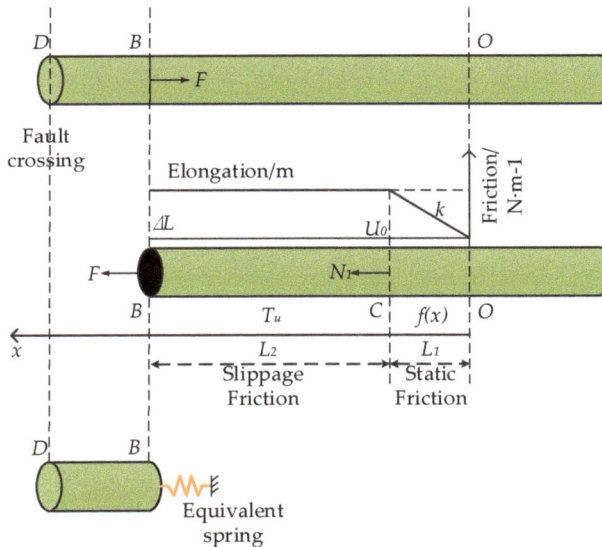

Figure 4. Constitutive model for the equivalent boundary spring.

3.4. Validation of Proposed Model

The full-scale experimental results for buried HDPE pipe subjected to strike-slip fault movements conducted by O'Rourke and a corresponding numerical model established by Xie et al. were used here to validate our established numerical model [11,18]. Material characteristics of HDPE material are similar with high-strength pipe strength, which has a roundhouse-type stress-strain curve but with a much smaller yield strength. The experimental pipe is a 10.56 m long pipe with 0.4 m diameter and 0.0024 m wall thickness, and the fault trace is located at the axial center of the pipe with a 65°

intersection angle with the pipe axial axis. As the tested pipe is smaller than the steel pipes mentioned in Section 3.2, dimensions of the pipe was revised to be the same as the experiment pipe. The soil spring parameters in the validation model were also directly obtained from the experimental results. Furthermore, as the test PE pipe has smaller pipe diameter, the pipe was re-meshed into 48 elements in pipe's circumferential direction, which is the same as Xie's numerical model. The calculation process is the same as the one described in Section 3.2. Figure 5 shows comparison results of longitudinal and circumferential strains in the internal and external surfaces of the buckling section in the pipe under 0.61 m fault displacement. The results contain the proposed numerical results (Num-INT and Num EXT in Figure 5) and the numerical results derived by Xie et al. through a refined 3D finite element model (Xie-INT and Xie EXT in Figure 5) [18] and the measured results (measured in Figure 5).

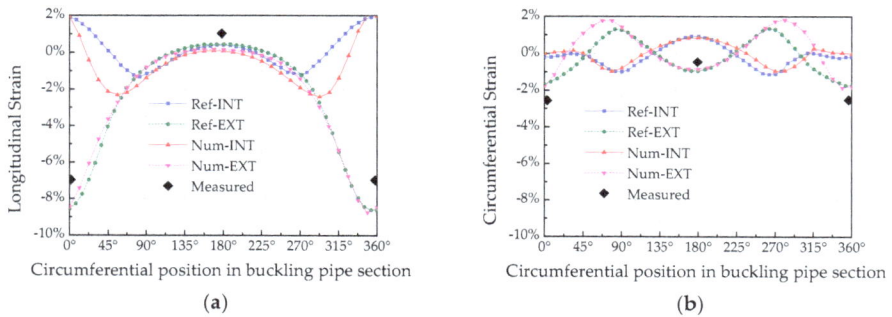

Figure 5. Comparison of strain distributions at cross-section where buckling occurs when $\delta = 0.61$ m. (**a**) Longitudinal strain distribution; (**b**) Circumferential strain distribution.

It can be readily obtained that the maximum longitudinal compressive strain equals -8.5% numerically, locating at $0°$ position of the pipe section. While the maximum longitudinal tensile strain is less than 1%, located at the opposite direction of the pipe. The circumferential strain values are much smaller than the longitudinal ones. This is mainly induced by the local buckling, which itself is induced by axial compression of pipe. Generally, it can be found that strain results of the presented numerical model matches quite well with measured experimental results and the numerical results derived by Xie et al. [18].

4. Results and Discussion

4.1. Baseline Analysis for the Local Buckling Failure of X80 Pipe Steel

In the section, a baseline analysis was presented for plastic buckling phenomenon of X80 pipe subjected to a strike-slip fault with an intersection angle to be $150°$. The pipe diameter and wall thickness are 1.219 and 0.0184 m, respectively. The operation pressure of pipe is 12 MPa. The pipe is buried in cohesive soil with buried depth of 2 m. The internal friction angle of soil is $33°$.

As one limit state of pipe, local buckling decreased pipe bearing capacity both longitudinally and circumferentially. For numerical investigation, the section axial force in the buckled area can be monitored for identification of onset of local buckling of pipe (Liu et al. [33,49]). The section axial force in this study represents the local longitudinal internal force of pipe segments in the buckled area. When its value decreases with the increasing of applied fault displacement, the partial stiffness of pipe in this area starts decreasing representing the initiating of local buckling. Relationship of the section axial force with the axial membrane stress of the shell element is as follows:

$$SF = \sigma_{axial} S_{SectionS4R} \tag{8}$$

where σ_{axial} is the axial membrane stress of the shell element; $S_{SectionS4R}$ is the section area of the shell element. As the pipe was discretized into 54 elements circumferentially, $S_{SectionS4R} = \pi Dt/54$.

Figure 6 shows the relationships of section axial force in buckled area SF and peak axial compressive strain ε_{axial} with fault displacement δ. If δ is relatively small, SF increases almost linearly with the δ. If $\delta > 0.2$ m, gradient of SF decreases obviously until SF reaches its peak value, where it drops suddenly representing the local collapse behavior of pipe. The critical pipe section axial force in the buckled area SF_{crit} and the relative critical fault displacement δ_{crit} are 5500 KN and 0.465 m, respectively. Compared with the trend of ε_{axial} with δ, it can be found that before δ reaches δ_{crit}, ε_{axial} increases gradually with the increase of δ; once δ reaches δ_{crit}, ε_{axial} exhibits an abrupt increase. The critical axial compressive strain ε_{crit} for the onset of local buckling can then be derived for the curve as -1.88%.

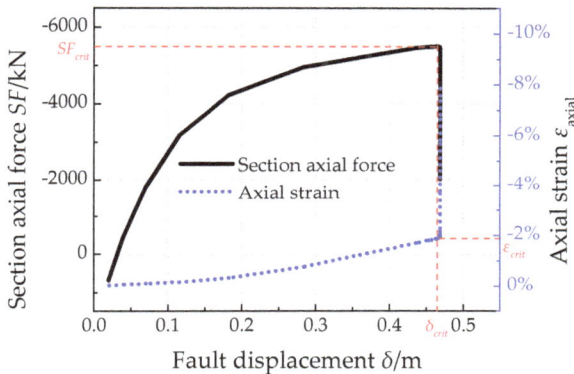

Figure 6. Trends of section axial force and axial strain in buckled area with fault displacement.

Figure 7 plots the contours of SF and ε_{axial} in pipe with various fault displacements, as well as the detail distribution of ε_{axial} in buckled area. Results show that, distribution of SF have almost no variations when δ increases from 0.182 m to the critical value 0.465 m. However, after that, a small increase of δ ($\delta = 0.468$ m) induces wavy distribution of SF with small amplitude in buckled area. Immediately after this, an obvious bulging pattern occurs in the pipe as well as a much more severe wavy distribution of SF. Compared with the section axial force results, variations of axial strain in the buckled area during this process are more obvious. At the critical fault displacement, an obvious strain concentration appears in the buckled area. After that, during the post buckling stage, severe wavy distribution of axial strain occurs when $\delta = 0.468$ m, which quickly concentrates to one wave in the center of the bucked area. Above all, results show that, when local buckling occurs, pipe loses its partial bearing capacity. The section axial force drops immediately after the onset of local buckling, with axial strain increasing abruptly induced by the large deformation.

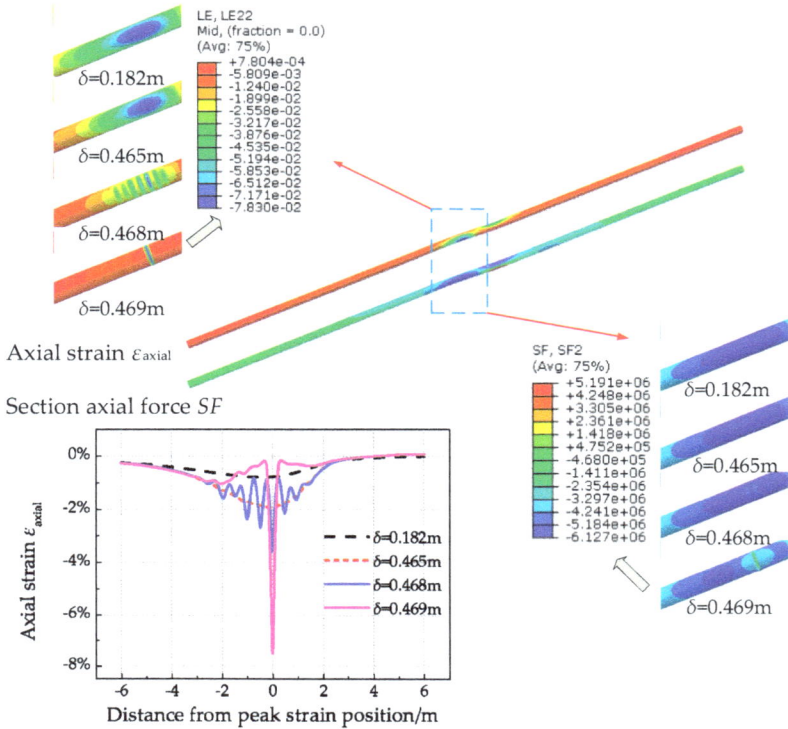

Figure 7. Section axial force and axial strain distribution in pipe at various fault displacements.

4.2. Effects of Pipe-Fault Intersection Angle

When pipes crossing compression strike-slip faults with pipe-fault intersection angle larger than 90°, they are subjected to combined compression and bending load induced by the axial and lateral soil displacement components. With a large intersection angle (i.e., 175°), the pipe is subjected to severe axial compression and relatively small bending. While with a small intersection angle (i.e., 105°), the pipe is subjected to severe bending and relatively small axial compression.

In this section, five typical intersection angles, i.e., 105°, 120°, 135°, 150°, 175°, are considered to investigate their effects on critical fault displacement and critical axial compressive strain for local buckling, as shown in Figure 8. When δ is smaller than 0.45 m, trends of *SF* with δ are quite similar for the cases with different intersection angles. When δ reaches 0.4612 m, *SF* for the condition that $\beta = 135°$ first drops, representing the local buckling of pipe.

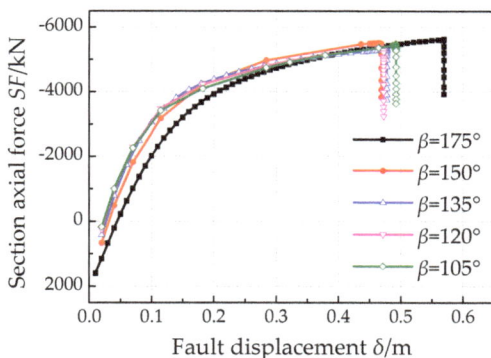

Figure 8. Relationship between section axial force and fault displacement at different intersection angles.

Relationships between the critical fault displacements δ_{crit} and pipe fault intersection angles β are also illustrated in Figure 9. It can be derived that for compression strike-slip fault, if $\beta < 135°$ δ_{crit} decreases with β, else δ_{crit} increases with β. Thus, when pipe fault intersection angle equals 135°, the pipe is more likely to be buckled and fail. This is mainly because when pipe fault intersection angle equals 135°, the combined effect of bending induced by lateral fault displacement $\delta_s \cos\beta$ and compression induced by the axial fault displacement $\delta_s \sin\beta$ is the severest, which makes the pipe more likely to be buckled.

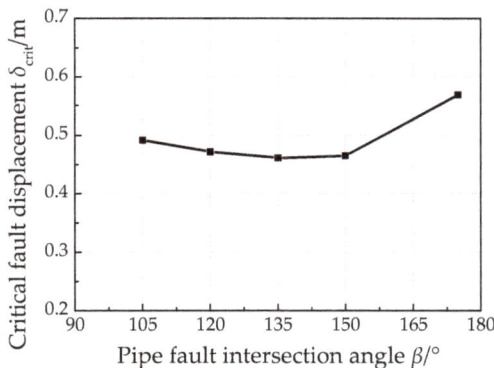

Figure 9. Relationships of the critical fault displacement and the intersection angle.

The critical axial compressive strain ε_{axial} for the five cases were captured and compared with the commonly used failure criteria introduced in Section 2. It should be mentioned that as a perfect pipe was used in numerical models in this study, the height of the geometry imperfection was set to be 1%t, a minimum value recommended by CRES (Liu et al. [42]) to calculate the critical compressive strain values. As shown in Figure 10, the derived critical strain values are all near 2%, with a small variation when the intersection angle changes. While for all the three models in Section 2 only consider effects of the pipe geometry parameters and operation pressures on ε_{axial}, their calculated critical strain results will keep the same for these cases with various intersection angles.

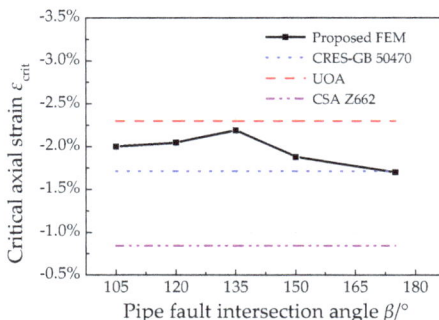

Figure 10. Relationships of the critical compressive strain and the intersection angles.

It can be further observed that for the three considered models, the CSA Z662 model results are mostly conservative, which is almost less than half of the results of the UOA model, CRES-GB 50470 model and FE model. The UOA model results are larger than derived FE model results, representing un-conservative prediction of pipe's strain capacity in actual loading conditions, which may limit its application. As for the CRES-GB 50470 model, its results are a little conservative and in good agreement with the FE model results.

The post buckling behaviors of pipelines with different fault pipe intersection angle cases from a lateral view are also illustrated in Figure 11. Pipe buckles in a similar shape, with a main wrinkling (elephant's foot buckling) near the fault trace for all conditions. In addition, with increase of β, the buckled position becomes a little further away from the fault trace. This is mainly because a smaller β induces a larger bending in pipe.

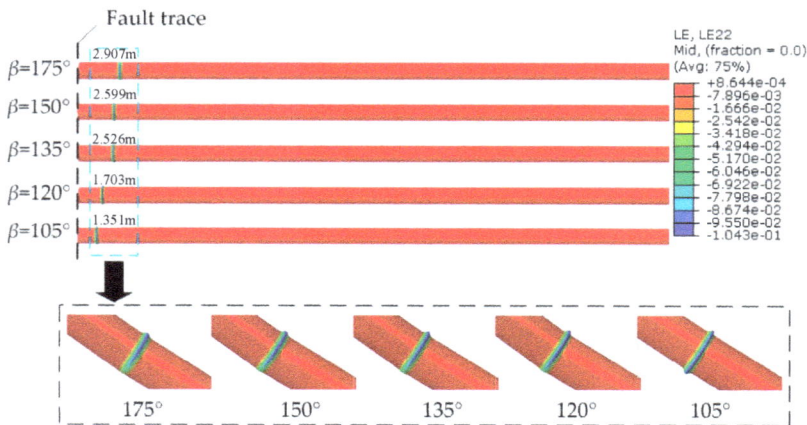

Figure 11. Post buckling behavior (axial strain contour) of pipeline at various intersection angles.

4.3. Effects of Pipe Operation Pressure

Pipe operation pressure is a typical working parameters for oil and gas pipelines in services [55,56]. Effects of operation pressure on pipe's mechanical behaviors are investigated in this section. Six values are considered, i.e., 0, 4, 6, 8, 10, 12 MPa, in which 12 MPa is the designed pressure for common high-strength natural gas pipelines [21]. Figure 12 illustrates trends of *SF* with δ with various pipe operation pressures. The section axial forces for smaller operation pressures are larger. This variation becomes more obvious with increase of fault displacement. When the operation pressure P is 12 MPa, SF reaches its peak value (the onset of local buckling) with the smallest fault displacement (0.472 m). In addition, generally, the critical fault displacement δ_{crit} decreases with the increase of P, except the conditions that $P = 0$ MPa, as shown in Figure 13. This phenomenon is induced by the different pipe buckling shape when the considered pipe is unpressurized, because a pressurized pipe has stronger anti-collapse capacity than non-pressurized pipe circumferentially, as an initial hoop stress exists in pressurized pipes, which makes the bulging type of deformation occur.

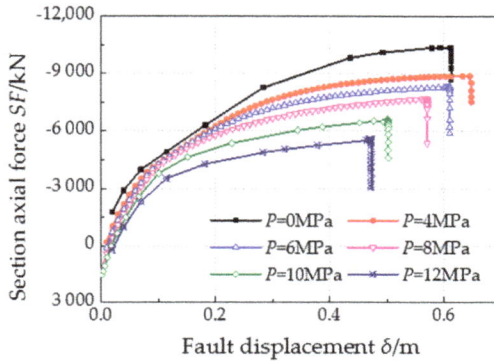

Figure 12. Relationship between section axial force and fault displacement with different operation pressure.

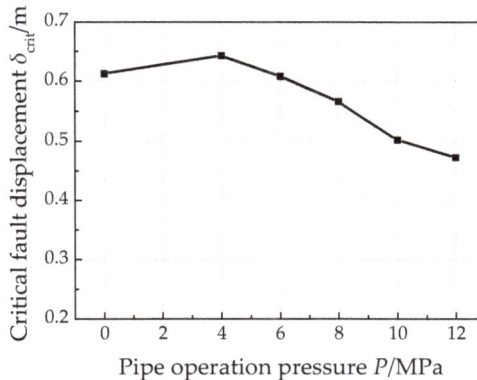

Figure 13. Relationships of the critical fault displacements with the operation pressure.

The critical axial compressive strains with different operation pressured derived from numerical analysis were also compared with the analytical models. A general increasing trend from about 1% to 2% can be found for all the results except those of CSA Z662 model. As concluded in the previous section, CSA Z662 model provides quite conservative results. In addition, both UOA model and the CRES model predicts rather good results when operation pressure P is no more than 10 MPa. Comparing Figures 14 with 13, a valuable conclusion can be drawn, although the pipe has higher compressive strain capacity with a higher operation pressure, but it is also more like to fail due to local buckling.

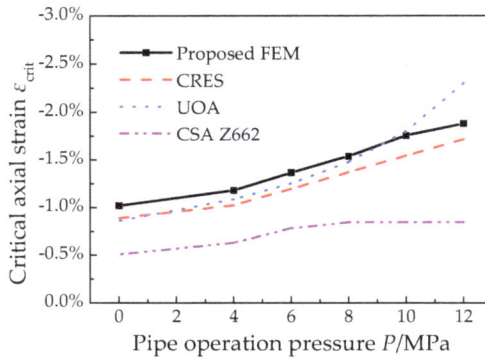

Figure 14. Relationships of the critical compressive strains with the operation pressure.

The post-buckling deformation as well as the axial strain distribution of pipe with different operation pressures were plotted in Figure 15. With an increase of operation pressure, the failure position is located a little nearer the fault trace generally. When *P* is larger than 6 MPa, the pipe buckles with almost a same shape of elephant's foot buckling. When *P* is 4 MPa, the pipe also performs an elephant's foot buckling but with a more abrupt local bulge. When *P* is 0 MPa, the pipe exhibits severe inward and outward deformation, also commonly known as "diamond buckling".

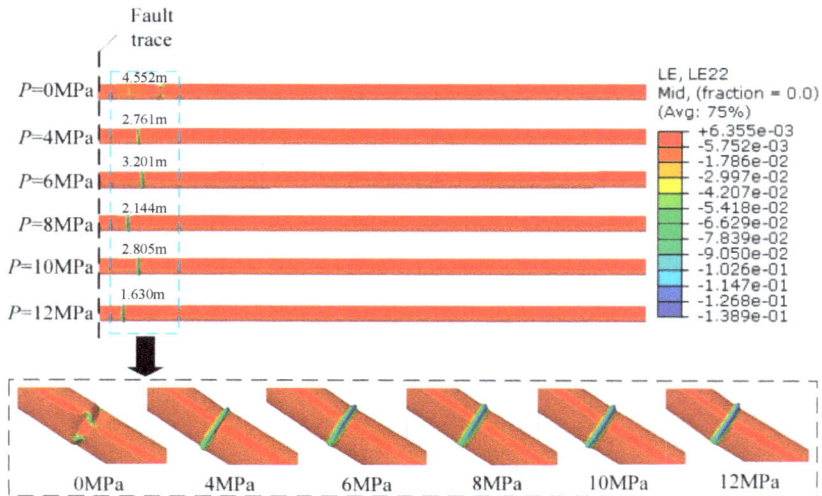

Figure 15. Post buckling behavior (axial strain contour) of pipeline with different operation pressure.

4.4. Effects of Pipe Wall Thickness

Due to various safety factors used in the design of pipelines for different regions, different pipe wall thicknesses will appear for pipes with the same outer diameter. In this section, five common values of X80 pipeline, i.e., 18.4, 19.1, 22.0, 26.4, 27.5 mm, were included to investigate effects of pipe wall thickness on the local buckling behavior for X80 pipe. Figure 16 illustrates trends of *SF* with *δ* for pipes with different pipe wall thicknesses. As a larger wall thickness induces a larger pipe stiffness, the pipe with a larger pipe wall thickness has a larger section axial force at the same fault displacement.

When δ reaches 0.47 m, the thinnest pipe (t = 18.4 mm) first buckled with a smallest critical section axial force SF_{crit}. In addition, with the increase of t, δ_{crit} and SF_{crit} both increase. A detailed relationship between δ_{crit} and t is plotted in Figure 17, showing δ_{crit} almost increases linearly with the increase of t, which is in good agreement of the theory results for critical buckling stresses of axial compressed steel cylinders as follows:

$$\sigma_{crit} = K\frac{2Et}{D} \tag{9}$$

where σ_{crit} is the critical buckling stress, E is the elastic modulus, K is the coefficient.

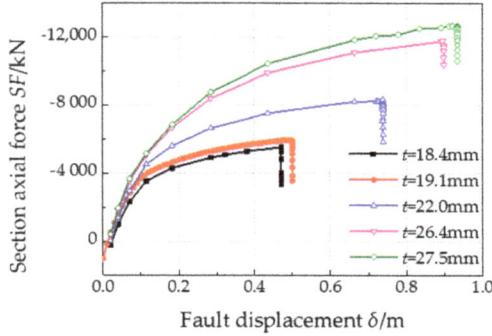

Figure 16. Relationship between section axial force and fault displacement with different pipe wall thickness.

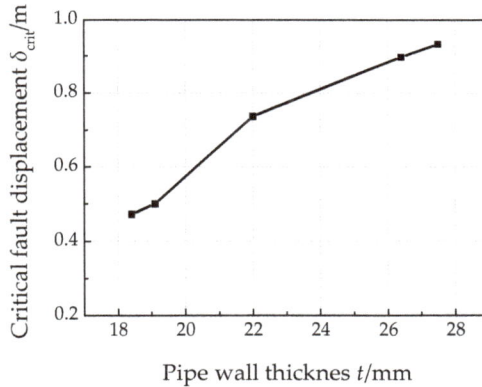

Figure 17. Relationships of the critical fault displacements with the pipe wall thickness.

Results of the critical compressive strains of various pipe wall thicknesses derived by different methods are further compared in Figure 18. With increase of t, critical compressive strains increase monotonically. CSA Z662 model predicted results are also quite conservative. If t > 22 mm, the UOA model results has better agreement with the proposed numerical results. While if t ≤ 19.1 mm, the CRES-GB 50470 model has better agreement with the proposed numerical results. However, it should also be noted that, in generally, both the UOA model and the CRES-GB 50470 model have rather good prediction results for a large range of pipe wall thickness.

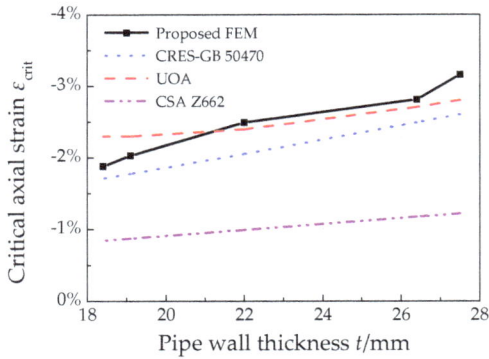

Figure 18. Relationships of the critical compressive strains with the pipe wall thickness.

Figure 19 illustrates the post buckling results for pipes with different pipe wall thickness. For the pipes with $t \leq 22$ mm, one wrinkle near the fault trace occurs after strain concentration induced by local buckling. For the pipe with $t \geq 26.4$ mm, two wrinkles will first appear in the pipe after local buckling, and finally concentrate to the major wrinkle. The local failure position is located further away from the fault trace with the increase of pipe wall thickness.

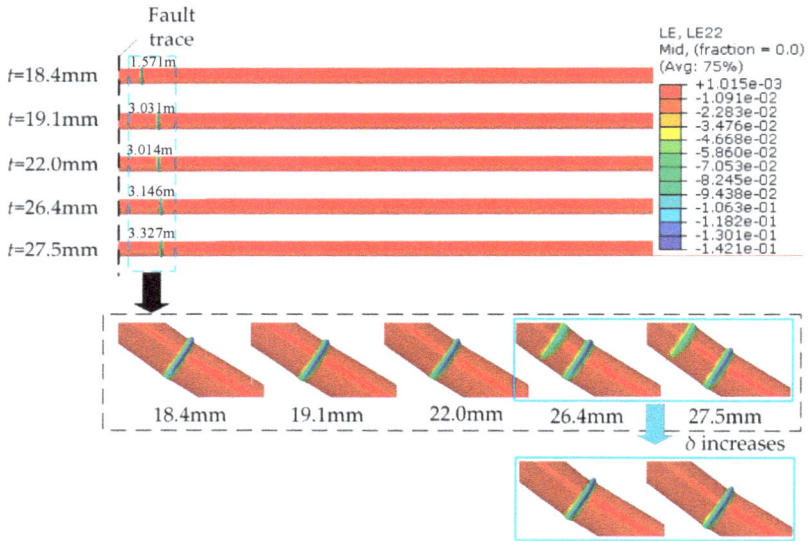

Figure 19. Post buckling behavior (axial strain contour) of pipeline with different wall thickness.

4.5. Effects of Soil Parameters

Buried pipelines are surrounded with soil. Constraints of soil on a pipe are directly related to the soil parameters as internal friction angle, cohesion, effective unit weight as well as the pipe soil friction coefficient according to ASCE-ALA Guideline [39]. In this section, five different kinds of clay with various parameters listed in Table 1 were considered to investigate the effects of soil constraints. The soil performed harder and has a larger resistance from Soil A to Soil E in sequence.

Table 1. Soil parameters of different types of soil sites in numerical investigation used for pipe soil interaction.

Soil Sites Type	Internal Friction Angle ψ (°)	Cohesion c (kPa)	Effective Unit Weight of Soil γ (kN/m³)	Pipe-Soil Friction Coefficient f
Soil A	30	30	22	0.6
Soil B	30	40	22	0.6
Soil C	33	50	22	0.6
Soil D	33	60	22	0.6
Soil E	33	70	22	0.6

Figure 20 shows the trends of section axial force *SF* in the buckled area with fault displacement δ in various soil types. When $\delta < 0.05$ m, the five curves are almost the same. After $\delta = 0.1$ m, *SF* for pipe buried in Soil E is the largest one, and it first drops when δ reaches 0.39 m. The relationships of the critical fault displacements derived from Figure 20 with the soil type are further illustrated in Figure 21, which shows that a harder soil stiffness induces a relatively smaller critical fault displacement. Thus, pipes crossing active faults locating hard soil sites have higher failure risk of plastic local buckling.

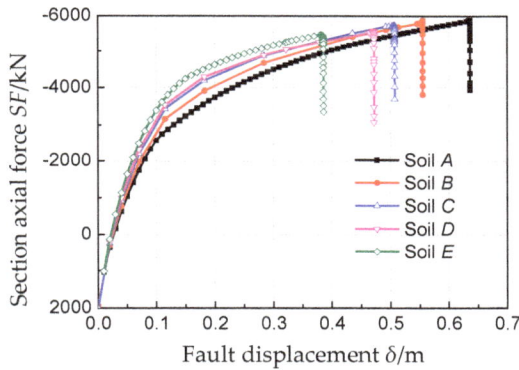

Figure 20. Relationship between section axial force and fault displacement with different soil parameters.

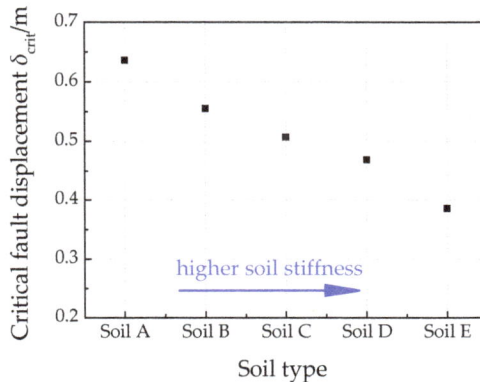

Figure 21. Relationships of the critical fault displacements with different soil parameters.

The critical compressive strains derived by the presented model for different soil conditions are plotted in Figure 22, comparing with common failure criteria based on the case parameters. The results of the CSA Z662 model, UOA model and CRES-GB 50470 model are all constants, similar to Figure 10, because the presented critical compressive strain prediction models are all based on pipe's pure bending test, which cannot consider the influences of external environment loads encountered by actual pipes in field. In addition, similar to the conclusions derived in the previous sections, the CSA model predicts rather conservative results. The UOA model and CRES-GB50470 model both have a good prediction, but the UOA model result is a little un-conservative. Furthermore, although ε_{crit} derived by the numerical model is around 2%, it should be noted that with the increase of soil stiffness surrounding pipe, ε_{crit} has a general decreasing tendency, which reflects that external load conditions should also affect some of pipe's strain capacity.

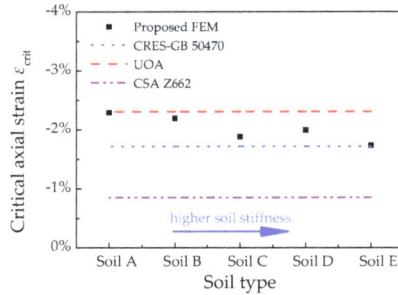

Figure 22. Relationships of the critical compressive strains for different soil parameters.

Post buckling axial strain contours of pipes buried in these five kinds of clay are compared in Figure 23. The pipes have almost same plastic section deformation with a bulge pattern induced by the operation pressure considered for all cases. However, variations exist between these five cases on the failure location. If the soil is relatively soft (i.e., Soil A–C), the local buckling failure position is located nearer the fault trace with a stiffer surrounding soil. However, if the soil is hard enough, harder than Soil C of the cases in this study, the local buckling failure position will have negligible variations for different soil types.

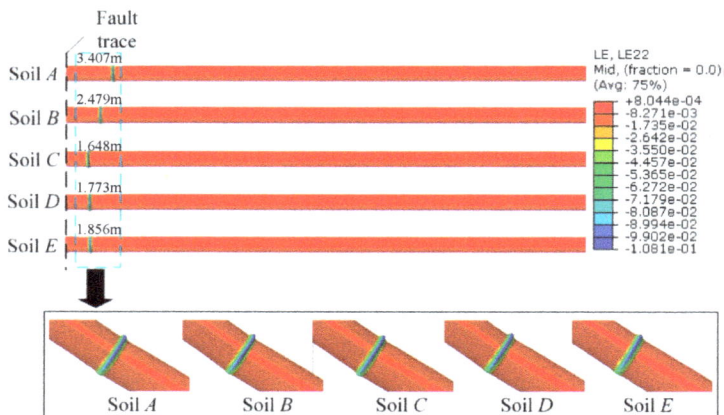

Figure 23. Post buckling behavior (axial strain contour) of pipeline with different soil parameters.

4.6. Effects of Pipe Buried Depth

Pipes are commonly buried in various depths according to the environment or pipe safety concern. In this section, five buried depth values, i.e., 0.5, 1.0, 1.5, 2.0, 3.0 m, possibly encountered in true engineering cases were adopted for parametric investigation. As illustrated in Figure 24, trends of SF with δ for pipes buried in different depth are the same, until the pipe buried in 3.0 m reaches its critical fault displacement (0.39 m), where its section axial force first drops. After that, pipe buckles in sequence from the deepest buried depth to shallowest buried depth. The quantitative relationship of the critical fault displacement δ_{crit} with pipe buried depth H is displayed in Figure 25, which shows that δ_{crit} decreases almost linearly with H, from 0.51 to 0.39 m, when H increases from 0.5 to 3 m. Thus, when subjected to fault displacement, a deeper buried depth of pipe enhances soil constraint on pipe leading to easier buckling failure in pipe.

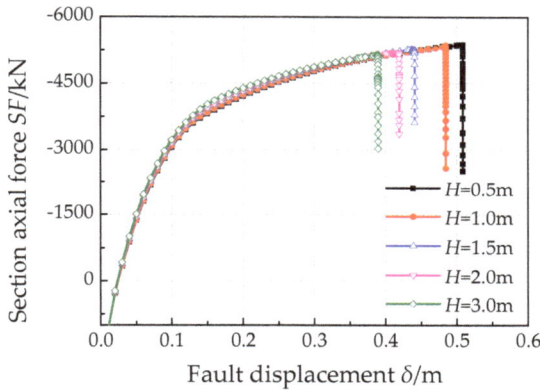

Figure 24. Relationship between section axial force and fault displacement with different buried depths.

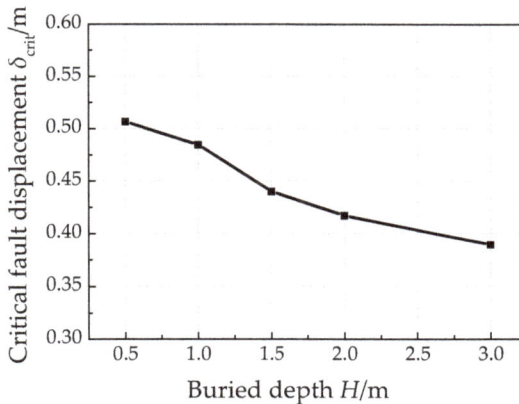

Figure 25. Relationship of critical fault displacements with buried depth.

Variations of the critical compressive strain in the pipe with buried depth are shown in Figure 26, which is quite similar with Figure 22. This is mainly because both increase of soil stiffness and increase of buried depth increase the soil resistance on pipe, which causes larger axial compression and lateral bending with the same fault displacement.

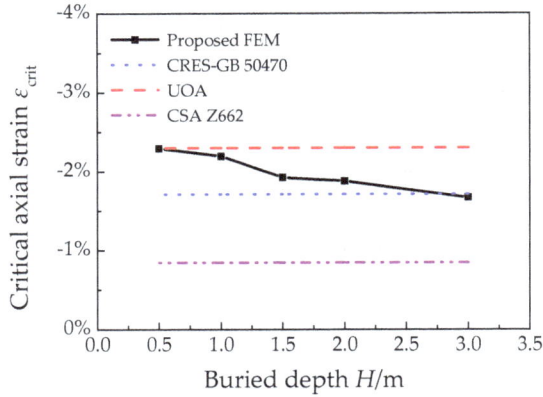

Figure 26. Relationships of critical compressive strains with buried depth.

The plastic deformation of pipes in the post buckling stage with the considered five buried depths are plotted in Figure 27. The same elephant foot buckling phenomenon appears in all cases. In addition, with the increase of pipe buried depth, the local buckling failure position is located nearer the fault trace as the trends found in Figure 23 for relatively soft soil types.

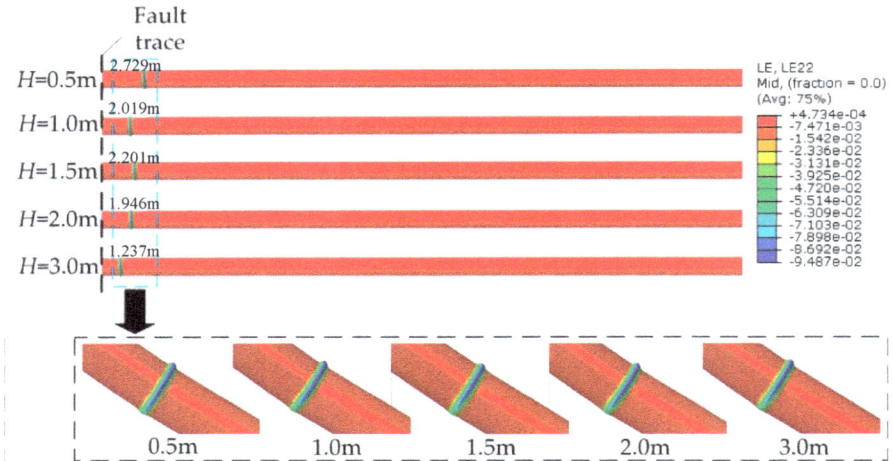

Figure 27. Post buckling behavior (axial strain contour) of pipeline with different buried depth.

5. Conclusions

Local buckling is a major failure type for buried steel pipelines subjected to active fault movements. Based on refined numerical model using finite element methods, it has been found that pipe subjected to strike-slip fault displacements incurs local buckling when it reaches the compressive limit state, which can be identified by the sudden drop of the section axial force in the buckled area. After buckling, wavy axial strain distribution appears initially and changes into strain concentration with wrinkling immediately. Effects of the pipe-fault intersection angle, pipe operation pressure, pipe wall thickness, soil parameters and pipe buried depth on the local buckling behavior of high-strength X80 pipes were further investigated through a parametric analysis. Some conclusions can be drawn:

1. For compressive strike-slip fault, the critical fault displacement increases initially and decrease afterwards with pipe fault intersection angle. When pipe fault intersection angle equals 135°, the pipe is most likely to be buckled and fail. The failure section of pipe locates further with increase of pipe fault intersection angle.

2. Pipe operation pressure decreases pipe's anti-buckling capacity generally, although a positive correlation between the critical compressive strain and pipe operation pressure was also found. In addition, operation pressure affects pipe's plastic deformation shape in post buckling stage. For the pipe investigated, when P is larger than 4 MPa, the pipe performs an elephant's foot buckling, and when P is 0 MPa, the pipe exhibits diamond buckling.

3. Pipe wall thickness has a positive relationship with both the critical fault displacement and critical axial strain for pipe. The failure pipe section of thicker pipes locates further away from fault trace.

4. Both increasing soil stiffness and pipe buried depth increase soil constraints on pipe, which will lead to a smaller critical fault displacement. In addition, pipe failure locations are nearer to fault trace with stronger constraints by surrounding soil.

5. For pure pipes without considering the geometry imperfections, the CSA Z662 model predicts too conservative results for compressive strain capacity. Generally, both the CRES-GB 50470 model and the UOA model have rather good critical compressive strain results compared with the numerical derived results for various conditions. Among them, the CRES-GB 50470 model is recommended for pipe failure assessment, because it is relatively conservative.

6. Numerical results show that not only the pipe parameters (i.e., geometrical parameters and internal pressure) but also external loads (pipe fault intersection angle, soil constraints on pipe) also have some minor effects on pipe's critical compressive strain, which is not considered by the commonly available buckling failure criteria.

Acknowledgments: This research has been co-financed by China National Key Research and Development Project under (Grant No. 2016YFC0802105), National Natural Science Foundation of China (Grant No. 51309236).

Author Contributions: Hong Zhang conceived and designed the physical model. Xiaoben Liu established the numerical model and wrote the paper. Baodong Wang, Kai Wu and Qian Zheng performed the numerical analysis. Mengying Xia and Yinshan Han analyzed the data.

Conflicts of Interest: The authors declare no conflict of interest.

Nomenclature

D	the pipe diameter (m)
t	the pipe wall thickness (m)
P	the pipe operation pressure (MPa)
ε_c^{crit}	the compressive strain capacity of steel pipes
σ_y	the yield strength of pipe steel (MPa)
E	the initial elastic modulus (MPa)
h_g	the height of the geometry imperfection (mm)
σ_u	the tensile strength of pipe material (MPa)
σ_a	the applied net-section stress in the longitudinal direction of pipe (MPa)
T_u	the axial peak resistant force per unit length of soil springs (kN/m)
P_u	the lateral peak resistant force per unit length of soil springs (kN/m)
Q_u	the vertical uplift peak resistant force per unit length of soil springs (kN/m)
Q_d	the vertical bearing peak resistant force per unit length of soil springs (kN/m)
Δt	the yield displacement in the axial direction (m)
Δp	the yield displacement in the lateral direction (m)
Δq_u	the yield displacement in the vertical uplift direction (m)
Δq_d	the yield displacement in the vertical bearing direction (m)
σ_{true}	the true stress (MPa)
ε_{true}	the true strain

r	the strain hardening exponent
α	the yield offset parameter
A	the pipe cross section area (m^2)
ΔL	the pipe extension at Point B (m)
δ	the fault displacement (m)
ΔX	the fault displacement in the pipe axial direction (m)
ΔY	the fault displacement in the pipe perpendicular direction (m)
β	the pipe fault intersection angle (°)
ε_{crit}	the critical compressive strain
SF	the section axial force in buckled area (kN)
SF_{crit}	the critical section axial force in buckled area (kN)
K	the coefficient of theory critical axial force
ε_{axial}	the axial compressive strain
δ_{crit}	the critical fault displacements (m)
ϕ	the internal friction angle of the soil (°)
c	the soil cohesion representative (kPa)
γ	the effective unit weight of soil (kN/m3)
f	the pipe–soil friction coefficient
H	the pipe buried depth (m)

References

1. Lower, M.D. Strain-Based Design Methodology of Large Diameter Grade X80 Linepipe. Ph.D. Thesis, University of Tennessee, Knoxville, TN, USA, 2014.
2. Liu, X.B.; Zhang, H.; Wu, K.; Xia, M.Y.; Chen, Y.F.; Li, M. Buckling failure mode analysis of buried X80 steel gas pipeline under reverse fault displacement. *Eng. Fail. Anal.* **2017**, *77*, 50–64. [CrossRef]
3. Newmark, N.M.; Hall, W.J. Pipeline design to resist large fault displacement. In Proceedings of the U.S. National Conference on Earthquake Engineering, Ann Arbor, MI, USA, 18–20 June 1975; pp. 416–425.
4. Kennedy, R.P.; Chow, A.W.; Williamson, R.A. Fault movement effects on buried oil pipeline. *Transp. Eng. J.* **1977**, *103*, 617–633.
5. Wang, R.L.; Yeh, Y.H. A refined seismic analysis and design of buried pipeline for fault movement. *Earthq. Eng. Struct. Dyn.* **1985**, *13*, 75–96. [CrossRef]
6. Karamitros, D.K.; Bouckovalas, G.D.; Kouretzis, G.P. Stress analysis of buried steel pipelines at strike-slip fault crossings. *Soil Dyn. Earthq. Eng.* **2007**, *27*, 200–211. [CrossRef]
7. Trifonov, O.V.; Cherniy, V.P. A semi-analytical approach to a nonlinear stress-strain analysis of buried steel pipelines crossing active faults. *Soil Dyn. Earthq. Eng.* **2010**, *30*, 1298–1308. [CrossRef]
8. Zhang, L.; Zhao, X.; Yan, X.; Yang, X. Elastoplastic analysis of mechanical response of buried pipelines under strike-slip faults. *Int. J. Geomech.* **2016**, *17*. [CrossRef]
9. Ha, D.; Abdoun, T.H.; O'Rourke, M.J.; Symans, M.D.; O'Rourke, T.D.; Palmer, M.C.; Steward, H.E. Buried high-density polyethylene pipelines subjected to normal and strike-slip faulting-a centrifuge investigation. *Can. Geotech. J.* **2008**, *45*, 1733–1742. [CrossRef]
10. Ha, D.; Abdoun, T.H.; O'Rourke, M.J.; Symans, M.D. Centrifuge modeling of earthquake effects on buried high-density polyethylene (HDPE) pipelines crossing fault zones. *J. Geotech. Geoenviron. Eng.* **2008**, *134*, 1501–1515. [CrossRef]
11. O'Rourke, T.D.; Jung, J.K.; Argyrou, C. Underground pipeline response to earthquake-induced ground deformation. *Soil Dyn. Earthq. Eng.* **2016**, *91*, 272–283. [CrossRef]
12. Jung, J.K.; O'Rourke, T.D.; Argyrou, C. Multi-directional force–displacement response of underground pipe in sand. *Can. Geotech. J.* **2016**, *53*, 1763–1781. [CrossRef]
13. Jalali, H.H.; Rofooei, F.R.; Attari, N.K.A. Performance of Buried Gas Distribution Pipelines Subjected to Reverse Fault Movement. *J. Earthq. Eng.* **2017**, *10*, 1–24. [CrossRef]
14. Jalali, H.H.; Rofooei, F.R.; Attari, N.K.A.; Samadian, M. Experimental and finite element study of the reverse faulting effects on buried continuous steel gas pipelines. *Soil Dyn. Earthq. Eng.* **2016**, *86*, 1–14. [CrossRef]

15. Rofooei, F.R.; Jalali, H.H.; Attari, N.K.A.; Kenarangi, H.; Samadian, M. Parametric study of buried steel and high density polyethylene gas pipelines due to oblique-reverse faulting. *Can. J. Civ. Eng.* **2015**, *42*, 178–189. [CrossRef]
16. Trifonov, O.V. Numerical stress-strain analysis of buried steel pipelines crossing active strike-slip faults with an emphasis on fault modeling aspects. *J. Pipeline Syst. Eng. Pract.* **2015**, *6*. [CrossRef]
17. Xie, X.J.; Symans, M.D.; O'Rourke, M.J.; Abdoun, T.H.; O'Rourke, T.D.; Palmer, M.C.; Stewart, H.E. Numerical modeling of buried HDPE pipelines subjected to normal faulting: A case study. *Earthq. Spectra* **2013**, *29*, 609–632. [CrossRef]
18. Xie, X.J.; Symans, M.D.; O'Rourke, M.J.; Abdoun, T.H.; O'Rourke, T.D.; Palmer, M.C.; Stewart, H.E. Numerical modeling of buried HDPE pipelines subjected to strike-slip faulting. *J. Earthq. Eng.* **2011**, *15*, 1273–1296. [CrossRef]
19. Joshi, S.; Prashant, A.; Deb, A.; Jain, S.K. Analysis of buried pipelines subjected to reverse fault motion. *Soil Dyn. Earthq. Eng.* **2011**, *31*, 930–940. [CrossRef]
20. Uckan, E.; Akbas, B.; Shen, J.; Rou, W.; Paolacci, W.; O'Rourke, M. A simplified analysis model for determining the seismic response of buried steel pipes at strike-slip fault crossings. *Soil Dyn. Earthq. Eng.* **2015**, *75*, 55–65. [CrossRef]
21. Liu, X.B.; Zhang, H.; Han, Y.S.; Xia, M.Y.; Zheng, W. A semi-empirical model for peak strain prediction of buried X80 steel pipelines under compression and bending at strike-slip fault crossings. *J. Nat. Gas Sci. Eng.* **2016**, *32*, 465–475. [CrossRef]
22. Melissianos, V.E.; Vamvatsikos, D.; Gantes, C.J. Performance assessment of buried pipelines at fault crossings. *Earthq. Spectra* **2017**, *33*, 201–218. [CrossRef]
23. Melissianos, V.E.; Vamvatsikos, D.; Gantes, C.J. Performance-based assessment of protection measures for buried pipes at strike-slip fault crossings. *Soil Dyn. Earthq. Eng.* **2017**, *101*, 1–11. [CrossRef]
24. Liu, X.B.; Zhang, H.; Gu, X.T.; Chen, Y.F.; Xia, M.Y.; Wu, K. Strain demand prediction method for buried X80 steel pipelines crossing oblique-reverse faults. *Earthq. Struct.* **2017**, *12*, 321–332. [CrossRef]
25. Xu, L.; Lin, M. Analysis of buried pipelines subjected to reverse fault motion using the vector form intrinsic finite element method. *Soil Dyn. Earthq. Eng.* **2017**, *93*, 61–83. [CrossRef]
26. Kaya, E.S.; Uckan, E.; O'Rourke, M.J.; Karamanos, S.A.; Akbas, B.; Cakir, F.; Cheng, Y. Failure analysis of a welded steel pipe at Kullar fault crossing. *Eng. Fail. Anal.* **2016**, *71*, 43–62. [CrossRef]
27. Zhang, J.; Liang, Z.; Han, C.J. Buckling behavior analysis of buried gas pipeline under strike-slip fault displacement. *J. Nat. Gas Sci. Eng.* **2014**, *21*, 921–928. [CrossRef]
28. Zhang, J.; Liang, Z.; Han, C.J.; Zhang, H. Numerical simulation of buckling behavior of the buried steel pipeline under reverse fault displacement. *Mech. Sci.* **2015**, *6*, 203–210. [CrossRef]
29. Vazouras, P.; Karamanos, S.A.; Dakoulas, P. Finite element analysis of buried steel pipelines under strike-slip fault displacements. *Soil Dyn. Earthq. Eng.* **2010**, *30*, 1361–1376. [CrossRef]
30. Vazouras, P.; Karamanos, S.A.; Dakoulas, P. Mechanical behavior of buried steel pipes crossing active strike-slip faults. *Soil Dyn. Earthq. Eng.* **2012**, *41*, 164–180. [CrossRef]
31. Vazouras, P.; Dakoulas, P.; Karamanos, S.A. Pipe-soil interaction and pipeline performance under strike-slip fault movements. *Soil Dyn. Earthq. Eng.* **2015**, *72*, 48–65. [CrossRef]
32. Vazouras, P.; Karamanos, S.A. Structural behavior of buried pipe bends and their effect on pipeline response in fault crossing areas. *Bull. Earthq. Eng.* **2017**, *4*, 1–26. [CrossRef]
33. Liu, X.B.; Zhang, H.; Li, M.; Xia, M.Y.; Zheng, W.; Wu, K.; Han, Y.S. Effects of steel properties on the local buckling response of high strength pipelines subjected to reverse faulting. *J. Nat. Gas Sci. Eng.* **2016**, *33*, 378–387. [CrossRef]
34. Kainat, M.; Lin, M.; Cheng, J.R.; Martens, M.; Adeeb, S. Effects of the initial geometric imperfections on the buckling behavior of high-strength UOE manufactured steel pipes. *J. Press. Vessel Technol.* **2016**, *138*. [CrossRef]
35. Neupane, S.; Adeeb, S.; Cheng, R.; Ferguson, J.; Martens, M. Modeling the deformation response of high strength steel pipelines—Part I: Material characterization to model the plastic anisotropy. *J. Appl. Mech.* **2012**, *136*, 272–275. [CrossRef]
36. Neupane, S.; Adeeb, S.; Cheng, R.; Ferguson, J.; Martens, M. Modeling the deformation response of high strength steel pipelines—Part II: Effects of material characterization on the deformation response of pipes. *J. Appl. Mech.* **2012**, *79*, 051003. [CrossRef]

37. Canadian Standard Association (CSA). *Oil and Gas Pipeline Systems*; CSA Standard; CSA Z662-11; Canadian Standard Association: Mississauga, ON, Canada, 2015.

38. Gresnigt, A.M. *Plastic Design of Buried Steel Pipelines in Settlement Areas*; STEVIN-laboratory of the Department of Civil Engineering, Delft University of Technology: Delft, The Netherlands, 1987.

39. American Lifelines Alliance. *American Society of Civil Engineers, Guidelines for the Design of Buried Steel Pipe (with Addenda through February 2005)*; American Lifelines Alliance: Reston, VA, USA, 2005.

40. Indian Institute of Technology Kanpur. *IITK-GSDMA Guidelines for Seismic Design of Buried Pipelines, Gandhinagar: Gujarat State Disaster Management Authority*; Indian Institute of Technology Kanpur: Kalyanpur, India, 2007.

41. Dorey, A.B.; Murray, D.W.; Cheng, J.J.R. Critical buckling strain equations for energy pipelines—A parametric study. *J. Offshore Mech. Arct. Eng.* **2006**, *128*, 248–255. [CrossRef]

42. Liu, M.; Wang, Y.Y.; Zhang, F.; Kotian, K. *Realistic Strain Capacity Models for Pipeline Construction and Maintenance*; US DOT PHMSA Other Transaction Agreement #DTPH56-10-T-000016, Draft Final Report; Center For Reliable Energy Systems: Dublin, OH, USA, 2013.

43. Codeofchina Inc. *GB 50470-2008 Seismic Technical Code for Oil and Gas Transmission Pipeline Engineering*; Codeofchina Inc.: Beijing, China, 2017.

44. American Petroleum Institute (API). *Design, Construction, Operation, and Maintenance of Offshore Hydrocarbon Pipelines (Limit State Design)*; API Recommended Practice, API RP 1111; American Petroleum Institute: Washington, DC, USA, 1991.

45. Det Norske Veritas (DNV). *Submarine Pipeline Systems*; DNV Offshore Standard, DNV-OS-F101; Det Norske Veritas: Oslo, Norway, 2010.

46. Committee on Gas and Liquid Fuel Lifelines of the American Society of Civil Engineers Technical Council on Lifeline Earthquake Engineering. *Guidelines for the Seismic Design of Oil and Gas Pipeline Systems*; ASCE: Reston, VA, USA, 1984; pp. 10–12.

47. Hibbitt, D.; Karlsson, B.; Sorensen, P. *ABAQUS Standard User's and Reference Manuals*; Version 6.14; Hibbitt, Karlsson & Sorensen, Inc.: Johnston, RI, USA, 2007.

48. Ramberg, W.; Osgood, W.R. *Description of Stress-Strain Curves by Three Parameters*; NACA Technical Note; No. 902; National Advisory Committee for Aeronautics (NACA): Washington, DC, USA, 1943.

49. Liu, X.B.; Zhang, H.; Xia, M.Y. Buckling behavior of buried steel pipeline under compression strike-slip fault. In Proceedings of the 2017 ASME Pressure Vessels & Piping Conference, Waikoloa, HI, USA, 16–20 July 2017.

50. Martini, A.; Troncossi, M.; Rivola, A. Leak Detection in Water-Filled Small-Diameter Polyethylene Pipes by Means of Acoustic Emission Measurements. *Appl. Sci.* **2017**, *7*, 2. [CrossRef]

51. Juliano, T.M.; Meegoda, J.N.; Watts, D.J. Acoustic Emission Leak Detection on a Metal Pipeline Buried in Sandy Soil. *J. Pipeline Syst. Eng. Pract.* **2013**, *4*, 149–155. [CrossRef]

52. Martini, A.; Troncossi, M.; Rivola, A. Vibroacoustic Measurements for Detecting Water Leaks in Buried Small-Diameter Plastic Pipes. *J. Pipeline Syst. Eng. Pract.* **2017**, *8*, 1–10. [CrossRef]

53. Yazdekhasti, S.; Piratla, K.R.; Atamturktur, S.; Khan, A. Experimental evaluation of a vibration-based leak detection technique for water pipelines. *Struct. Infrastruct. Eng.* **2017**, *14*, 46–55. [CrossRef]

54. Liu, A.; Hu, Y.; Zhao, F.; Li, X.; Takada, S.; Zhao, L. An equivalent-boundary method for the shell analysis of buried pipelines under fault movement. *Acta Seismol. Sin.* **2004**, *17*, 150–156. [CrossRef]

55. American Society of Mechanical Engineers. *Pipeline Transportation Systems for Liquid Hydro Carbons and Other Liquids*; ANSI/ASME2016, B31:4; American Society of Mechanical Engineers: New York, NY, USA, 2016.

56. American Society of Mechanical Engineers. *Gas Transmission and Distribution Piping Systems*; ANSI/ASME2016, B31:8; American Society of Mechanical Engineers: New York, NY, USA; 2016.

![metals logo] *metals*

MDPI

Article

Evaluation of Failure Pressure for Gas Pipelines with Combined Defects

Tadas Vilkys [1], Vitalijus Rudzinskas [2], Olegas Prentkovskis [3,*], Jurijus Tretjakovas [4], Nikolaj Višniakov [2] and Pavlo Maruschak [5]

[1] Vilpros Pramonė, UAB, Vilniaus g. 11, Izabelinė LT-14200, Lithuania; tadas.vilkys@gmail.com

[2] Department of Mechanical and Material Engineering, Faculty of Mechanics, Vilnius Gediminas Technical University, J. Basanavičiaus g. 28, Vilnius LT-03224, Lithuania; vitalijus.rudzinskas@vgtu.lt (V.R.), nikolaj.visniakov@vgtu.lt (N.V.)

[3] Department of Mobile Machinery and Railway Transport, Faculty of Transport Engineering, Vilnius Gediminas Technical University, Plytinės g. 27, Vilnius LT-10105, Lithuania

[4] Department of Applied Mechanics, Faculty of Civil Engineering, Vilnius Gediminas Technical University, Saulėtekio al. 11, Vilnius LT-10223, Lithuania; jurijus.tretjakovas@vgtu.lt

[5] Department of Industrial Automation, Ternopil National Ivan Pul'uj Technical University, Rus'ka str. 56, Ternopil 46001, Ukraine; maruschak.tu.edu@gmail.com

* Correspondence: olegas.prentkovskis@vgtu.lt; Tel.: +370-5-2744784

Received: 11 March 2018; Accepted: 8 May 2018; Published: 11 May 2018

Abstract: The paper presents the study of the influence of mechanical damage on the safe operation of gas transmission pipelines. The main types of pipeline damage with the actual parameters and their influence on the operational parameters are analysed. The damaged fractures of the section of the pipeline Kaunas (Lithuania)–Kaliningrad (Russia) were investigated in the laboratory. The main operational characteristics and the structure of the pipeline's metal after the period of long-term operation were determined using various research and experimental methods. The influence of the pipeline's damage was modelled by using the Finite Element Method and the ANSYS code. The predictions of the failure pressure were made, taking into consideration the actual properties of the pipeline's metal. Techniques including the hardness and microhardness measurement, chemical analysis, the impact strength test, and metallography analysis with an optical microscope, were used in the experimental study.

Keywords: combined defect; gas pipeline; mechanical damage; failure pressure; finite element method

---

1. Introduction

Lithuania has a well-developed gas pipeline network, which is continuously being improved. However, most of the gas transmission pipelines were laid when Lithuania was a part of the Soviet Union. In the European Union, some of the trunks were laid underground. The total length of the gas pipelines in the Soviet Union in 1970 was 70 thousand kilometres, while in 1986, it reached 174 thousand kilometres [1,2]. Most of the pipes used for the trunks in the Soviet Union were the first- generation pipes made of normalised steel 17G1S GOST 19281-73 (1970s) [3]. The second-generation pipes were made of the improved (carbide-hardened) steel of grades 17G2SF or 17G2SAF GOST 19281-73 (since 1975) [3]. The steel of grades 17G2SF and 7G2SAF were used to produce spiral-welded pipes from 530 to 1220 mm in diameter according to TU 14-3-731-78 [4]. The Minsk–Vilnius–Kaunas–Kaliningrad (Belorussia–Lithuania–Russia) gas pipeline (ø 530 mm), which is currently operating on the territory of the Republic of Lithuania, was put into operation in 1985. These pipelines have been operated for more than 30 years [1]. This period of pipeline operation was marked by the increased likelihood of accidents due to metal aging, fatigue, and corrosion.

Thirty or forty years ago, the building technologies were imperfect and the quality of work was very low. Therefore, pipelines with various mechanical damages were buried and put into operation [5–8]. The operating time of these pipelines was inevitably very short. The analysis of the practical use of gas pipelines [9–11] allows us to state that the main causes of pipeline accidents and leakages are as follows:

- Corrosion;
- Gouges;
- Plain and kinked dents;
- Smooth dents on welds;
- Smooth dents with other types of defects;
- Manufacturing defects in the pipe body;
- Girth and seam weld defects;
- Cracking;
- Environmental cracking.

Mechanical damage is the most common causes of pipeline accidents: the majority being external damages (49.6%). Manufacturing defects reach 16.5%, corrosion makes 15.3%, and ground displacement makes 7.3%, while other causes contribute 11.3% [12–15].

Because of the high level (49.6%) of accidents caused by mechanical damages, the companies maintaining the buried pipelines are often reluctant to apply expensive diagnostics, monitoring, and repairing techniques. As a rule, the buried pipelines are checked by using the intelligent pigging and inspection methods. The application of these technical solutions helped to detect many dangerous defects during the pipelines' operation [16–19]. Sometimes, the shape defects were so huge that the intelligent pigging robot could not move forward.

The shape defects, which are characterised by large plastic deformations with (or without) reduced wall thickness, are very dangerous because they can initiate (or can be combined with) various types of corrosion defects or cracks [20–23]. Therefore, these defects should also be examined to decide whether the pipelines of this age can be safely used further, or whether these defects must be removed.

For this reason, it is important to model the influence and evolution of local shape defects in order to assess the pipeline's reliability. The traditional analytical methods for evaluating the pipeline's failure pressure and probability of leaks are based on the determination of the defect's depth and length. Well-known mathematical methods, codes, and mathematical expressions can be used to predict the failure pressure of steel pipelines with various corrosion defects. The Finite Element Method allows for predicting failure pressure and evaluating complex defects for obtaining better results than the traditional methods can yield [23].

2. The Object of Research

The research object is the fragment of the pipeline ø 530 mm with mechanical surface defects, which has been operated since 1985. The pipeline's steel 17G2SF (GOST 19281-73) [3], which contains alloying elements such as Nb, V, and Ti, was investigated (Table 1) [24]. Its standard yield strength R_e is 380 MPa and the tensile strength R_m is 550 MPa (Table 2) [3]. The class of the steel strength is K55 according to GOST 19281-73 [3], while the pipe's grade is Ch55 according to API 5L [25].

Table 1. Chemical composition of 17G2SF steel according to GOST 19281-73 [3].

Element	C	Si	Mn	P	S	N	Cu	Cr	Ni	Nb	Ti	V
wt. %	0.15–0.2	0.4–0.7	1.3–1.7	Max 0.035	Max 0.04	Max 0.012	Max 0.3	Max 0.3	Max 0.3	0.02–0.05	0.01–0.03	0.05–0.1
					Fe residue							

Table 2. Typical mechanical properties of 17G2SF steel according to GOST 19281-73 [3].

Yield Strength R_a, MPa	Tensile Strength R_m, MPa	Elongation A, %	Impact Strength at Ambient Temperature KCU, J/cm²	Impact Strength at Ambient Temperature KCV, J/cm²
380–470	550–650	23–30	44–65	Min 40

The geometric parameters of the pipe according to TU 14-3-731-78 [4] and the specification GOST 20295-85 [26] are as follows: the external diameter is ø 530 ± 3 mm and the wall thickness is 8 mm. The working pressure of the gas pipelines, which are from 530 mm to 1020 mm in diameter, is 5.45 MPa during the operation period [24]. The surface of the fragment of the investigated pipe had some combined defects such as dents, gouges, and corroded areas. The detected defects differed in size, but the depth of the most dangerous geometric defects reached 4 mm (about 50% of the total wall thickness) (Table 3). The geometry of the combined defects is presented in Figure 1.

Table 3. Characteristics of the combined defects.

Vessel	Sample 1	Sample 2
Defect type	Dent and gouge	Dent and corrosion
Dent depth b, mm	17.4 mm (3.5% 2·r)	17.4 mm (3.5% 2·r)
Gouge length 2·a, mm	8 mm (t)	
Gouge width 2·c, mm	4 mm ($t/2$)	
Gouge depth d, mm	4 mm ($t/2$)	
Gauge radius δ, mm	2 mm ($t/4$)	
Pitting diameter e, mm		4 mm ($t/2$)
Depth of pitting corrosion d, mm	4 mm ($t/2$)	4 mm ($t/2$)
Failure location	Gouge	Area of pitting corrosion

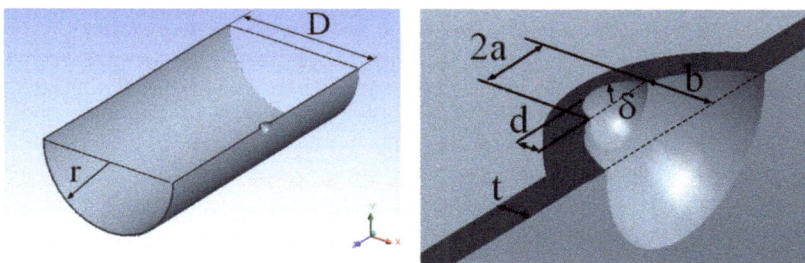

Figure 1. The geometry of the combined defect: a denotes a half-length of the gouge, mm; b is the depth of the dent, mm; c is the width of the gouge, mm; d is the depth of the gauge, mm; t is wall thickness, mm; r is the internal radius of the pipe, mm, and D is the external diameter of the pipe, mm.

3. Methodology of the Experiment

The chemical analysis of a metal sample was performed to identify the metal grade and to make sure that it meets the requirements of the project documentation and GOST 19281-73 [3] and to assess the metal equivalent carbon content.

The chemical analysis was performed using the optical spectroscopy BELEC Compact Lab N according to GOST 7565-81 [27]. The carbon equivalent of steel was calculated by the equation as follows [3]:

$$CE = C + \frac{Mn}{6} + \frac{Si}{24} + \frac{Cr}{5} + \frac{Ni}{40} + \frac{Cu}{13} + \frac{V}{14} + \frac{P}{2} \qquad (1)$$

where CE is carbon equivalent, %, and C, Mn, Si, Cr, Ni, Cu, V, and P denote the concentration of the elements, wt. %.

The main mechanical properties of the pipe's metal were determined to assess the changes in the metal properties during the elaboration period and to determine their state at the present moment. The Charpy pendulum impact test was performed according to the specification EN ISO 148-1:2016 [28]. Since the tube's wall thickness is 8 mm, the standard-size sample could not be used. Therefore, the V-notch test pieces of the square section with 7.5 mm sides and 55 mm length were used. The pendulum impact tester 2130KM-03 (Soviet Union) and an 8 mm strike were used in testing. The tests were made at the ambient temperature of 20 °C. The test pieces for the impact test were taken from various zones of the tube (Figure 2). Three specimens were taken from the undamaged zone and one from the dented zone. The specimen from the dented zone was aligned prior to impact testing because the shell of the pipe in this area was deformed.

Figure 2. A schematic view of the preparation of Charpy V-notch test pieces.

Metallographic tests were performed to assess the potential level of degradation of the metal in the vat. The samples were taken from the undamaged and deformed sections of the pipe.

The Buehler Beta 2 (Buehler, Lake Bluff, IL, USA) grinding and polishing machine was used for preparing metallographic samples (Figure 3). After performing the abrasive grinding and polishing of the specimens, the polished surface was etched with the most commonly used metallographic etchant, 2% Nital. The microstructure of the metal in the pipe bend was examined by using a microscope MA200 (Nikon, Tokyo, Japan) with optical magnification of ×50 to ×500. The strengthening of steel in the deformation (dent) area was investigated by performing a microhardness test according to the requirement of EN ISO 9015-2:2016 [29]. The automatic microhardness tester ZHμ (Zwick Roell, Ulm, Germany) with a load of 25 g and 10 s of storage was used in the experiment. The measurements were made on the surfaces of metallographic samples taken from the deformed and undamaged sections of the pipe.

The basic mechanical properties of the pipe's metal were determined by performing the hardness and stretching tests. Metal hardness tests were made on the outer surface of the pipe according to the specifications EN ISO 6507-1:2005 [30]. The Vickers' hardness was measured in the dented zone and compared to the hardness in the non-dented zone.

The universal hardness tester ZHU (Zwick Roell, Ulm, Germany) with a diamond pyramid indenter was used for the test. Since hardness tests do not allow for determining the important mechanical properties of the metal of the pipe, destructive tensile tests were also made. Only the tensile testing of the samples taken from the undamaged pipe's area was performed, since the tensile testing of the metal taken from the area with defects could not be made because of the fragment's small dimensions.

Tensile tests were performed according to EN ISO 6892-1:2016 [31]. The universal testing machine TIRA Test 2300 (up to 100 kN) (TIRA GmbH, Schalkau, Germany) with a multi-channel electronic PC data acquisition system SPIDER8 (HBM, Darmstadt, Germany)was used for the tensile test.

The machined longitudinal test pieces were of E type, with cross section $So = 112.5$ mm^2, original gauge length $Lo = 60$ mm, and parallel length $Lc = 68$ mm, were used for testing at room temperature.

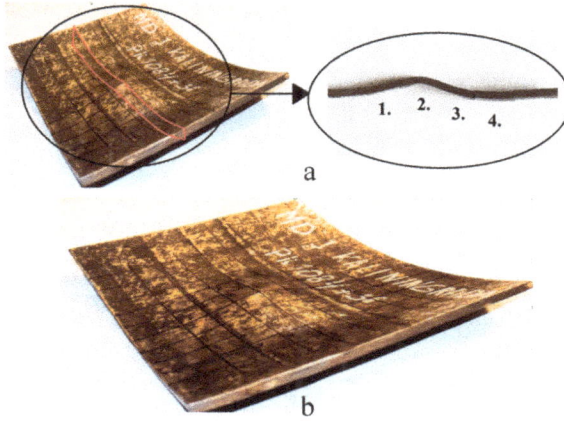

Figure 3. A schematic view of the preparation of test pieces: (**a**) denotes the disposition of metallographic test pieces; (**b**) denotes the areas of hardness testing.

4. Experimental Results and Discussion

The chemical composition of the metal of the tested pipe's fragment meets the 17G2SF steel grade (Table 4). This steel is analogous to 17G1S, but is alloyed with Ti and V.

Table 4. Chemical composition of 17G2SF grade steel.

Element	C	Si	Mn	P	S	N	Cu	Cr	Ni	Nb	Ti	V
wt. %	0.18	0.3	1.09	0.029	0.028	0.01	0.1	0.11	0.07	0.004	0.01	0.05
					Fe residue							

The carbon equivalent of the carbon steel calculated by Equation (1) was 0.423%. Steel with $CE <$ 0.45% is welded appropriately, and its additional heating is optional.

According to the results of the impact-tensile test, it is evident that the impact load characteristics are similar to the characteristics of the deformed and undamaged pipe sections. There was no significant change in the impact durability of the investigated steel (Table 5). The impact strength of the samples is practically equivalent (the average KCV 65.27 J/cm^2) to the ordinary 17G2SF steel strength, according to GOST 19281-73 [3].

Table 5. The results of Charpy pendulum impact test.

Specimen	h, mm	b, mm	S_0, cm^2	KV_8, J	KCV, J/cm^2
1	7.5	7.5	0.41	26.7	64.8
2	7.5	7.5	0.41	28.3	68.7
3	7.5	7.5	0.41	27.5	66.7
4	7.5	7.5	0.41	25.1	60.9

The obtained results guarantee the plastic fracture of this type of steel after long-term operation.

A typical steel 17G2SF microstructure consists of ferrites, pearlite, and dispersive carbides. The welded spiral pipes are made of these steel sheets, which have been normalised. The microscopic microstructure of the investigated samples is typical of this steel grade. The microstructure consists of the ferrite and pearlite phases (Figure 4).

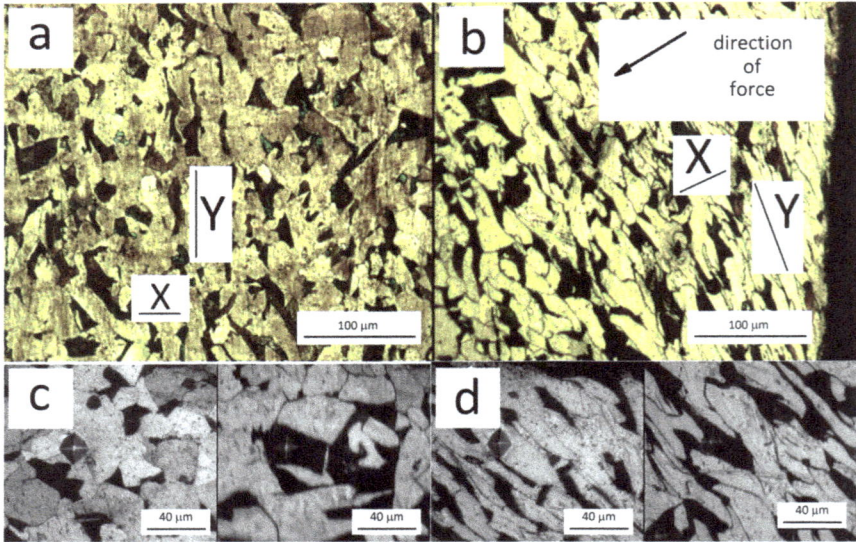

Figure 4. Steel microstructure: (**a**) denotes the undeformed area of the pipe; (**b**) is the area of the dent; (**c**) is an enlarged view of the microstructure of the steel with the microhardness indentations of the undamaged section of the pipe; (**d**) is an enlarged view of the microstructure of the steel with the microhardness indentations of the deformed tube section; ferrite—white grains; pearlite—dark grains.

Metallographic evaluation of the steel grain size was performed according to the specification EN ISO 643:2013 [32]. The 17G2SF steel pipe has a fine-grained structure with a typical ferrite grain size G 7–9 (with the average diameter of 0.015–0.031 mm) [24]. The metal grain size of the analysed tube fragments corresponds to G 6.5 (when the mean diameter of the ferrite grain is 0.035 mm). However, in the deformed tube section, there is a significant deformation of the grain (Table 6). The deformation of the grain due to the pipe's deflection was evaluated according to equation [33]:

$$K = \frac{D_y}{D_x} \qquad (2)$$

where K is the deformation coefficient; D_x is the diameter of the grain along x-axis, mm; D_y is the diameter of the grain along y-axis, mm.

Table 6. The average grain size and grain deformation levels.

Parameter	Ferrite		Pearlite	
	Undeformed Section	Dent Area	Undeformed Section	Dent Area
The diameter of the grain along x-axis, mm	0.031	0.013	0.012	0.011
The diameter of the grain along y-axis, mm	0.040	0.046	0.033	0.036
Deformation coefficient K	1.29	3.53	2.75	3.27

In the undeformed section of the pipeline, ferrite grains have a regular shape. In the area of defects, the deformation of ferrite grains in the direction along the x-axis is 2.7 times as large. Pearlite dimensions along the X and Y directions have not changed. The orientation of the grains has changed in the dent area. This is logical because pearlite grains are harder than the ferrite ones.

According to the orientation of the stretched asymmetric grains, it is possible to determine the direction of the external loads in the deformed area (Figure 4b). These structural changes increase the material's anisotropy because the deformed grains have different mechanical properties in different directions. This was confirmed by the results of the microhardness testing of this pipe. In the undeformed section of the pipe, the impression diagonals on the ferrite and pearlite grains are symmetrical (Figure 4c). Meanwhile, in the deformed part of the tube, the impressions are asymmetrical (Figure 4d). As shown in Figure 4d, the ferrite deformation level is higher than that of pearlite. In the area of the undeformed ferrite grains, the difference in the diagonal length of the impression is 1.4 times. As shown in Figure 5, the changes in hardness can be observed only in the deformed part of the pipe (the dent area). The hardness of HV in the undeformed section was 152 HV10, while in the deformed part it was 197 HV10. The degree of hardening in the dent area, when compared to that of the undeformed section of the tube, was approximatively 1.3.

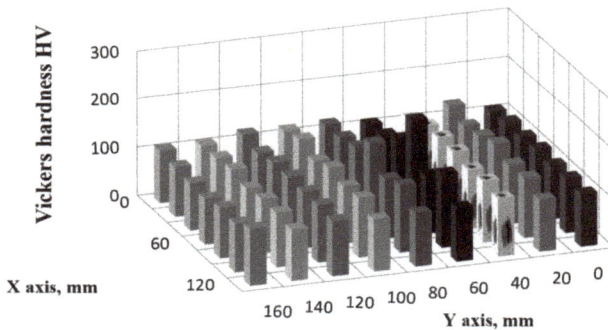

Figure 5. The differences in Vicker's hardness on the sample's surface.

The microstructure and hardness measurements provided important information about the anisotropy of the metal of the pipe in the deformed tube's section. Theoretically, based on the Vickers' hardness value, it is possible to determine the approximate tensile strength of the metal [34] as follows:

$$R_m = HV \cdot \frac{10}{3} \tag{3}$$

where R_m is tensile strength, MPa; HV is the Vickers hardness, MPa.

According to the above equation, the calculated strength limit for an undeformed pipe is 506 MPa. This value is very close (with the difference not exceeding 9%) to the typical steel tensile strength of 550 MPa. In the deformed section of the pipe, the strength limit calculated according to the given equation is 656 MPa. The results of the tensile tests confirm that the metal of the pipe remains sufficiently plastic ($A = 23\%$). Its relative elongation still meets the requirements of GOST 19281-73 [3]. However, the tensile strength of the metal pipe tested is about 10% lower than the tensile strength of similar new pipes (Table 7).

Table 7. The mechanical properties of the pipe's metal.

Properties	Yield Strength R_a, MPa	Tensile Strength R_m, MPa	Elongation A, %
The initial characteristics of steel according to the certificate	380	550	25
Service-exposed steel	350	502	23

5. The Calculation of Failure Pressure

Now, there is a number of methodologies for calculating the critical pressure of a pipeline with defects. The critical pressure *P* is close to the limit pressure given by a simple equation [14,15]:

$$P = \frac{R_m \cdot t}{r} \tag{4}$$

where R_m is the ultimate strength determined by the tension test (502 MPa); *t* is the wall thickness (8 mm); *r* is the outer radius of the pipe (257 mm).

The burst pressure of a ø 530 mm vessel made of 17G2SF steel without any defects is equal to 15.5 MPa. When there are deep gouges in the pipeline, causing the decrease in the wall thickness by half or more, the burst pressure, according to this equation, can reach the operating pressure of 5.45 MPa. Following the considered equation, this can occur when a gouge's depth reaches about 5 mm.

6. The Numerical Modelling of the Pipe

The ANSYS code [35] was used for calculation and simulation by employing the finite elements method. A model of the 3D FEA (Finite Element Analysis) of the pipe with the combined defects was developed. The 8-node brick finite elements for the volumetric model of the pipe with defects were used. The main geometrical characteristics of the pipe (Figure 6) were as follows: the external diameter of the tube was d_e = 530 mm and the thickness of the pipe's wall was *t* = 8.00 mm. The internal pressure *p* of the fluid pipeline can vary from 2.5 MPa to 10 MPa [36], but in practice, these pipelines operate at a pressure of up to 5.4 MPa [37]. Since the length of the pipe was infinite, for modelling, it was considered a constant equal to 1.0 m. The material of the pipe was assumed to be elastoplastic, with the nonlinear stress–strain relation, as well as homogeneous and isotropic. However, in the course of the experimental investigation, the anisotropy of the metal was found to be in the field of the pipe's defects. The experimentally obtained mechanical characteristics of the steel were applied in testing. The stress–strain curve of the material modelled by BKIN (Bilinear Kinematic Hardening Model) was used. The plastic behaviour of the material characterised by a non-recoverable strain could be observed when the internal stresses exceeded the material's yield point. Static stress analysis of the pipe with defects was performed by simulating three pressure modes (minimum pressure of 3.6 MPa, medium pressure of 4.4 MPa, and maximum pressure of 5.4 MPa).

Figure 6 shows a classical model of an internally pressurised pipe with the external radius d_e and thickness *t*. The plastic behaviour of the material, characterised by the non-recoverable strain, can be observed when the stresses exceed the yield point of the material.

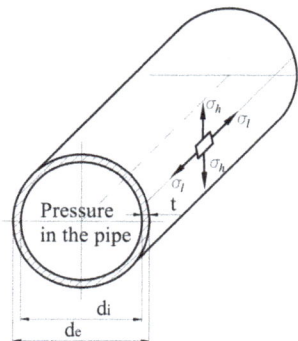

Figure 6. The hoop and the longitudinal stresses of the pipe.

The hoop stresses σ_h and longitudinal stresses σ_l induced by the internal pressure are given by the closed-form equations:

$$\sigma_h = \frac{p \cdot d_i}{2 \cdot t} \tag{5}$$

$$\sigma_l = \frac{p \cdot d_i}{4 \cdot t} \tag{6}$$

The hoop stresses of the pipeline without defects under normal operating conditions (at a pressure of 3.6 MPa) could reach 115.4 MPa (Table 8).

Table 8. The results of hoop stress testing.

Stresses	Analytical	Numerical	Error, %
σ_h, MPa	115.4	117.5	1.8

Over the course of the research, several defect combinations were modelled: (1) only a gouge or a pitting corrosion defect was modelled; (2) a combination of a dent and a gouge was modelled (Figure 7). Their depth could vary. Thus, in the model of the pipe, the defects' depth increased in 0.5 mm increments from zero to 4.0 mm. The influence of the defect's depth on the distribution of principal stresses, σ_1, of the pipe was investigated numerically. The principal stresses of the pipe with a gouge defect under 3.6 MPa pressure are presented in Table 9 and Figure 8. Table 9 shows that in the pipeline with a 4-mm deep gouge at a working pressure of 3.6 MPa, the values of the maximum principal stresses are 1.96 times as large as these values obtained for a pipe without any defects.

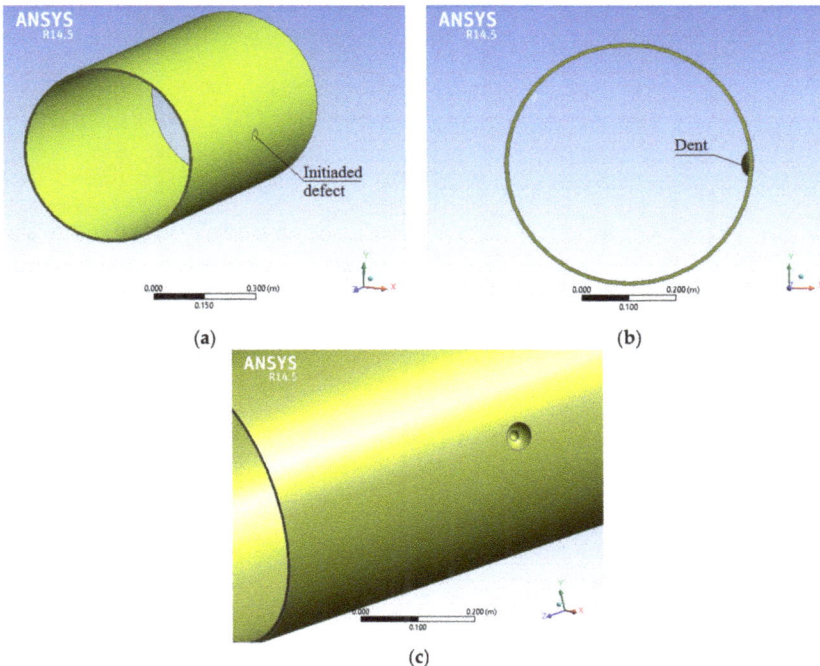

Figure 7. The model of the pipe with the initiated combined defects: (**a**) denotes a pipe with a gouge defect; (**b**) is a pipe with a dent; (**c**) is a pipe with a combined defect.

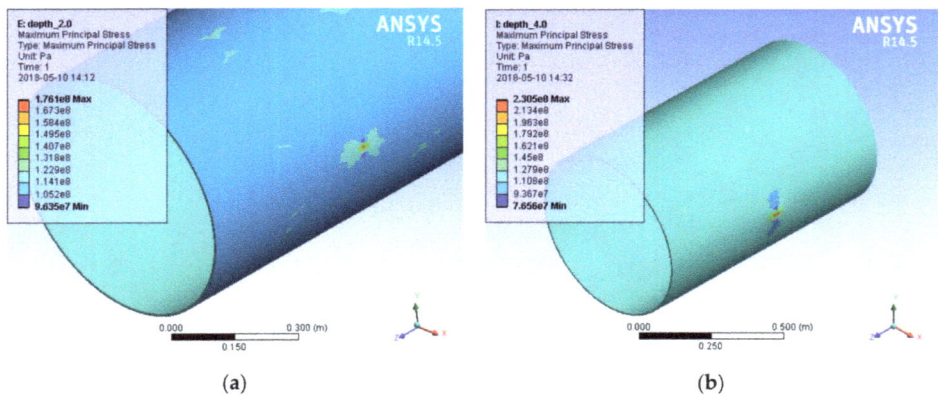

Figure 8. Principal stresses in the pipe with a gouge of various depths: (**a**) 2.0 mm; (**b**) 4.0 mm.

Table 9. Principal stresses in the pipe with a gouge defect of various depths.

Depth, mm	σ_1, MPa
0	117.4
0.5	130.2
1.0	147.0
1.5	172.4
2.0	176.1
2.5	191.3
3.0	201.9
3.5	211.6
4.0	230.5

Similarly, the principal stresses in the pipe with a gouge defect under the pressures of 4.4 MPa and 5.4 MPa were simulated (Figure 9). In the pipeline with a 4-mm deep gauge under the working pressure of 4.4 MPa, the values of the maximum principal stresses reached 282 MPa (Figure 9a). Then, the elastic-plastic geometrical linear model with a 4-mm deep gouge was solved, using the load with the maximum value $p = 5.4$ MPa. Stress of 354 MPa (Figure 9b) developed in the defect area. The yield stress of the pipe's steel was 350 MPa, therefore, the plastic deformation of metal could be initiated in the defect area. If the defect is larger than 4 mm, the internal stress of 354 MPa in the defect area significantly exceeds the metal's yield strength. Therefore, the working pressure of the pipe at 5.4 MPa becomes dangerous for its safe operation because of the risk of the pipe bursting.

Another case of the defect's simulation was based on the combination of a dent and a gouge. A case of a 4-mm deep gouge combined with a dent defect was modelled. The dent surface was modelled as a spherical surface and the depth of this defect was about 17.4 mm. Under the same working pressure conditions, the internal stress in the field of the compound defect of 5.4 MPa was significantly higher than the yield strength of steel and higher than that obtained by simulating the defect of the same depth gouge. The stress in the area of the combined defects increased to 396 MPa (when the gouge depth was 4 mm, Figure 10a) and to 417 MPa (when the gouge depth was 4.5 mm), respectively (Figure 10b). The dent's factor changed the maximal stress location from the defect's centre to the dent's edge. Therefore, the simulation of the dent of variable depth and several gouges in the centre and on the perimeter of the dent's edge was the next step of investigation.

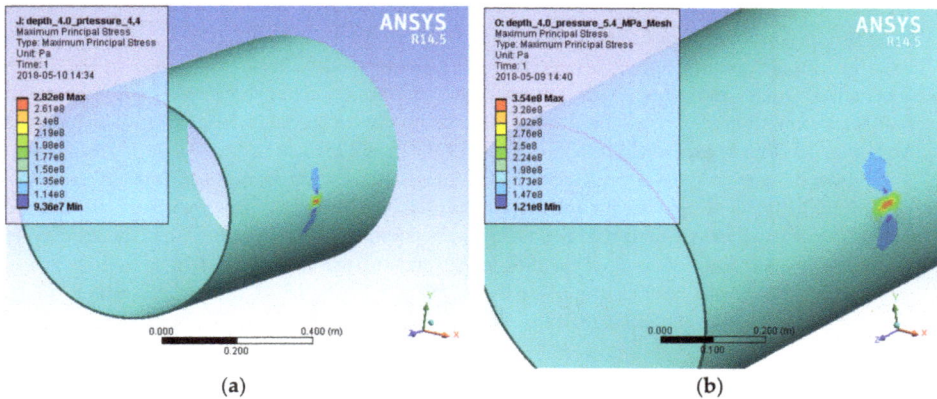

Figure 9. Maximal stresses in the pipe with a 4-mm deep gouge: (**a**) at a pressure of 4.4 MPa; (**b**) at a pressure of 5.4 MPa.

As demonstrated above, using FEM modelling of the combined defects allows for evaluation of the danger and accident risk for the pipeline by determining the values of the local stresses at the location of these defects under various working pressures and the dimensions of actual defects. However, experimental validation of the model should be performed. In this way, modelling of the considered defects, as well as evaluating anisotropy of the pipeline's metal, the effect of cycling loading under the pulsating pressure and the change in the temperature of the pipe could be performed. The authors intend to address this problem and to perform the validation of the model in further research.

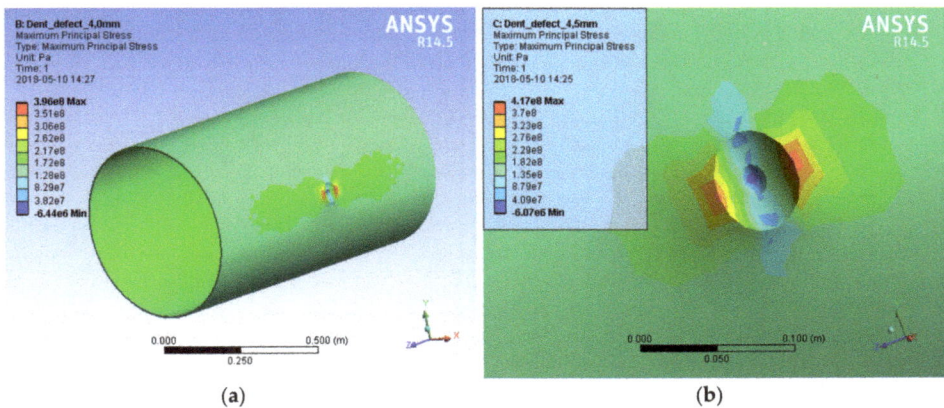

Figure 10. Stresses in the pipe with dent and gouge defects: (**a**) 4.0 mm deep; (**b**) 4.5 mm deep.

7. Conclusions

Based on the results of the experimental research, it can be stated that the mechanical properties and the structure of 17G2SF steel of the pipelines remained sufficiently stable after 33 years of operation. The mechanical strength of steel decreased by about 10%. However, in practice, due to the inseparably linked shell and corrosion defects, the operation of these pipelines under normal pressure of 5.4 MPa is unsafe. The use of the FEA modelling technique for predicting the failure pressure allowed for assessment of the critical sizes of the combined defects, the localization of the internal stresses and plastic deformation zones around these defects, and the critical pressure.

The simulation performed by using the finite element method has shown that, in the case of the combined defects under the operating pressure of 5.4 MPa, the internal stress in the metal alloy exceeds the permissible metal yield limit (350 MPa). The situation becomes dangerous when the depth of the combined defects exceeds half the thickness of the pipe's wall.

The most dangerous cases are associated with combined defects, because maximal stresses make up 83% of the ultimate stress and change their location in the direction from the centre to the dent's edge. Therefore, further research will be made based on using the available numerical model and the formation of new gouge defects of various depths on the perimeter of the dent's edge.

The most rational way for reducing the risk of accidents in pipelines operating for a long-time, is the reduction of their working pressure or the repair of defects of the described types by welding them after detection.

Author Contributions: All authors contributed equally to this work.

Conflicts of Interest: The authors declare no conflict of interests.

References

1. Maruschak, P.; Panin, S.; Vlasov, I.; Prentkovskis, O.; Danyliuk, I. Structural levels of the nucleation and growth of fatigue crack in 17Mn1Si steel pipeline after long-term service. *Transport* **2015**, *30*, 15–23. [CrossRef]
2. Maruschak, P.; Poberezhny, L.; Pyrig, T. Fatigue and brittle fracture of carbon steel of gas and oil pipelines. *Transport* **2013**, *28*, 270–275. [CrossRef]
3. GOST 19281-73. *Low-Alloy Steel Bars and Shapes*; Russian Standard: Moscow, Russia, 1973.
4. TU 14-3-731-78. In *Hot-Deformed and Cold-Deformed Seamless Pipes from Corrosion-Resistant Steel Grade 12Ch18N1OT*; Russian Technical Rules: Moscow, Russia, 1978.
5. Hrabovs'kyi, R.S. Determination of the resource abilities of oil and gas pipelines working for a long time. *Mater. Sci.* **2009**, *45*, 309–317. [CrossRef]
6. Kishawy, H.A.; Gabbar, H.A. Review of pipeline integrity management practices. *Int. J. Press. Vessels Pip.* **2010**, *87*, 373–380. [CrossRef]
7. Vianello, C.; Maschio, G. Risk analysis of natural gas pipeline: Case study of a generic pipeline. *Chem. Eng. Trans.* **2011**, *24*, 1309–1314. [CrossRef]
8. Wahab, M.A.; Sabapathy, P.N.; Painter, M.J. The onset of pipewall failure during "in-service" welding of gas pipelines. *J. Mater. Process. Technol.* **2005**, *168*, 414–422. [CrossRef]
9. Maruschak, P.; Danyliuk, I.; Prentkovskis, O.; Bishchak, R.; Pylypenko, A.; Sorochak, A. Degradation of the main gas pipeline material and mechanisms of its fracture. *J. Civ. Eng. Manag.* **2014**, *20*, 864–872. [CrossRef]
10. Cosham, A.; Hopkins, P. The Pipeline Defect Assessment Manual. In Proceedings of the 4th International Pipeline Conference, Calgary, AB, Canada, 29 September–3 October 2002; The American Society of Mechanical Engineers: New York, NY, USA, 2003; pp. 1565–27067. [CrossRef]
11. Alexander, C.R.; Kiefner, J.F. *Effects of Smooth and Rock Dents on Liquid Petroleum Pipelines*; American Petroleum Institute (API): Washington, DC, USA, 1997; Available online: http://www.chrisalexander.com/assetmanager/assets/1998%20Apr%20-%20API%20Pipeline%20Conf%20Rocks%20on%20dents.pdf (accessed on 10 January 2018).
12. Otegui, J.L. Challenges to the integrity of old pipelines buried in stable ground. *Eng. Fail. Anal.* **2014**, *42*, 311–323. [CrossRef]
13. Cosham, A.; Hopkins, P. The effect of dents in pipelines–guidance in the pipeline defect assessment manual. *Int. J. Press. Vessels Pip.* **2004**, *8*, 127–139. [CrossRef]
14. Allouti, M.; Schmitt, C.; Pluvinage, G. Assessment of a gouge and dent defect in a pipeline by a combined criterion. *Eng. Fail. Anal.* **2014**, *36*, 1–13. [CrossRef]
15. Allouti, M.; Schmitt, C.; Pluvinage, G.; Gilgert, J.; Hariri, S. Study of the influence of dent depth on the critical pressure of pipeline. *Eng. Fail. Anal.* **2012**, *21*, 40–51. [CrossRef]

16. Maruschak, P.; Konovalenko, I.; Prentkovskis, O.; Tsyrulnyk, O. Digital analysis of shape and size of dimples of ductile tearing on fracture surface of long-operated steel. *Procedia Eng.* **2016**, *134*, 437–442. [CrossRef]

17. Zhang, Y.M.; Fan, M.; Xiao, Z.M.; Zhang, W.G. Fatigue analysis on offshore pipelines with embedded cracks. *Ocean Eng.* **2016**, *117*, 45–56. [CrossRef]

18. Poberezhnyi, L.; Maruschak, P.; Prentkovskis, O.; Danyliuk, I.; Pyrig, T.; Brezinová, J. Fatigue and failure of steel of offshore gas pipeline after the laying operation. *Arch. Civ. Mech. Eng.* **2016**, *16*, 524–536. [CrossRef]

19. Hood, J.E. Fracture of steel pipelines. *Int. J. Press. Vessels Pip.* **1974**, *2*, 165–178. [CrossRef]

20. Maruschak, P.; Prentkovskis, O.; Bishchak, R. Defectiveness of external and internal surfaces of the main oil and gas pipelines after long-term operation. *J. Civ. Eng. Manag.* **2016**, *22*, 279–286. [CrossRef]

21. Ghajar, R.; Mirone, G.; Keshavarz, A. Ductile failure of X100 pipeline steel—Experiments and fractography. *Mater. Des.* **2013**, *43*, 513–525. [CrossRef]

22. Azevedo, C.R.F.; Sinatora, A. Failure analysis of a gas pipeline. *Eng. Fail. Anal.* **2004**, *11*, 387–400. [CrossRef]

23. Terán, G.; Capula-Colindres, S.; Velázquez, J.C.; Fernández-Cueto, M.J.; Angeles-Herrera, D.; Herrera-Hernández, H. Failure pressure estimations for pipes with combined corrosion defects on the external surface: A comparative study. *Int. J. Electrochem. Sci.* **2017**, *12*, 10152–10176. [CrossRef]

24. Matrosov, Y.I.; Litvinenko, D.A.; Golovanenko, S.A. Steel for pipelines. *Metallurgy 288*, 56–67. (In Russian)

25. *API 5L: Specification for Line Pipe*, 45th ed.; American Petroleum Institute (API): Washington, DC, USA, 2013; p. 192.

26. GOST 20295-85. *Steel Welded Pipes for Main Gas-and-Pil Pipelines*; Russian Standard: Moscow, Russia, 1985.

27. GOST 7565-81. In *Iron, Steel and Alloys, Sampling for Determination of Chemical Composition*; Russian Standard: Moscow, Russia, 1981.

28. EN ISO 148-1:2016. *Metallic Materials—Charpy Pendulum Impact Test—Part 1: Test Method*; Italian Standards: Milano, Italy, 2016.

29. EN ISO 9015-2:2016. *Destructive Tests on Welds in Metallic Materials—Hardness Testing—Part 2: Microhardness Testing of Welded Joints*; ISO: Geneva, Switzerland, 2016.

30. EN ISO 6507-1:2005. *Metallic Materials—Vickers Hardness Test—Part 1: Test Method*; CENELEC: Brussel, Belgium, 2005.

31. EN ISO 6892-1:2016. *Metallic Materials—Tensile Testing—Part 1: Method of Test at Room Temperature*; CENELEC: Brussel, Belgium, 2016.

32. EN ISO 643:2013. *Steels-Micrographic Determination of the Apparent Grain Size*; CENELEC: Brussel, Belgium, 2013.

33. Pashinskaya, E.G.; Zavdoveev, A.V.; Metlov, L.S.; Nepochatikh, Y.I.; Maksakova, A.A.; Tkachenko, V.M. Non-trivial changes in physical and mechanical properties and structure of low carbon steel wire, produced by rolling with shear and cold drawing. *Mater. Phys. Mech.* **2015**, *24*, 163–177.

34. Ashby, M.F. *Materials Selection in Mechanical Design*; Butterworth-Heinemann: Oxford, UK, 2005; ISBN 978-0-08-100599-6.

35. Moaveni, S. *Finite Element Analysis Theory and Application with ANSYS*; Prentice Hall: Upper Saddle River, NJ, USA, 2007; ISBN 978-0131890800.

36. SNIP 2.0506-85. *Pipelines. Russian Building Codes and Rules*; SNIP: Gosstroj, Russia, 1985.

37. Bobrickij, N.V.; Jufin, V.A. *Fundamentals of Gas Industry Petroleum and Gas Industry*; Nedra: Moscow, Russia, 1988. (In Russian)

metals

MDPI

Article

Characterization of Corrosion Products on Weathering Steel Bridges Influenced by Chloride Deposition

Vit Krivy [1], Monika Kubzova [1,*], Katerina Kreislova [2] and Viktor Urban [1]

[1] Department of Building Structures, Faculty of Civil Engineering, VSB-Technical University of Ostrava, L. Podeste 1875, 708 00 Ostrava, Czech Republic; vit.krivy@vsb.cz (V.K.); viktor.urban@vsb.cz (V.U.)
[2] SVUOM Ltd., U Mestanskeho Pivovaru 934/4, 170 00 Prague 7, Czech Republic; kreislova@svuom.cz
* Correspondence: monika.kubzova@vsb.cz; Tel.: +420-734-206-860

Received: 15 August 2017; Accepted: 26 August 2017; Published: 31 August 2017

Abstract: The article presents the results of experimental testing of corrosion processes on weathering steel bridges. Two bridge structures spanning various obstacles were selected for the experimental measurement. The tested bridges are situated in the same location and structural solution of these bridges is similar. Differences in development of corrosion products are mainly affected by the microclimate below the bridge structure. Special attention is paid to a bridge over the motorway which is strongly affected by the deposition of chlorides. The dependences between the measured deposition of chlorides and parameters of corrosion layers (thickness of corrosion products, corrosion rates, and chemical composition) are discussed and evaluated in this article.

Keywords: atmospheric corrosion; weathering steel; experimental tests; corrosion losses; deposition rate of chlorides

1. Introduction

Steel structures located in the outdoor environment are exposed to factors of atmospheric corrosion causing the formation of corrosion products on the metal surface and consequently a decrease in the thickness of the material. Among the main factors of atmospheric corrosion belong mainly humidity, temperature, aggressive stimulants present in the atmosphere such as SO_2, Cl^-, NO_x, solid particulate matter, and others [1]. The inappropriate use of structural carbon steel without corrosion protection in the outdoor environment and others reasons has led to the development of new types of low-alloy steels [2]. The first steel with increased resistance to atmospheric corrosion, the so-called weathering steel, was patented in the USA in 1933. In 1964, weathering steel was first used for the design of the main supporting structure of a bridge [3,4]. The alloying elements contained in weathering steel remain within 2 wt. % and are primarily represented by Cu, Cr, P, and Ni. What matters is the balance of individual alloying elements, especially the Cu–P–Cr combination.

Weathering steel is mainly used in designing bridge structures, lattice towers from sections and tubes, but also as roofing and cladding materials or decorative material. The increased resistance to corrosion in the atmosphere results from the development of a compact adhesive corrosion layer on the surface that reduces corrosion rate to a technically permissible limit. The corrosion layer with protective properties, the so-called "patina", forms after 3 to 7 years of exposure under suitable environmental conditions with respect to the concentration of SO_2 in the atmosphere. The necessary atmospheric conditions include, in particular, regular wetting and drying of the steel surface and also an environment remaining within the permissible limits of aggressive corrosion stimulators. In order to ensure a favourable development of the protective corrosion layer, it is necessary to ensure regular maintenance of the structure as well as regular inspections, mainly in the first years after construction, revealing places prone to accumulation of dust particles, impurities, or moisture. When designing structures increased attention must be paid to the layout design of the bridge structure, the structural details, and the design of the drainage system for the bridge [5].

As mentioned above, the most important aggressive corrosion stimulators occurring in the atmosphere include sulphur dioxide (SO_2) and chloride ions (Cl^-). Sulphur dioxide is brought to the air primarily as a product of fossil fuel combustion. It had the greatest impact on steel corrosion rate in the Czech Republic between 1970 and 1980. The subsequent introduction of desulphurization units brought a substantial reduction of SO_2 concentration in the atmosphere and the corrosion losses of carbon steel in the Czech Republic, current values stay below 10 µg/m^3 [6–8], see Figure 1.

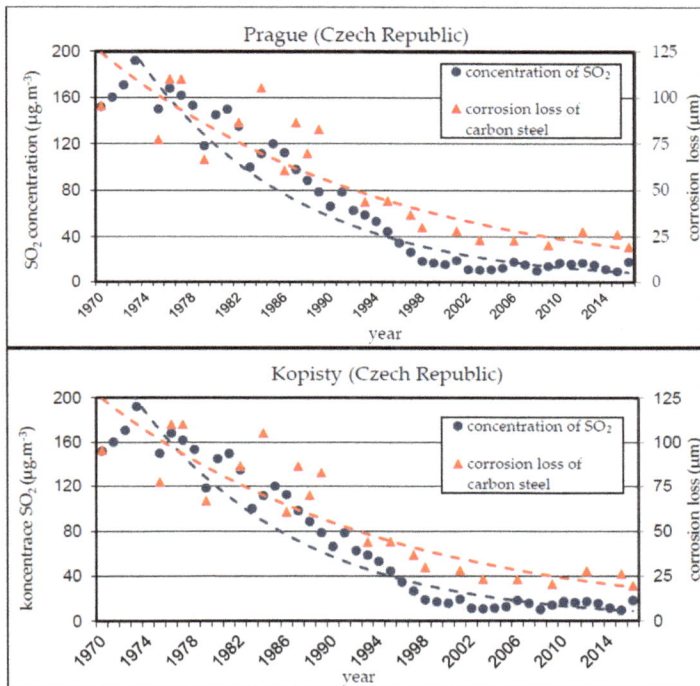

Figure 1. Decreasing concentration of SO_2 and one year corrosion losses of carbon steel (CS) in atmospheric test sites Prague and Kopisty (Czech Republic).

Due to the notable decline in SO_2 concentrations, chlorine ions (Cl^-) originating from de-icing salt used for winter road maintenance have become much more important factors of corrosive damage. The action of the chlorides leads to the formation of a corrosion layer at an increased rate, the resulting layer being thick and layered with numerous pores limiting the protective function of the corrosion layer on the surface of the weathering steel [8,9]. In coastal areas, the deposition rate of chlorides ranges from 250 mg·m^{-2}·day^{-1} to 2000 mg·m^{-2}·day^{-1}, which is mainly sea salt aerosol in the air [8]. In inland areas, chlorides are spreading from de-icing chemicals used during the winter. In the Czech Republic, road maintenance mainly uses sodium chloride (NaCl). In the vicinity of roads so treated, the chloride deposition rate can then be expected to reach values otherwise typical for coastal areas. Significant deposition can be expected during the winter period, although certain studies indicate that higher deposition rates can still be measured for up to two and a half months after the last application of de-icing agents [9,10]. It is thus currently needed to concentrate more intensively on the issue of chlorides spreading around bridge structures. Increased chloride deposition can cause significant corrosive damage and reduce the formation of the protective layer on the weathering steel surface.

In view of the above, this article presents an experimental monitoring of chloride deposition on a supporting steel structure of selected bridges designed from weathering steel. The measurement results are evaluated in relation to the results of atmospheric corrosion tests carried out in parallel with the chloride deposition measurements. The main objective of the research is to monitor the influence of the environment under the bridge structure (especially the impact of increased chloride deposition from road transport under the bridge) on the development of the corrosion layer and its protective properties.

2. Materials and Methods

2.1. Bridge Structures for Experimental Tests

To verify experimentally the influence of road transport under the bridge structure on the amount of chloride deposition and the subsequent development of corrosion products on typical surfaces of bridges designed from weathering steel, two bridge structures located on the same road in Ostrava, Czech Republic (distance from coastal area is about 500 km) were selected (see Figure 2):

- Bridge No. 1: Bridge on road No. 479 over highway D1 (built in 2001);
- Bridge No. 2: Bridge on road No. 479 over a railway (built in 1983).

Figure 2. Location of the bridge structures.

Bridge No. 1 takes the road No. 479 across the busy highway D1. The composite bridge is designed with a reinforced concrete slab and I-shaped main girders. Bridge No. 2 brings the same road across a railway line and is designed with an orthotropic deck and main box girders. Both bridge constructions are designed with Atmofix B [11] weathering steel and are located in Ostrava-Svinov, Czech Republic. The site can be characterized as an urban environment with a degree of atmospheric corrosivity categories C2 to C3 for steel complying with EN ISO 9223 [12]. The distance between the centers of both bridges is approximately 200 m. Figure 3 shows a view of both bridges.

Figure 3. View at selected bridges for experimental testing: (**a**) Bridge No. 1 and (**b**) bridge No. 2.

The general conditions for the development of corrosion products on typical surfaces of the bridge constructions are very similar—the bridges are situated in the same location, with just minimal differences in orientation. The design solutions are not too different from one another, either—both bridge constructions are designed as girder road bridges with upper deck. Any significant variations in the development of corrosion products on a particular typical surface of each individual bridge are thus conditioned primarily by the specific local microclimate surrounding the bridge. The decisive factor within those microclimates is the impact of road traffic under the bridge. Only bridge No. 1 is exposed to intense road traffic underneath. In addition, the abutments of bridge No. 1 stand in close proximity to the highway, which makes is somewhat similar to the so-called "tunnel-like conditions" and the associated increased deposition of chlorides on the structure [13–15], see Figure 4.

Figure 4. Bridge No. 1—Environment under the bridge is similar to the environment in the tunnels.

2.2. Experimental Tests for Monitoring the Development of Corrosion Products on Weathering Steel

To monitor the development of corrosion layers on the surface of the weathering steel, corrosion samples for modified atmospheric corrosion tests were installed on both bridges. Corrosion samples are attached to selected typical surfaces of the bridge structure in such a way that the conditions for development of protective corrosion layers on the surface of the samples correspond as closely as possible to the conditions on the adjacent surface of the bridge structure, see Figure 5. Three corrosion samples have been fastened to each selected surface, to be withdrawn later after 1, 3, and 10 years of exposure. For more detailed information on installing corrosion samples, see [5]. The typical surfaces common to both bridges being compared include:

- Surface P1: External wall of the main girder (north orientation);

- Surface P2: External wall of the main girder (south orientation);
- Surface P3: Internal wall of the main girder;
- Surface P4: Upper flange of the main girder—bottom surface;
- Surface P5: Bottom flange of the main girder—external upper surface;
- Surface P6: Bottom flange of the main girder—internal upper surface;
- Surface P7: Bottom flange of the main girder—bottom surface.

The following experimentally measured properties of corrosion layers are used to evaluate the course of corrosion processes on both selected bridge structures:

- Average thickness of corrosion products after long-term exposure;
- Average thickness of corrosion products after one year of exposure of corrosion samples;
- Corrosion loss after one year of exposure of corrosion samples;
- Chemical composition of corrosion products;
- Dry deposition of chlorides.

(a)   (b)

Figure 5. Corrosion samples on the upper surface of the bottom flange at the time of installation (**a**) and after 1 year of the exposure (**b**).

The thickness of the corrosion products is measured by magnetic-induction method. For each surface under evaluation, an average of 30 measurements is determined. The measurements are made directly on the bridges' surfaces (long-term exposure) and also on the surfaces of the corrosion samples (one-year exposure). Corrosion rates r_{corr} (i.e., corrosion losses) after one year of exposure of the corrosion samples are evaluated according to EN ISO 9226 [16] and EN ISO 9223 [12].

A laboratory analysis of the withdrawn corrosion layers determines the weight percentage of individual elements in the corrosion products. The element demonstrating the exposure to an environment affected by deposits of de-icing salts is chlorine (Cl). An increased proportion of dust deposits is reflected in the concentrations of silicon (Si) and aluminium (Al).

In addition to the chemical composition of the corrosion products, a phase composition analysis was also performed. The representation of individual phases is determined by an X-ray diffraction analysis of the corrosion products. The most stable phase is goethite α-FeOOH. The least stable one is lepidocrocite γ-FeOOH. In the corrosion products, compounds like magnetite Fe_3O_4 or maghemite γ-Fe_2O_3 are produced; in environments containing chlorine, it is also akaganeite β-FeOOH phase. The relative ratio of the main compounds and phases in the corrosion product layer can be used to determine the Protective Ability Index [17,18].

2.3. Experimental Tests for Monitoring the Deposition Rate of Chlorides

A sampling device was installed on both bridge structures to monitor the deposition rate of chlorides. Two basic methods according to EN ISO 9225 [19] are used to measure the dry deposition of chlorides: Wet candle method and dry plate method.

When measuring using the dry plate method, standard flat samples are installed on the bridge structure together with a frame with gauze stretched across. The perforation of the gauze used for the measurement allows dust deposits from the atmosphere to be collected without risking their loss by being blown away again. This method was applied to measure the deposition rate of chlorides from chemical de-icing agents used for winter road maintenance in Japan [9].

Wet candle method means installing bottles on the bridge structure, containing a solution of glycerol and a wick made of inert material and wrapped in surgical gauze. The chlorides are gradually deposited on the surface of the gauze wick and, after dissolution, they mix with the solution in the bottle. The chloride deposition measuring equipment was placed on the following selected positions of the bridges under evaluation:

- L1—bridge No. 1—external girder, north orientation (corresponds to surfaces P1 and P5);
- L2—bridge No. 1—internal girder, south orientation (corresponds to surfaces P3 and P6);
- L3—bridge No. 1—external girder, south orientation (corresponds to surface P2);
- L4—bridge No. 2—external girder, north orientation (corresponds to surfaces P1 and P5).

The location of the sampling device is shown in Figure 6. Each position includes a sampling device both for wet candle method and dry plate method in a horizontal and a vertical direction, simulating the deposition of chlorides on the vertical and horizontal parts of the structure's surface. The individual positions are shown in Figure 7. The sampling devices were installed on 1 December 2016. Samples are replaced after one month of exposure; the sample is removed from the sampling device and an analysis is performed according to the procedure set out in [19] to determine the amount of chlorides stored in the wet candle solution and on the dry plate. The wind direction and velocity were not monitored within the research.

(a) (b)

Figure 6. Position of the measuring devices on the selected bridge structures—bridge No. 1 (**a**) and bridge No. 2 (**b**).

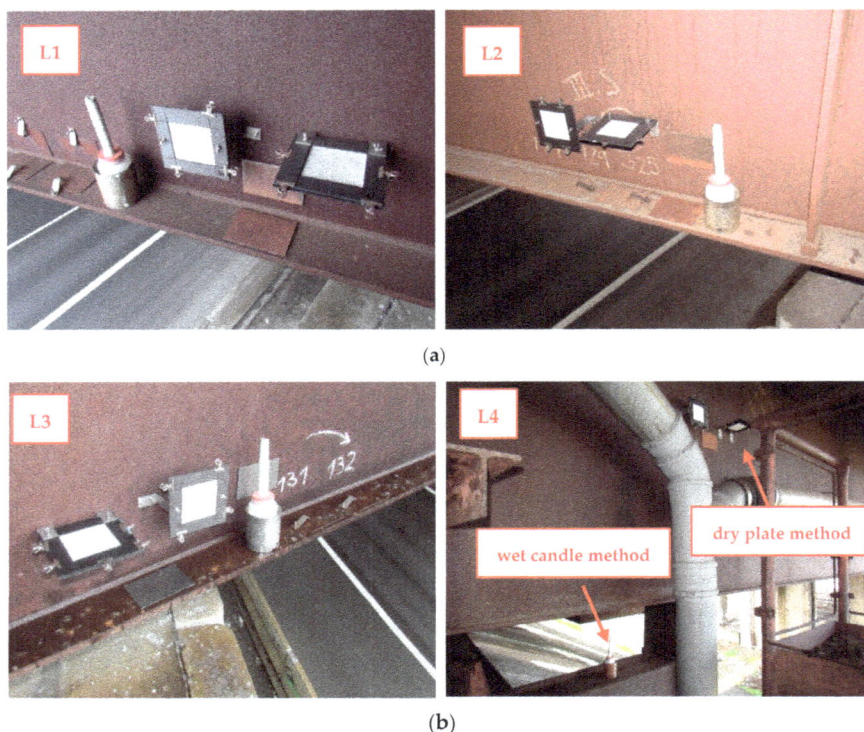

Figure 7. (a) Individual positions of the measuring devices on the selected bridges; (b) Individual positions of the measuring devices on the selected bridges.

3. Results

Comparison of the results from the experimental measurements carried out on the bridge structures as described above can be used to evaluate the effects of road transport under bridges. The following part of the paper presents the results of experimental measurements of the corrosion products' average thickness, corrosion loss in corrosion samples after one year of exposure, the chemical and phase composition of the corrosion layers, and the measured values of chloride deposition rate.

3.1. Average Thickness of Corrosion Products After Long-Term and 1-Year Exposure

The thickness of corrosion products t_{corr} was measured on typical surfaces of bridge structures near the position of corrosion samples (long-term exposure: 16 years for the bridge No. 1 and 33 years for the bridge No. 2), as well as on corrosion samples themselves after one year of exposure. The results of the measurement of corrosion thicknesses on typical surfaces of the bridge structures under evaluation are given in Table 1 (average values from 30 measurements are listed in the table).

Table 1. Average thickness of corrosion products t_{corr} after long-term and 1-year exposure (μm).

Tested Surface	Tested Bridge Structures			
	No. 1	No. 2	No. 1	No. 2
	After Long-Term Exposure		After 1-Year Exposure	
P1—external wall of the main girder (north)	127.1	90.9	54.4	27.0
P2—external wall of the main girder (south)	89.9	82.4	37.5	27.0
P3—internal wall of the main girder	144.3	108.9	48.5	14.2
P4—upper flange of the main girder (bottom surface)	134.1	-	45.6	-
P5—bottom flange of the main girder (external upper surface)	272.0	-	156.0	-
P6—bottom flange of the main girder (internal upper surface)	700.0 [1]	-	118.1	-
P7—bottom flange of the main girder (bottom surface)	158.4	168.2	63.9	35.5

[1] The internal surfaces of the bottom flanges of the main girders have not developed a sufficiently protective corrosion layer. Non-adherent corrosion products are formed which gradually fall off the surface of the flange, see Figure 8.

The values of thickness of corrosion layer measured on bridge No. 1 exceed the values of thickness measured on bridge No. 2. The impact of intensive road transport under the bridge and of the design creating a partial "tunnel-like effect" was most pronounced at the upper surfaces of the bottom flanges. The increased occurrence of dust deposits (including chlorides) has had a very negative influence, especially so in the case of non-ventilated internal surfaces of the bottom flanges, which do not develop a sufficiently protective corrosion layer, see Figure 8.

Looking at the values given in Table 1, both bridges under evaluation show a significant influence of location and orientation of the area on the development of corrosion products. The highest corrosion product thickness values were found on the upper surfaces of the bottom flanges of the main girders. More details on the differences in the development of corrosion products on typical surfaces of bridge structures can be found in [20,21].

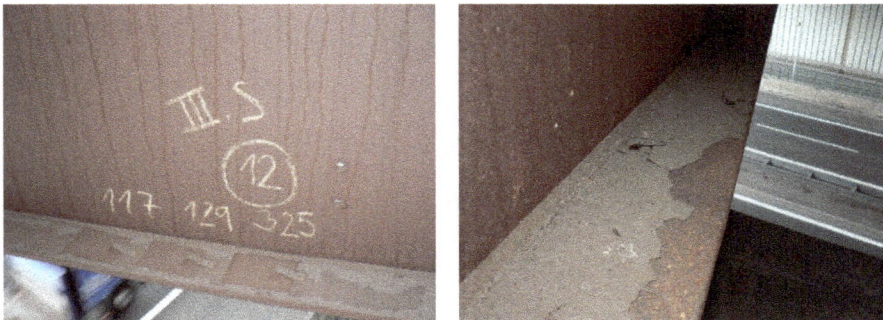

Figure 8. Bridge No. 1—unfavorable development of corrosion products on the internal upper surface of the bottom flange (surface P6).

3.2. Corrosion Losses After 1 Year of Exposure

The determined values of corrosion loss after the first year of exposure are shown in Table 2. The values of corrosion loss are determined in accordance with EN ISO 9226 [16].

There is a strong correlation between the values of corrosion losses and thicknesses of corrosion layers measured on corrosion samples after one year of exposure [20,21]. The microclimate affected by the intense transport under bridge No. 1 was thus also significantly reflected in the evaluation of corrosion loss values, where as much as a three-fold increase can be observed compared to the reference bridge over the highway.

Table 2. Corrosion losses after 1 year of exposure r_{corr} ($\mu\text{m}\cdot\text{year}^{-1}$).

Tested Surface	Tested Bridge Structures	
	No. 1	No. 2
P1—external wall of the main girder (north)	14.6	5.3
P2—external wall of the main girder (south)	10.5	5.3
P3—internal wall of the main girder	12.0	4.0
P4—upper flange of the main girder—bottom surface	8.5	-
P5—bottom flange of the main girder—external upper surface	35.2	-
P6—bottom flange of the main girder—internal upper surface	24.1	-
P7—bottom flange of the main girder—bottom surface	10.8	8.8

3.3. Chemical Composition of Corrosion Products

Four samples of corrosion products for composition and phase analysis were collected from the surface of the steel supporting structure of bridge No. 1 (bridge over highway D1). The samples were withdrawn from the external wall of the main girder (surface P1), the internal wall of the main girder (surface P3), the upper external surface of the bottom flange of the main girder (surface P5), and the upper internal surface of the bottom flange of the main girder (surface P6).

The results of the composition analysis are summarized in Table 3. The table only lists concentrations of chlorine (Cl) identifying the amount of de-icing salts deposited and then silicon (Si) and aluminium (Al) concentrations indicating the amount of dust deposits.

Table 3. Composition analysis—samples withdrawn from surfaces of bridge No. 1.

Tested Surface	Concentration (wt. %)		
	Cl	Si	Al
P1—external wall of the main girder (north)	1.297	2.436	0.624
P3—internal wall of the main girder	1.072	0.645	0.159
P5—bottom flange of the main girder—external upper surface	1.152	1.811	0.498
P6—bottom flange of the main girder—internal upper surface	2.513	6.334	1.644

The highest concentrations of chloride and also elements pointing to an increased proportion of dust deposits (silicon and aluminium) have been identified on surface P6, i.e., the internal upper surface of the bottom flange of the main girder. It is logical that horizontal surfaces are exposed to increased deposition of dust and chlorides. However, the internal girders do not get cleaned by rain and wind as much as the external surfaces, and non-protective corrosion products have thus developed on the bottom flanges.

Representation ratios of individual compounds of the corrosion layer of weathering steel can be obtained by X-ray diffraction analysis. Protective functionality expressed by indexes PA (protective ability indexes) can be assessed by comparing these ratios [17,18]. An X-ray diffraction analysis was also performed on the withdrawn samples of corrosion products and PA Index values were determined from the obtained phase composition; see Table 4. The phase analysis clearly shows that the corrosion products collected from surfaces P1, P3, and P5 correspond to the already relatively steady state of the corrosion layer—goethite is the dominant phase there. The akaganeite, indicative of the action of chlorides, was detected at high concentrations on surface P6; increased values were also identified on surface P1. This finding corresponds to the composition analysis and the identified weight percentages of chloride in the collected corrosion products. A higher proportion of akaganeite was manifested in the values of PAI_α and PAI_β. The increased proportion of dust deposits on surface P6 was signaled by the heavy occurrence of aluminium and certain compounds.

Table 4. X-ray diffraction analysis—samples withdrawn from structures of bridge No. 1.

Tested Surface	X-ray Diffraction Analysis	PAI_α	PAI_β
P1	strongly goethite, akaganeite and quartz, weakly lepidocrocite	0.47	0.86
P3	very strongly goethite, weakly lepidocrocite and akaganeite	0.83	0.60
P5	very strongly goethite, weakly lepidocrocite and akaganeite	1.09	0.60
P6	very strongly quartz, strongly goethite and akaganeite, very weakly lepidocrocite, weakly limestone $CaSO_4 \cdot 2H_2O$, very weakly muscovite $KAl_2(Si, Al)_4O_{10}(OH)_2$	0.25	0.90

Both bridges also underwent a composition analysis of corrosion products developed after 1 year of exposure on the surface of corrosion samples. The main girders of bridge No. 2 are made of box sections with a small overlap of the bottom flange; that surface thus cannot be used to compare both bridges. That is why surface P1, i.e., the external wall of the main girder, was selected for comparison.

The results of the composition analysis of the corrosion products for both bridges and the reference surface P1 are given in Table 5. The concentrations of all monitored elements (Cl, Si, and Al) are approximately three times higher at bridge No. 1 (bridge over the highway) when compared to bridge No. 2 (bridge above the railway). Again, this finding points to the adverse corrosive effects of intense road transport beneath the bridge structure.

Table 5. Composition analysis of corrosion products after one-year exposure—comparison of both bridges.

Tested Surface	Concentration (wt. %)		
	Cl	Si	Al
bridge No. 1—surface P1 (external wall of the main girder)	0.311	0.329	0.168
bridge No. 2—surface P1 (external wall of the main girder)	0.119	0.143	0.021

3.4. Deposition Rate of Chlorides

At present, experimentally measured chloride deposition rates are available for the period from December 2016 to June 2017, see Figures 9–11. Chloride deposition values are measured using wet candle method and dry plate method. The results include three months (December 2016 to February 2017) during which the winter maintenance with de-icing salts was performed on the roads concerning the bridges under evaluation.

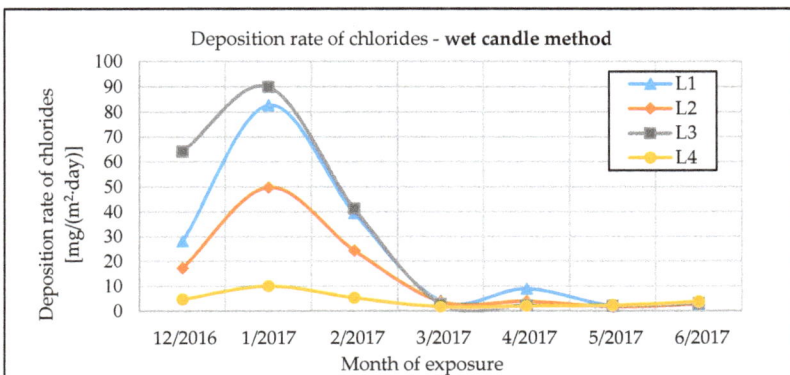

Figure 9. The deposition rates of chlorides determined by the wet candle method.

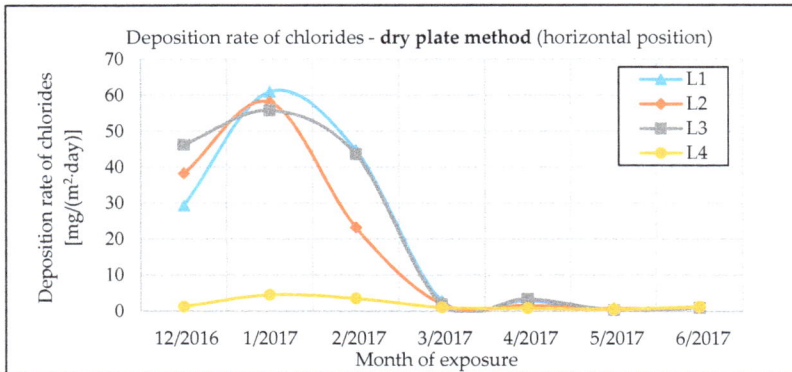

Figure 10. The chloride deposition rates determined by the dry horizontal plate method.

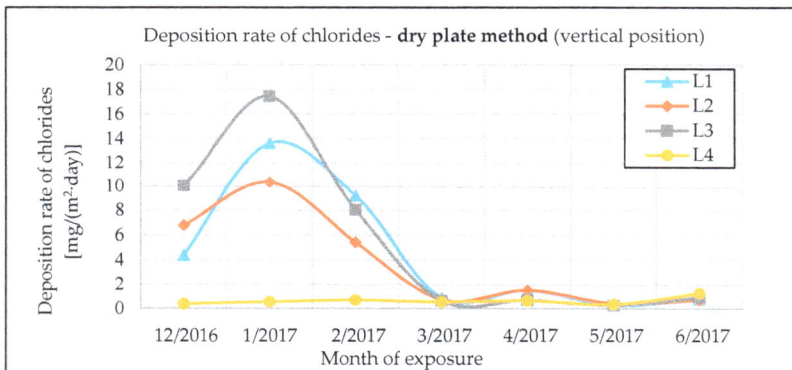

Figure 11. The chloride deposition rates determined by the dry vertical plate method.

Several interesting results and dependencies can be observed when looking at the chloride deposition rates measured:

- There is a clear link between the amount of deposited chlorides and the winter maintenance using de-icing salt. The chloride deposition rates during the winter maintenance period (December 2016 to February 2017) significantly exceed the values measured during the spring months. The difference is evident especially with bridge No. 1, which is affected by the intense traffic on the D1 highway underneath. The drop in deposition rate at the end of the winter maintenance period is rather steep. The chloride deposition rate during the winter period amounts to 5 mg·m^{-2}·day^{-1}. The gradual decline in the deposition rate over approximately two months after the end of winter maintenance, reported in the literature [9,10], has not been observed.
- Comparing the results from sampling devices located at bridge No. 1 (positions L1, L2, and L3) with results from bridge No. 2 (position L4) shows the significant influence of the specific microclimate found under the bridge structure. It is quite evident that the road traffic under the bridge structure is the main source of chloride deposition upon the bridge's supporting elements. The results from bridge No. 2 show, on the other hand, that a suitable design of the bridge (girder bridge with an upper deck and sufficient overhang over the external main

girders), significantly reduces the amount of chloride deposits originating from the road traffic on the bridge.

- The method used for measuring the dry deposition of chlorides significantly affects the resulting values. For most measurements, the highest chloride deposition rates are obtained using the wet candle method (maximum deposition rate measured was 90 mg·m^{-2}·day^{-1}). In some cases, however, the highest chloride deposition rate was measured with a dry plate method in a horizontal position. Dry plate vertical measurements provided in all cases significantly lower values when compared with other measurement methods (approximately four times less than the dry plate in the horizontal position).

- The results indicate that the external surfaces of the supporting steel structure are exposed to higher levels of chloride deposition when compared to internal surfaces (this observation is in line with the results presented in [9,10]). However, dust deposition (including chlorides) on the external surfaces of bridge No. 1 is regularly washed by rainfall or wind. Non-protective corrosion products only form on the internal flanges, despite the fact that the chloride deposition rate on the internal surfaces is lower compared to the external ones.

4. Discussion

All experimentally determined data clearly show that the development of corrosion products largely depends on the local microclimate under the bridge structure. Intense road traffic under the bridge becomes the source of increased dust and chloride deposition on the bridge bearing elements. The most affected areas of the load-bearing steel structure are the internal surfaces of the bottom flanges of the I-shaped cross-sections, unlike the external ones, these horizontal surfaces do not benefit from sufficient washing from regular rainfall and blowing of the dust deposits by the wind.

Experimental measurements have also shown that dust and chloride deposition on the load-bearing steel structure coming from road traffic on the bridge is significantly lower than the deposition resulting from the traffic underneath. For that, however, some basic constructional principles must be observed regarding the design of bridges with an upper deck. The supporting structure under the deck is then protected against the corrosive effects of the traffic on the bridge.

To evaluate the development of corrosion products on the typical surfaces of bridge structures, various methods can be used. A basic visual assessment should always be supplemented by a measurement of the corrosion products' average thickness. Measurements are easily taken with portable thickness gauges and the results are readily available. Average thickness of corrosion products of 350 to 400 μm is usually considered as the limit value of a favorable development of the protective corrosion layer for long-exposed bridges [19,22]. The values of corrosion product thickness measured during the initial bridge's exposure can be used for an accurate estimation of the corrosion rate, as there is a strong correlation between thickness of corrosion products and corrosion loss [21].

The atmospheric corrosion tests using corrosion samples installed on the surfaces of the bridge structure provides the most valuable data on the development of corrosion processes. By applying a sufficiently large number of corrosion samples to the surface of the structure, it is possible to obtain data on the corrosion product development in time. The progression of corrosion loss, changes in the thickness of the corrosion layer and the chemical composition of the corrosion products can be observed. An important finding resulting from the atmospheric corrosion tests is the fact that the development of corrosion products is significantly affected by the location of the evaluated surface within the structure [21]. For example, on bridge No. 1, corrosion losses after one year of exposure to the bottom external surface of the flanges are approximately four times higher compared to the bottom surface of the upper flanges:

- Bridge No. 1, surface P4, 1 year exposure: $r_{corr} = 8.5$ μm·year^{-1};
- Bridge No. 1, surface P5, 1 year exposure: $r_{corr} = 35.2$ μm·year^{-1}.

The composition analysis of corrosion products can be used to identify the decisive stimulators of corrosion. The increased deposition of road salt results in a higher mass ratio of chlorine (Cl) in the corrosion products. Higher ratios of silicon (Si) and aluminium (Al) in corrosion products indicate increased deposition of dust. There is not yet enough statistical data to estimate the permissible limit values for chlorine concentrations in the corrosion product layer. The results obtained from bridge No. 1 indicate that during the development of corrosion layer on a surface suffering from the deposition of chlorides, the concentration of chloride in the corrosion products is gradually increasing:

- Bridge No. 1, surface P1, 1 year exposure: wt. % Cl = 0.311 (see Table 5);
- Bridge No. 1, surface P1, 14 years exposure: wt. % Cl = 1.297 (see Table 3).

The increased chloride deposition rate also reflected in the phase composition of corrosion products determined by X-ray diffraction analysis, see Table 4. Corrosion products developed in combination with an increased deposition of chlorides also show high ratios of akaganeite. A greater presence of akaganeite was reflected in the values of the protective ability indexes PAI_α and PAI_β. The literature states [17,18] that a fully protective corrosion layer with a corrosion rate $r_{corr} < 10$ μm·year^{-1} correspond to $PAI_\alpha > 1$. If the PAI_α ratio < 1, calculation of the PAI_β is necessary to determine another factor of the layer's protective ability ($PAI_\beta < 0.5$—a corrosion product layer without protective ability; or $PAI_\beta > 0.5$—partially protective corrosion layer). Interpreting the detected values of PA Indexes is still rather complicated. As an example, it can be compared the values of the corrosion product layer average thickness t_{corr} measured on the surface of the structure, the initial r_{corr} corrosion rates and the PAI_α indexes of the corrosion layer on surfaces P1, P3, P5, and P6; see Table 6. The highest PAI_α value of corrosion layer protective ability was identified on surface P5, but that surface also shows the highest initial corrosion rate r_{corr} and a high value of average corrosion thickness t_{corr}. The results indicate that there may be no positive correlation between corrosion loss and the patina protective ability index.

Table 6. Comparison of the average thicknesses of the corrosion layer t_{corr} measured on the surface of the structure, corrosion rates r_{corr} and PAI_α indexes.

Tested Surface	t_{corr} (μm)	r_{corr} (μm·year^{-1})	PAI_α
P1—external wall of the main girder (north)	127.1	14.6	0.47
P3—internal wall of the main girder	144.3	12.0	0.83
P5—bottom flange of the main girder—external upper surface	272.0	35.2	1.09
P6—bottom flange of the main girder—internal upper surface	700.0	24.1	0.25

The monitoring of chloride deposition rate explicitly showed the potential negative effects of microclimatic conditions under the bridge structure. Winter road maintenance using de-icing salts combined with intense road traffic under the bridge structure are a significant source of chloride and dust deposition on the bridge's load-bearing elements. The maximum measured value of chloride deposition rate on bridge No. 1 was almost 90 mg·m^{-2}·day^{-1}. The effect can be largely intensified by an inappropriate design of the bridge, creating tunnel-like conditions. Measurements of chloride deposition rates also showed that with a proper bridge design, there are very low amounts of chloride deposits from road traffic on the bridge, compared to the potential effects of traffic going on under the same bridge. The maximum measured value of chloride deposition rate on bridge No. 2, which is not affected by road traffic underneath, was only 10 mg·m^{-2}·day^{-1}.

The chloride deposition rates are measured using the wet candle and the dry plate methods in compliance with the EN ISO 9225 standard [19]. The standard sets the relationship between chloride deposition rates measured by wet candle $S_{d,c}$ and on the dry plate $S_{d,p}$ (the dry plate is assumed to be in a vertical position):

$$S_{d,c} = 2.4 \times S_{d,p}. \tag{1}$$

The results of the measurements differ slightly from the assumption in the standard. The average value of the conversion factor for all measurements is m = 3.99; it is thus higher than the 2.4 value assumed by the standard. In addition, there was substantial variation among individual measurements: Coefficient of variation v = 0.29, minimum value min = 1.40, maximum value max = 6.06.

5. Conclusions

As a result of the significant decrease of SO_2 concentration in the Czech Republic, the influence of other corrosion stimulators is increasing. Roads and their surrounding areas are exposed to a significant source of corrosive damage in the form of chloride deposition. The chlorides come from de-icing salt used during winter maintenance. The space around the bridge structure is specific. Measuring of chloride deposition on surfaces of the bridge structure should be monitored as they can influence the development of the corrosion protection layer. The evaluation of corrosion processes on road bridges designed from weathering steel in the Czech Republic shows that most surfaces of the steel supporting structures under evaluation have developed a sufficiently protective corrosion layer. At some bridges, though, faults in the protective corrosion layer development have been identified—which in turn relates to the deposition of chlorides and dust deposits.

There are several cases where the development of protective corrosion layer on bridges located on the same road has been very different and affected by the deposition of chlorides and dust deposits. The bridge structures mentioned in this article can be used as a typical example of these cases: Bridge No. 1 is negatively affected by intense road traffic on the highway under the bridge, bridge No. 2 crosses a railway and the potential use of chlorides deposition is negligible compared to bridge No. 1.

A comparison of the negative impact of road traffic under the bridge construction has been demonstrated by several results of experimental testing mentioned in this article. The thicknesses of the protective corrosion layer after 1-year of exposure were compared for similar surfaces of these bridges. Thicknesses of the layer of corrosion products after 1-year of exposure for the bridge No. 1 are about two times higher than thicknesses measured for bridge No. 2. Also thicknesses of corrosion layer after long-term exposure are higher for the bridge No. 1. Moreover, for the upper surface of bottom flange of the main internal girder for bridge No. 1 a sufficiently protective corrosion layer is not yet created there, nor after 16 years of exposure. A second pointer showing the negative effect of the microclimate under the bridge structure on the development of a sufficiently protective layer of corrosion products on the surface of the weathering steel are corrosion losses obtained after 1 year of exposure of corrosion samples. Corrosion losses for bridge No. 1 are even up to three times higher, for some compared surfaces, than for bridge No. 2. The highest corrosion losses were again found on the upper surface of the bottom flange of the main internal girder. The measurements thus confirmed that the progression of corrosion processes on the surface of the bridge's load-bearing structure is significantly affected by the character of the obstacle that the bridge is designed to overcome.

The structural design of the bridge is another very important factor determining the amount of chlorides and dust deposited. This was confirmed by the measurement of the deposition rate of chlorides. Values of chloride deposition in the winter even reached a value of 90 $mg \cdot m^{-2} \cdot day^{-1}$ for bridge construction No. 1, while for bridge No. 2 above the railway reached chloride deposition had a maximum value of 10 $mg \cdot m^{-2} \cdot day^{-1}$. Measurements on bridge No. 2 show that a proper structural design of the bridge can eliminate the negative effects of road transport on the bridge itself to a large extent. The arrangement of abutments of bridge No. 1 is a good example of improper structural design from the point of view of chloride and dust deposition, as it creates tunnel-like conditions. In this case, the design contributes to the corrosive damage of the bridge structure, as confirmed by the results of the experimental tests.

Bridge structures must be designed in such a way that they can function reliably throughout their intended lifetime. A reliable estimation of the microclimatic conditions around the projected bridge structure will have a major impact on the future functionality of the selected system of protection against corrosion. For structures designed from unprotected weathering steel, real corrosion loss must not exceed the assumptions of the static calculation. The knowledge obtained from the measurements carried out on bridge structures in operation greatly enhances our understanding of the environment surrounding road infrastructure and its characteristics. Another valuable finding resulting from the experimental measurements is the correlation between the bridge's structural design and the progression of corrosion processes on individual surfaces of the structure.

Acknowledgments: This outcome has been achieved with funds of Conceptual development of science, research, and innovation assigned to VSB—Technical University of Ostrava by Ministry of Education Youth and Sports of the Czech Republic and the project MSMT-3375/2017-1 Durability and Sustainability of Engineering Constructions.

Author Contributions: Vit Krivy and Katerina Kreislova conceived and designed the experiments; Vit Krivy, Monika Kubzova and Viktor Urban performed the experiments; Vit Krivy, Katerina Kreislova and Monika Kubzova analyzed the data; Vit Krivy and Monika Kubzova wrote the paper.

Conflicts of Interest: The authors declare no conflict of interest.

References

1. Fontana, M.G. *Corrosion Engineering*, 3rd ed.; Tata McGraw-Hill: New Delhi, India, 2005.
2. Bhadeshia, H.K.D.H.; Honeycombe, R. *Steels Microstructure and Properties*, 3rd ed.; Elsevier: Burlington, ON, Canada, 2006; ISBN 9780080462929.
3. Morcillo, M.; Díaz, I.; Chico, B.; Cano, H.; de la Fuente, D. Weathering steels: From empirical development to scientific design. A review. *Corros. Sci.* **2014**, *83*, 6–31. [CrossRef]
4. Nickerson, R.L. *Performance of Weathering Steel in Highway Bridges a Third Phase Report*; American Iron and Steel Institute: Washington, DC, USA, 1995.
5. Krivy, V.; Urban, V.; Kreislova, K. Development and failures of corrosion layers on typical surfaces of weathering steel bridges. *Eng. Fail. Anal.* **2016**, *69*, 147–160. [CrossRef]
6. Kreislova, K.; Geiplova, H.; Bartak, Z.; Majtas, D. Atmospheric corrosion models. *Koroze Ochr. Mater.* **2017**, *61*, 59–66. [CrossRef]
7. Kreislova, K.; Knotkova, D. The Results of 45 Years of Atmospheric Corrosion Study in the Czech Republic. *Materials* **2017**, *10*, 394. [CrossRef] [PubMed]
8. Alcantara, J.; Fuente, D.; Chico, B.; Simancas, J.; Diaz, I.; Morcillo, M. Marine Atmospheric Corrosion of Carbon Steel: A Review. *Materials* **2017**, *10*, 406. [CrossRef] [PubMed]
9. Iwasaki, E. Scattering of Deicing Salt and Corrosion of Steel Bridges. In Proceedings of the Challenges in Design and Construction of an Innovative and Sustainable Built Environment, 19th IABSE Congress, Stockholm, Sweden, 21–23 September 2016; pp. 1436–1441.
10. Yamaguchi, E. Maintenance of weathering steel bridges. *Steel Constr. Today Tomorrow* **2015**, *45*, 12–15.
11. Krivy, V.; Konecny, P. Real material properties of weathering steels used in bridge structures. *Procedia Eng.* **2013**, *57*, 624–633. [CrossRef]
12. International Organization for Standardization. *Corrosion of Metals and Alloys—Corrosivity of Atmospheres—Classification, Determination and Estimation*; ISO 9223; International Organization for Standardization: Geneva, Switzerland, 2012.
13. Dolling, C.; Hudson, R. Weathering steel bridges in the U.K. *Rev. Métall.* **2003**, *100*, 1125–1133. [CrossRef]
14. Xanthakos, P.P. *Theory and Design of Bridges*; Consulting Engineer: Washington, DC, USA, 1994.
15. Kreislova, K.; Knotkova, G. *Atmosférická Koroze*; SVÚOM: Praha, Czechia, 2014; ISBN 978-80-87444-11-5. (In Czech)
16. International Organization for Standardization. *Corrosion of Metals and Alloys—Corrosivity of Atmospheres—Determination of Corrosion Rate of Standard Specimens for the Evaluation of Corrosivity*; ISO 9226; International Organization for Standardization: Geneva, Switzerland, 2012.
17. Kamimura, T.; Hara, S.; Miuyki, H.; Yamashita, M.; Uchida, M. Composition and protective ability of rust layer formed on weathering steel exposed to various environments. *Corros. Sci.* **2006**, *48*, 2799–2812. [CrossRef]
18. Hara, S.; Kamimura, T.; Miyuki, H.; Yamashita, M. Taxonomy for protective ability of rust layer using its composition formed on weathering steel bridge. *Corros. Sci.* **2007**, *49*, 1131–1142. [CrossRef]

19. International Organization for Standardization. *Corrosion of Metals and Alloys—Corrosivity of Atmospheres—Measurement of Environmental Parameters Affecting Corrosivity of Atmospheres*; ISO 9225; International Organization for Standardization: Geneva, Switzerland, 2012.
20. Urban, V.; Krivy, V.; Kreislova, K. The development of corrosion processes on weathering steel bridges. *Procedia Eng.* **2015**, *114*, 546–554. [CrossRef]
21. Krivy, V.; Urban, V.; Kubzova, M. Thickness of corrosion layers on typical surfaces of weathering steel bridges. *Procedia Eng.* **2016**, *96*, 56–62. [CrossRef]
22. Morcillo, M.; Chico, B.; Díaz, I.; Cano, H.; de la Fuente, D. Atmospheric corrosion data of weathering steels, a review. *Corros. Sci.* **2013**, *77*, 6–24. [CrossRef]

metals

Article

Bayesian Correlation Prediction Model between Hydrogen-Induced Cracking in Structural Members

Taejun Cho [1,†], David Joaquin Delgado-Hernandez [2], Kwan-Hyeong Lee [3], Byung-Jik Son [4] and Tae-Soo Kim [5,*]

1   Department of Civil Engineering, Daejin University, Hogookro 1007, Pocheon 11159, Korea; taejun@daejin.ac.kr
2   Postgraduate Engineering Department, Autonomous University of the State of Mexico, Toluca 50130, Mexico; delgadoh01@yahoo.com
3   Department of Robotics Engineering, Daejin University, Hogookro 1007, Pocheon 11159, Korea; khlee@daejin.ac.kr
4   Department of International Civil and Plant Engineering, Konyang University, Nonsan 35365, Korea; strustar@konyang.ac.kr
5   Department of Architectural Engineering, Hanbat National University, Daejeon 34158, Korea
*   Correspondence: tskim0709@daum.net; Tel.: +82-42-821-1121
†   Current address: Sinwondang 103–708, Sungsa, Goyang, Geyeongi 10455, Korea.

Academic Editors: Ricardo Branco and Filippo Berto
Received: 19 March 2017; Accepted: 31 May 2017; Published: 5 June 2017

Abstract: Background: A quantitative model was developed and applied for analyzing the correlation between hydrogen-induced corrosion cracking in both main cable wires and degraded stiffening of the girders of a cable suspension bridge, considering maintenance effects across time and space. Method: Bayesian inference is applied for predicting the correlations among the wires in the main cables owed to hydrogen-induced cracking (HIC) in the cable wires of a steel bridge, by using the improved hierarchical Bayesian models proposed here. Results: The simulated risk prediction under decreased strength of cable wires, due to the corrosion cracking, yields posterior distributions based on prior distributions and likelihoods. The Bayesian inference model can be applied to the design and maintenance of highly corroded and correlated components Data are updated through analyzed information from previous crack steps. A numerical example including not only reliability indices but also probabilities of failure for cable wires, damaged by HIC, is then presented. Compared with a conventional linear prediction model, the one herein developed provides highly improved convergence and closeness to the analyzed data. Conclusion: The proposed model can be used as a diagnostic or prognostic prediction tool for the performance of corroded bridge cable wires with crack propagation, allowing the development of maintenance plans for mechanical components and the overall structural system.

Keywords: corrosion; cable wires; hydrogen-induced cracking; hierarchical Bayesian inference; correlation model; maintenance interventions

1. Introduction

As a mixture of corrosion and crack growth deterioration in structural steel, Hydrogen-Induced Cracking (HIC) has been present in the cable wires of many suspension bridges [1–9]. Because it may become the primary cause underlying abrupt failure of structural systems, HIC of cable wires is a critical issue in cable-supported structures. However, very few studies have attempted to quantify the correlation between HIC and structural components. The quantification of this correlation is essential for developing an optimized maintenance plan in terms of budget management, leading to the design of appropriate rehabilitation methods, to the location of damaged members, and to the identification of maintenance tasks. Therefore, this piece of research introduces HIC and a method for quantifying the deterioration of steel members in structural systems.

Corrosion reactions often result in the formation of hydrogen gas. Hydrogen atoms can either diffuse or be absorbed into the lattice of the metal, leading to the deterioration of material properties. This, combined with the stress applied to the metal, can result in crack propagation. Diffused hydrogen atoms can recombine to form hydrogen molecules, which can exert pressure on the surrounding steel, resulting in the crack propagation in wires under high tensile stress [3,4].

Such corrosion can happen under exposure to carbonated and bicarbonated, caustic, nitrate, cyanide, anhydrous ammonia (liquid), mixture of nitric and sulfuric acid, mixture of magnesium chloride and sodium fluoride, or $CO/CO_2/H_2O$ mixtures, and under hydrogen attack. Some mechanisms have been proposed in the literature for explaining the phenomenon; these can be classified as either anodic Stress Corrosion Cracking (SCC) mechanisms or mechanical fracture processes. Anodic SCC involves the rupture of the protective oxide layer at the crack tip, anodic dissolution of the base metal, and crack growth under constant stress. Crack growth in turn can be intergranular or transgranular [10].

Another similar concept is that a film is formed on a metal surface, and brittle fracture follows due to dealloying or vacancy injection. The crack proceeds through the film and across the film/metal interface into the metal, where it propagates under the stress of the applied load. Once the crack propagation stops, the process restarts with the formation of a new film. It is worth noting that this mechanism is a combination of Hydrogen-Assisted Cracking (HAC) and SCC [11].

The differences between SCC and HAC have been investigated by Wen-Ta Tsai et al. [11]. Depending on the potential, the types of and reasons underlying, SCC vs. HAC are different. In their tests, when the potential was moved towards the cathodic direction (<-900 mVSCE; millivolts silver chloride and saturated calomel), HAC was observed. The loss of ductility, as indicated by a decrease in the reduction in the area of the specimen, as compared to that in an air test, was linked to crack initiation and propagation. In short, hydrogen plays a key role in many cracking mechanisms including SCC and HAC. HAC encompasses a number of different mechanisms and, in some cases, is considered to be interchangeable with HIC, hydrogen embrittlement and hydrogen damage.

The mechanism of HIC begins with the hydrogen atoms diffusing throughout the material. At elevated temperatures, hydrogen tends to have increased solubility, allowing it to disperse into the material. When such hydrogen atoms combine again in very small metal voids to build molecules of hydrogen, they produce pressure within the cavity. The pressure created from the buildup can further elevate, which makes the metal to lose its tensile strength and ductility, reaching the cracking point, or HIC [12].

Hydrogen-Induced Cracking (HIC) refers to the internal cracks brought about by the material trapped in budding hydrogen atoms. It involves atomic hydrogen, which is the smallest atom, diffusing into a metallic structure. In the case of a crystal lattice becoming saturated or coming into contact with atomic hydrogen, many alloys and metals may lose their mechanical properties.

If the buildup of molecular H is repressed, the emerging atomic H can disperse into the metal rather than forming a gaseous reaction. This, in turn, produces a crack in the metal. Certain chemical elements may contribute to this phenomenon, such as selenium, antimony, arsenic and cyanides. However, the main one is H_2S, or hydrogen sulfide.

In this study, the focus is on the reduction of ductility and tensile strength due to the hydrogen absorption in high-strength steel, which ultimately results in the brittle failure of a structural system. Moreover, the research allowed the authors to develop a correlation model for predicting the HIC in cable wires and steel members of target structural systems.

Although there are strong correlations among sources of corrosion and cracking in steel wires and other members in steel structures, a quantitative correlation model for modeling them has scarcely been investigated. In terms of the HIC of the cables of a suspension bridge, the local and global safety of bridge systems have been compared by the authors of this study and by other researchers [3,4,11,13]. With regard to the local safety analysis, a decoupling technique based on two-dimensional finite element models has been developed for evaluating hydrogen-diffusion-driven crack propagation in a wire section. A serviceability limit state based on the global responses of a stiffening girder, the ultimate limit state, and the reliability of time-dependent and crack depth-dependent HIC of a cable wire has been calculated for individual components. Parallel system reliability analyses have also been carried out. While the proposed solutions show local and global risks successfully and separately calculated based on observed differences in the global and local safety behaviors of the suspension bridge system, a quantitative correlation between the two behaviors has not been modeled [3,4].

The correlation quantification among risky elements is important, especially when the components of a complex system are highly correlated. Predicting future degradation through the application of stochastic regression models needs additional consideration given their synergistic or dependent causes, which may result in growing damages. Thus far, efforts to determine correlations among the causes of deterioration and results have focused on only a few variables. For example, HIC in the main cables of a suspension bridge could beget torsion stress in stiffening girders [14] or nuts being pulled off in hangers. Hence, it is necessary to identify the correlations among the causes of deterioration and their results for an appropriate bridge management system, followed by the prediction of crack propagation in cable wires due to HIC.

For analyzing these correlations and the resulting deterioration, conventional stochastic modeling of structural components and systems has been applied mainly on the basis of Event Tree Analysis (ETA) [15], Fault Tree Analysis (FTA) [16], and regression simulation in terms of the Response Surface Method (RSM) or adaptive RSM [17,18]. ETA is a method for illustrating the sequence of outcomes that may arise after the occurrence of a selected initial failure event and for ranking accidents, considering that the order of events needs permutation-based calculations, which produce a large number of cases for predicting failure scenarios.

Modeling uncertainties in environmental load components could be considered in a correlation analysis by conducting uncertainty analysis between the resultant stress ranges and the expected fracture lives, due to hydrogen-induced corrosion cracking based on Bayesian inference modeling. Compared with ETA or FTA, the Bayesian inference model could evaluate and detect the partial failure as a probabilistic quantification, which is allowed neither in ETA nor in FTA.

Bayesian analysis is based on posterior inference [19]. Parameter estimates are usually summary statistics of the marginal posterior distributions such as the posterior mean, median, mode, and standard deviation. In nonhierarchical Bayesian models, it is often easy to analytically derive the marginal posterior distributions and obtain the summary statistics. However, in highly correlated system models, specifically, given that the parameters are of multiple dimensions, it is often difficult to present the marginal distribution of each parameter analytically.

The results of this study could appreciably reduce the known limitations for predicting future degradation and enable the full modeling of field variables, which are highly correlated and show inelastic behavior. The limitations are as follows:

1. Most mechanical components affect future events because of the evaluation of joint probability distributions of current maintenance data, which are highly correlated among the components.
2. Modeling uncertainty, when multiple parameters can be regarded as related or connected in some manners due to hydrogen-induced corrosion cracking, implies that a joint probability model should reflect the dependence among these parameters, which could then be modeled in a hierarchical inference.
3. Prediction of future degradation of a component or a structural system is performed through the implementation of an updated stochastic model; reducing errors arising from corroded and correlated components, nonlinear behaviors, and including maintenance interventions as well.

With these ideas in mind, the objectives and scope of this paper are:

- To assess risk correlation between shear and bending deformation of steel members and HIC in high-strength steel of main cable wires.
- To simulate degradation of cable wires on the basis of the proposed Bayesian prediction models in terms of locally correlated cracking in cable wires just before the collapse of a structural system, considering maintenance interventions by implementing an inverse analysis of the proposed Bayesian inference network model.

2. Bayesian Hierarchical Model for Correlated Data

2.1. Hierarchical Modeling

Bayes' theorem gives the posterior distribution for the parameters of interest, in terms of the prior distribution, failure model, and the observed data, which in the general continuous form is written as [14]:

$$\pi_1(\theta|x) = \frac{f(x|\theta)\pi(\theta)}{\int f(x|\theta)\pi(\theta)d\theta}, \tag{1}$$

where, $\pi_1(\theta \mid x)$ is the posterior distribution for the parameter of interest, denoted as θ. The observed data enter via the likelihood function, $f(x \mid \theta)$, and $\pi(\theta)$ is the prior distribution of θ. $f(x)$ is called the marginal or unconditional distribution of x. The range of integration is over all possible values of θ, being the probability of seeing x events, referred to as the predictive distribution for x [15].

Hierarchical models have been introduced in population-based problems [16], describing efficiently complex datasets incorporating correlation. Hence, when multivariate or repeated responses are observed, correlation can be included in the model via random effect for all measurements referring to the same individual. Random effects and the corresponding hierarchical structure are applied to appropriately specify the marginal sampling distribution frequently referred to as data augmentation, which has been considerably simplified in the Markov Chain Monte Carlo methods (MCMC) scheme that can be used to estimate the posterior distributions of interest. One of the MCMCs, Metropolis Hastings (M-H) sampling [17] numerically and efficiently simulates posterior distribution of parameters, which has been adopted in this study.

2.2. Hierarchical Bayesian Inference Model to Predict Degradation of Cable Wires due to HIC

In order to reconstruct incompletely observed or missing data, imputation models are commonly selected, which include the regression method for monotone data, the non-parametric Propensity method [18], and the MCMC [16] for non-monotone cases. In this research, creating a predictive distribution has been applied based on predictive distributions, while the data are averaged over all possible parameter values for the maintenance data of a bridge structure. The validation of the suggested model has been compared with the measured maintenance data with the partially deleted input prior distribution data.

For this reason, when datum y has not been observed yet, predictions are based on the marginal likelihood:

$$f(y) = \int f(y|\theta)f(\theta)d\theta, \tag{2}$$

which is the likelihood averaged over all parameter values supported by prior beliefs, where, $f(y)$ is called prior predictive distribution.

The posterior predictive distribution is given by:

$$f(y'|y) = \int f(y'|\theta)f(\theta|y)d\theta, \tag{3}$$

which is the likelihood of the future data averaged over the posterior distribution $f(\theta|y)$.

This distribution is termed as the predictive distribution since prediction is usually attempted only after observation of a set of data y. Future observations y' can be alternatively viewed as additional parameters under estimation. From this perspective, the joint posterior distribution is now given by $f(y', \theta | y)$. The MCMC method is used to obtain this posterior distribution from which the imputed values for missing observations or future predicted data are drawn. Inference on the future observations y' can be based on the marginal posterior distribution $f(y' | y)$ by integrating out all nuisance parameters, one of which in this case, is the parameter vector θ. Hence, the predictive distribution is given by Equation (3), since past and future observables, y and y', are conditionally independent given the parameter vector θ [19].

Linear regression models are common in statistical sciences. In linear regression models, the response variable Y is considered to be a continuous random normally distributed variable defined in the whole set of real numbers. The following equation is selected:

$$Y_{ij} = \alpha_i + \beta_i (x_j - \bar{x}), \tag{4}$$

where $\bar{x}$ = mean value of maintenance (duration of service). Due to the absence of a parameter representing the correlation between α_i and β_i, standardizing the x_j's around their mean to reduce dependence between α_i and β_i in their likelihood is carried out, achieving complete independence.

The synergic effects have been evaluated in deterministic and probabilistic ways, which revealed worse deterioration than linearly superposed. Therefore, the following quadratic regression models are proposed, for which each dependent variable serves as the dependent variable and the other variables in the dataset serve as the independent variables:

$$Y_{ij} = \alpha_i + \beta_i (x_j - \bar{x}) + \gamma_i (x_j - \bar{x})^2, \tag{5}$$

The model parameter estimates are then used in making random draws from the multinomial distribution for each missing response on the dependent variable in the regression.

3. Correlation Model for HIC in the Main Cable Wires

3.1. Deterministic Global Analysis Model for HIC of High-Strength Steel Wires in Suspension Bridges

HIC propagation in cable wires of a suspension bridge has been analyzed previously [3,4]. The structural model of a cable suspension bridge has 336 frame elements, 242 catenary cable elements, and 461 node numbers (2766 degrees of freedom, Figure 1). The maximum response of a bridge system has been obtained for tensile stresses at two girder section locations in the center span (Figure 2). The maximum values of the shear stress and horizontal deformation are obtained when the live load is loaded at the side span.

There are two load cases for generating the maximum tensile and shear stresses. Case 1 is for generating the maximum tensile stress, which is the combination of dead load (DL), traffic or live load (LL), wind load (WL), and temperature load (TL). Case 2 is for generating the maximum shear stress, which is the combination of DL, WL, and TL, i.e., (DL + WL + TL). The stresses and deflections pertaining to Load Case 1 are evaluated at $L/4$ and $L/2$ of the central span (L refers to the length of the central span), i.e., elements 48 and 72, and nodes 199 and 223 (Figure 2). The stresses and deflections pertaining to Load Case 2 are evaluated at the end of the central span, i.e., element 2, and node 153 (Figure 2).

Experiments on hydrogen-induced corrosion cracking in cable wires of suspension bridges were conducted [20]. The prestressing tendon were tested by Toribio [21]. The experimental results were compared with the solutions yielded by the analytical equations of Forman and Shivakumar [22] and the author's FEM analysis program written for the ANSYS61 platform [3,4]. The test results of the experiments on the hydrogen-induced corrosion cracking of cable wires [20] and the reduced brittle failure strength of wires closely follow the constitutive relationship for undamaged new wires. Figure 3 shows that the cracks lead to sudden brittle failure for that crack depth for which the stress intensity factor at the crack tips was equal to the critical Stress Intensity Factor (SIF) or fracture toughness (K_c) of the material, which was determined using the aforementioned FE program [3], compared with Forman's theoretical equation for SIF [23].

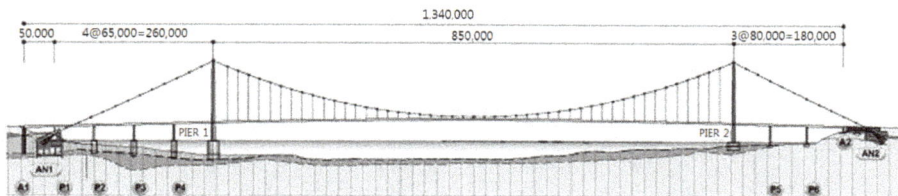

Figure 1. Side view of example suspension bridge (Unit: mm).

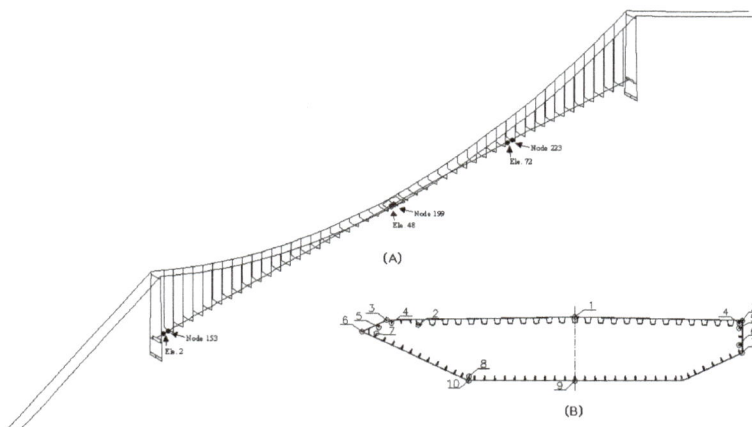

Figure 2. Observation positions of structural responses: (**A**) nodes and elements; and (**B**) cross section.

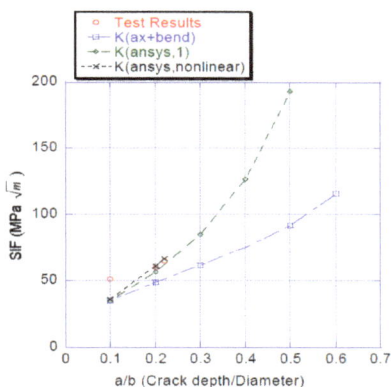

Figure 3. Constitutive relationship while the crack is growing in the main cables and the Calculated SIF is compared with the test results. Calculated SIF by elastic and elastic-plastic material behavior, compared with test results [3].

HIC of the main cable wires is directly related with the collapse of a suspension bridge. The safety and correlation of components with regard to the structural system has been determined using a decoupled FE analysis model and by conducting reliability analysis (Figure 4) in which only the local correlation of HIC in cable wires was measured.

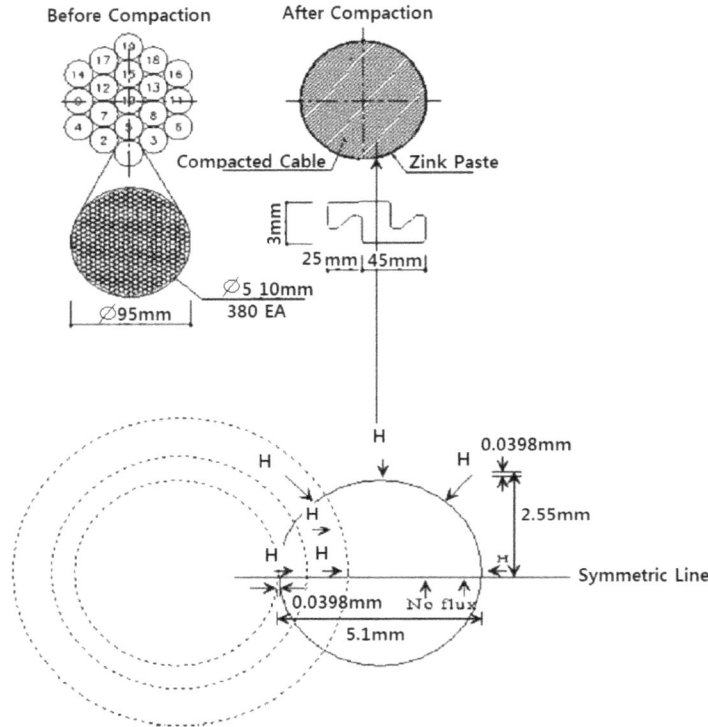

Figure 4. Finite Element Analysis model for crack propagation in main cable wires [3].

Considering periodical maintenance tasks, it would be helpful if the correlation between local and global components in terms of the HIC of cable wires and the structural responses of the bridge itself was quantified, which is described in the following section.

3.2. Correlation between Local and Global Component Reliabilities

Figure 1 presents the suspension bridge subjected to the evaluation of HIC and the overall dimension of the above stated bridge system. All stresses in the cross section are evaluated at the two positions, i.e., 1 and 2, as shown in Figure 2. The limit state functions for evaluating reliability indices and failure probabilities are composed of 10 cases, whereas the extent of corrosion of the main cable wires varies from 0 to 100%.

STEPs 1–6 mean 10–60% of the loss in the total 7220 wires, respectively. Hence, at Step 1, 722 wires are broken by HIC, and in Step 2, 1444 wires are fractured. For Step 6 in Figure 5, 4332 wires are removed from the active wires. While wires are broken, because the total loads are constant, the remaining wires will be more loaded. As tensile loads and stresses are increased in the wires, the diffusion of hydrogen atoms and the propagation rates of cracks into wires are accelerated. Consequently, the wires' lifetime is reduced. When the degree of corrosion is 60% in Step 6, the wires will reach the fracture toughness, K_c (45 MPa $\sqrt{m}$), immediately.

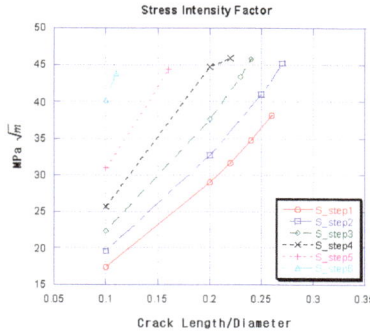

Figure 5. Stress Intensity Factor (*K*) while crack propagates in each degree of corrosion [3].

The reliability results for tensile and shear stresses are shown in Figures 6 and 7, respectively. As can be seen, for normally distributed random variables, the reliability index is related to the probability of failure by:

$$\beta = - \Phi(P_f) \text{ or } P_f = \Phi(-\beta), \tag{6}$$

where Φ^{-1} is the inverse standard normal distribution function, when P_f is less than 0, the probability of failure is larger than 50%. Table 1 compares several calculated reliability indices with the Probability of Safe, P_S, and the Probability of Failure, P_f.

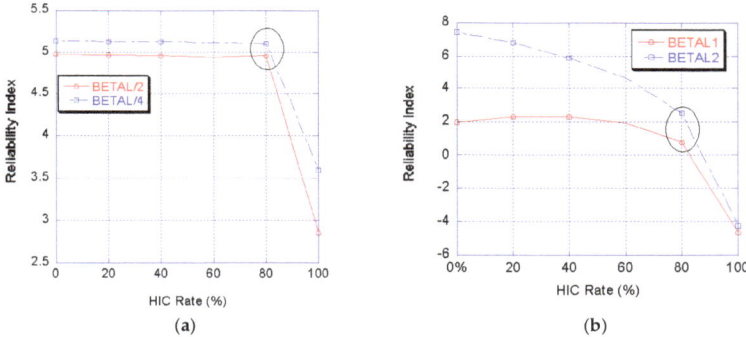

Figure 6. Reliability indices for tensile stresses and shear stresses in cable wires, where $L/4$ means span/4 (Node 223) and $L/2$ means span/2 (Node 199) (see Figure 2): (**a**) tensile stresses; and (**b**) shear stresses.

Table 1. Five Correlated cubic functional values with random effects.

X (Year)	Y_1	Y_2	Y_3	Y_4	Y_5
1	2.5	3.162463	2.393176	5.211623	3.625013
2	4	5.270988	3.154476	6.554857	6.958659
3	−0.5	−0.59202	−0.43391	−0.56402	−0.86762
4	−17	−24.256	−12.6095	−23.2487	−26.4017
5	−51.5	−68.6477	−42.3458	−89.3462	−96.6057
6	−110	−160.169	−87.4144	−184.485	−162.822
7	−199	−208.906	−141.385	−286.652	−269.2
8	−323	−359.445	−269.061	−364.23	−529.286
9	−490	−522.926	−374.941	−822.392	−604.255
10	−704	−716.606	−507.154	−989.295	−989.443

The following serviceability limit states are selected for the system reliability of a cable wire in two load cases and at two check points as a corrosion dependent function:

$$g(C) = \sigma_{cr} \times N_{\sigma_{cr}} - \left(a_0 + \sum_{i=1}^{n} a_i x_i + \sum_{i=1}^{n} a_i x_i^2 \right), \tag{7}$$

$$g(C) = \tau_{cr} \times N_{\tau_{cr}} - \left(a_0 + \sum_{i=1}^{n} a_i x_i + \sum_{i=1}^{n} a_i x_i^2 \right), \tag{8}$$

where σ_{cr} is the code specified allowable tensile stress (190 MPa), as a resistance value; τ_{cr} is the code specified allowable shear stress (110 MPa); and $N_{\sigma_{cr}}$ and $N_{\tau_{cr}}$ are the bias factors of tensile stress and shear stress respectively. The values in parenthesis are the corrosion dependent tensile and shear stresses, responses of the bridge system when the areas of the main cables and hangers are perturbed, as a load term.

The selected random variables are about three kinds of cable section area. Since the Bucher-Bourgund method [24] is employed, the three axial points have the distances of $\pm\sigma$ (standard deviation) between the center and axis points. The time-dependent safety of each response is evaluated by applying Equations (6) and (9) for the component and system reliability, respectively.

The upper and lower bounds of the probabilities of failure in the positively correlated parallel system occur as follows [17].

$$\prod_{i=1}^{n} F_{Ri(q)}(t) = \prod_{i=1}^{n} P_{fi}(t) \leq \Phi \left(-\beta_e \times \sqrt{\frac{n}{1 + (n-1)\bar{\rho}}} \right) \leq P_f(t) \leq \min(P_{fi}(t)), \tag{9}$$

where P_f is the probability of failure of the parallel-connected structural system; $F_{Ri(q)}(t)$ is the cumulative distribution function; q_i is the normalized base variables of load; and P_{fi} is the probability of failure of the ith component, which is calculated by standard normal distribution as $P_{fi} = \Phi(-\beta_i)$; n is the number of parallel elements, and $\bar{\rho}$ is the coefficient of partial correlations.

Figures 6 and 7 show that reliabilities decrease and the probabilities of failure increase when the HIC ratio is greater than 80%, as indicated with the black circles. If the degree of corrosion for wires due to HIC is greater than 60% (Figure 5, Fracture analysis result), the cable wires will fail immediately. Note that all wires are fully correlated via the hierarchical Bayesian prediction model.

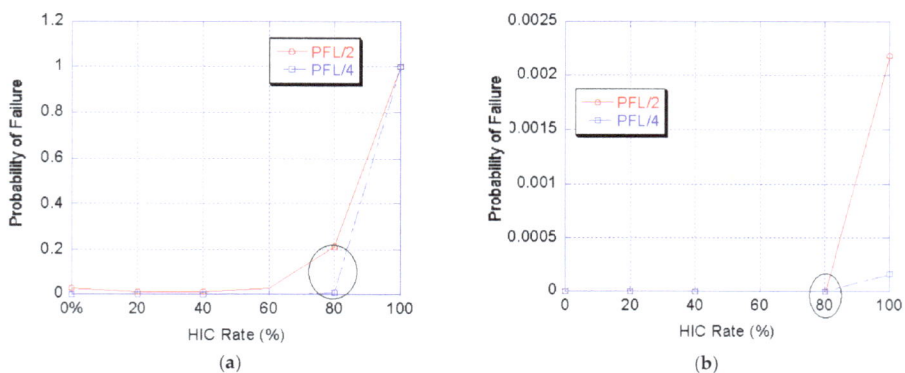

Figure 7. Failure probabilities for tensile stresses and shear stresses in cable wires, where $L/4$ means span/4 (Node 223) and $L/2$ means span/2 (Node 199) (see Figure 2): (**a**) tensile stresses; and (**b**) shear stresses.

As shown in Figure 8, although the expected mean values have slightly higher indices (less risky) than the calculated reliability indices, the confidence intervals or the predicted distributions of the predicted values incorporate the previous calculations shown in Figure 9 for the tensile and shear stresses' reliability indices.

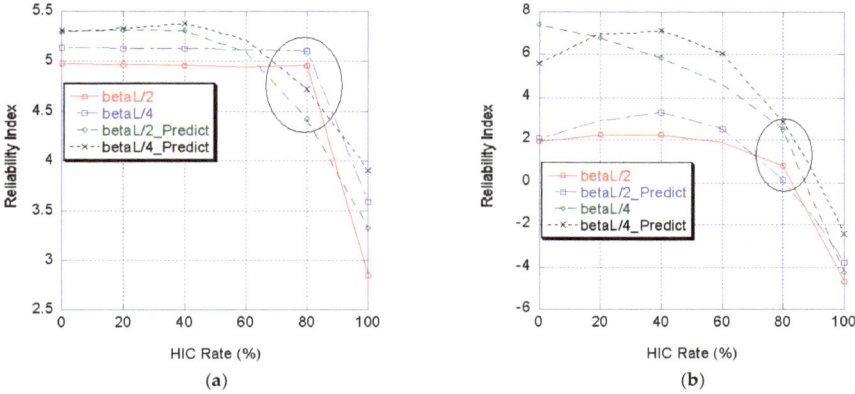

Figure 8. Predicted reliability indices for tensile and shear stresses in cable wires, where $L/4$ means span/4 (Node 223) and $L/2$ means span/2 (Node 199) (see Figure 2): (**a**) tensile stresses; and (**b**) shear stresses.

The correlation model between tensile stress and shear stress could be very important for identifying the correlation between the shear and tensile stress responses, in which data measured during daily monitoring show correlated responses from correlated structural behaviors in the target structure. In addition, shear stresses show sharper responses than other types of stresses, as shown in Figures 8 and 9. The correlation modeling and the resulting Bayesian belief network are discussed in the following section.

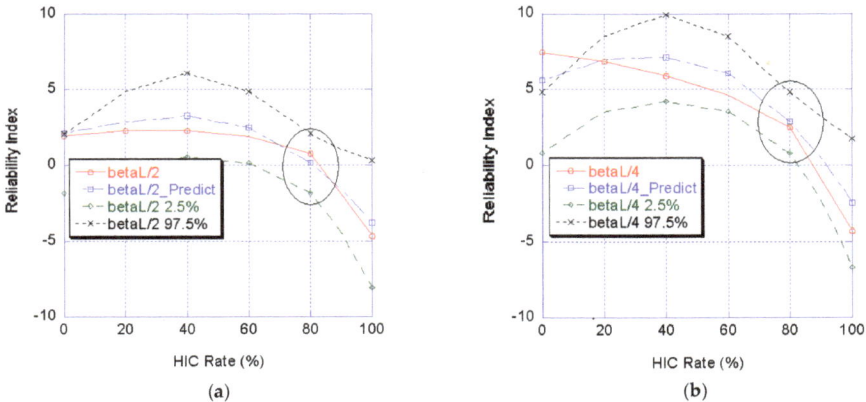

Figure 9. Confidence intervals of predicted reliability indices for shear stresses in cable wires, where $L/4$ means span/4 (Node 223) and $L/2$ means span/2 (Node 199) (see Figure 2): (**a**) tensile stresses; and (**b**) shear stresses.

3.3. Dynamically Correlated System Responses due to the Main Cable Crack Growing

While crack propagated in strands of main cable from 0 to 100%, tensile and shear stresses in stiffening girders increase. The proposed quadratic Bayesian prediction model is applied while strands have failed in terms of local modeling of HIC in wires of strands. After the local prediction modeling, correlation model could be applied to identify the correlation between shear and tensile stress responses, which can be important criteria of maintenance before determining appropriate management tasks.

The probability density function based on the joint probability, following the Markov condition is calculated as:

$$P(X_{1:T}) = \prod_{t=1}^{T} \prod_{i=1}^{n} P(X_t^i | \pi(X_t^i)),$$

(10)

where X_i represent nodes, and parents represent the stress responses at stiffening girders. T is the crack propagation time in unit of percentile from 0 to 100%.

The time dependent coefficients of correlation among nodes are:

$$\rho_{xy}^t = \frac{1}{n} \frac{\sum_{t=1}^{T} \sum_{i=1}^{n} x_i^t y_i^t - \sum_{t=1}^{T} \overline{xy}}{\sum_{t=1}^{T} S_x^t S_y^t},$$

(11)

where ρ_{xy}^t is the correlation coefficient between random variables, x_i^t and y_i^t denote the random variables, n represents the total number of random variables, S_x^t and S_y^t are the standard deviations of the variables, and $\overline{x}, \overline{y}$ are the mean values of the random variables at time t.

A Dynamic Bayesian Belief Network model (DBBN) was employed here, as shown in Figure 10. After 100,000 iterations of the Markov Chain Monte Carlo simulation, the following correlations between tensile and shear stresses in stiffening girders were observed in Figure 11. In the figure, correlations between tensile and shear stresses at $L/2$ and shear stress at $L/4$ are plotted while crack propagates from 0 to 100%.

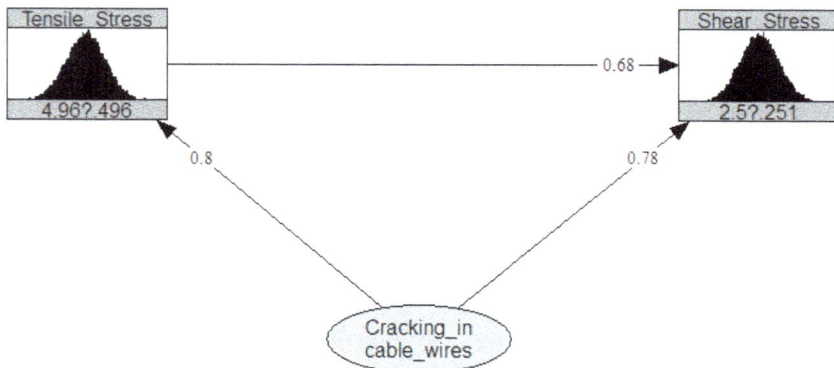

Figure 10. Random variables modeled with nodes and arcs using Uninet version 2.74 (LightTwist Software, Melbourne, Australia).

As shown in Figure 11, there are higher correlations between the tensile stresses at $L/2$ and $L/4$, which are the same for shear stresses at two girder locations because the value is on average higher than 0.3 between mu[1,*] and [2,*], where * means crack propagation steps. The percentile, mean, standard deviation, and error of MCMC is presented within the Appendix A in Table A1. The quantity reported in the MC error column gives an estimate of $\sigma/N1/2$, the Monte Carlo standard error of the mean.

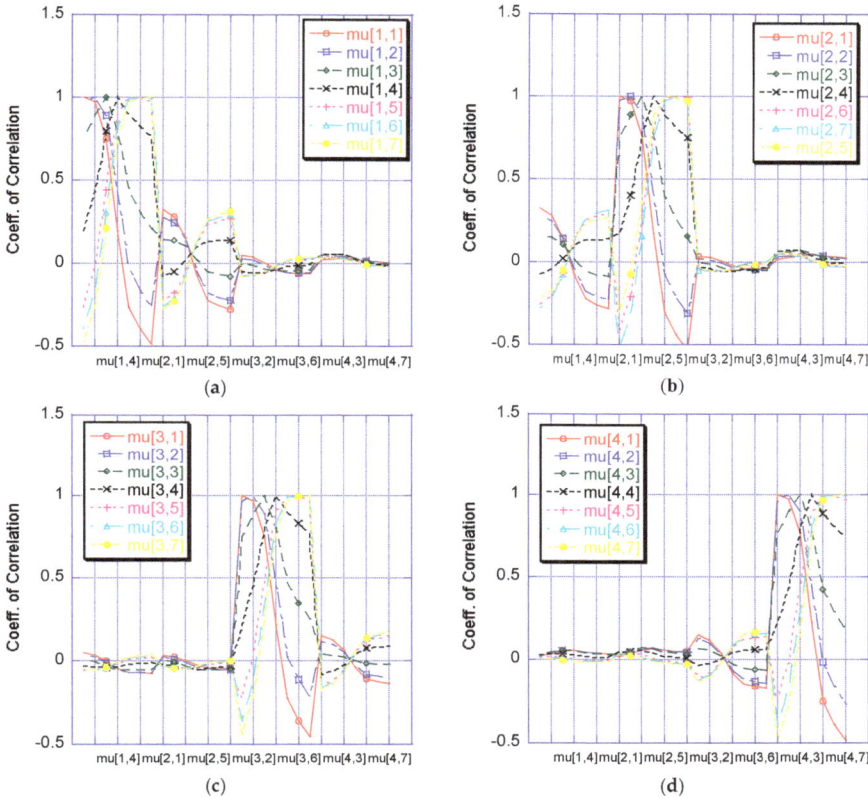

Figure 11. Correlations between tensile and shear stresses at $L/2$ and shear stress at $L/4$, where $L/4$ means span/4 (Node 223) and $L/2$ means span/2 (Node 199) (see Figure 7). Here, for mu[*x*, *y*], *x* = 1,2 implies tensile stress at $L/2$ and $L/4$; *x* = 3, 4 implies shear stress at $L/2$ and $L/4$; and *y* = 1 to 7 implies crack propagation from 0 to 20%, 40%, 60%, 80%, 90%, and 100%, respectively. (**a**) Tensile stress at $L/2$ and other variables. (**b**) Tensile stress at $L/4$ and other variables. (**c**) Shear stress at $L/2$ and other variables. (**d**) Shear stress at $L/4$ and other variables.

This value increases as the crack propagates through the main cable wires. The difference in the level of correlation at the two aforementioned locations indirectly proves that given the difference in these responses, there is little effect of the imposed direct load on the girder. This inference contradicts Betti-Maxwell's reciprocal theorem, but proves that they are dependent on the HIC in the main cable wires.

On the basis of these observations, the reference response to shear stresses at $L/4$ could be important in determining the risk stages for the considered structural system. However, two major problems can arise in the case of maintenance interventions for replacing the main cable wires or reinforcing the girder; for instance, increasing the sectional area or moment of inertia would lead to a change in the tensile or shear responses.

In those cases, a modified response should be determined for identifying the source of deformation, i.e., crack propagation in the main cable wires, which can be calculated based on the proposed inverse

DBBN model and Equation (11). Two cases of stiffening girder are considered, which is reinforced by widening or thickening the girder's flanges. Hence, the two reinforced girder cases are modeled as decreasing tensile stresses from 100% to 25% and 50% under 80% crack propagation in the wire, as summarized in Table 2.

Table 2. Reliability Index versus Probability of Failure.

Reliability Index	Reliability, P_S (=1 − P_f)	Probability of Failure, P_f
0	0.5	0.5
0.5	0.691	0.309
1	0.841	0.159
1.5	0.9332	6.68×10^{-2}
2	0.9772	2.28×10^{-2}
2.5	0.99379	6.21×10^{-3}
3	0.99865	1.35×10^{-3}
3.5	0.999767	2.33×10^{-4}
4	0.9999683	3.17×10^{-5}
4.5	0.9999966	3.40×10^{-6}
5	0.999999713	2.87×10^{-7}
5.5	0.999999981	1.90×10^{-8}
6	0.999999999	9.87×10^{-10}
7	1	1.28×10^{-12}
8	1	6.11×10^{-16}

It would be important to vary the responses while changing the extent of crack propagation in the main cable wires from the viewpoint of maintenance tasks, involving replacing the main cable wires. Therefore, two cases of the main cable wire rehabilitation are considered. The reliability indices of the reinforced cable wires are assumed to increase from 0.5 to 3.0 and 5.0 under 80% crack propagation in the wires, as referenced from the replacement of cable wires in real structures [25]

The inverse Bayesian belief models are shown in Figure 12, in which the increased reliability indices of 3.0 and 5.0 are shown. All types of probabilistic distributions of the reliability indices of stresses and cracking in wires are assumed as normally distributed with a standard deviation of 0.1.

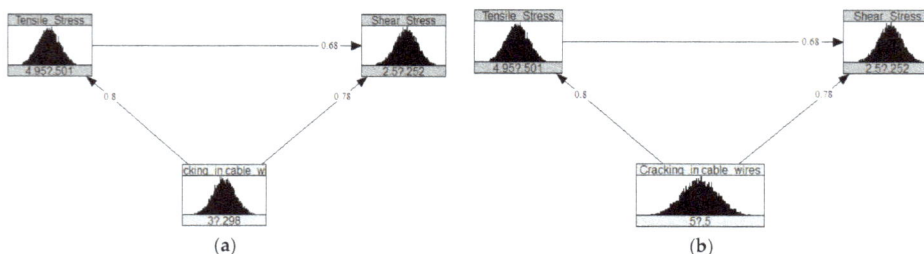

(a)
(b)

Figure 12. Inverse analysis of Bayesian belief network under 80% crack propagation in wire and reinforced as reliability indices increased from 0.5 to: (**a**) 3.0; and (**b**) 5.0. (**a**) Reliability index = 3.0 in cable. (**b**) Reliability index = 5.0 in cable.

Figure 13 shows the results of the inverse Bayesian belief network simulations in terms of the reliability indices for tensile and shear stress responses, while cracking in wire in 80% reinforced as the reliability indices increased from 0.5 to 3.0 and 5.0. Compared with the originally predicted indices, the reinforced main cable wires, having indices of 3.0 and 5.0, show reduced tensile and shear stresses.

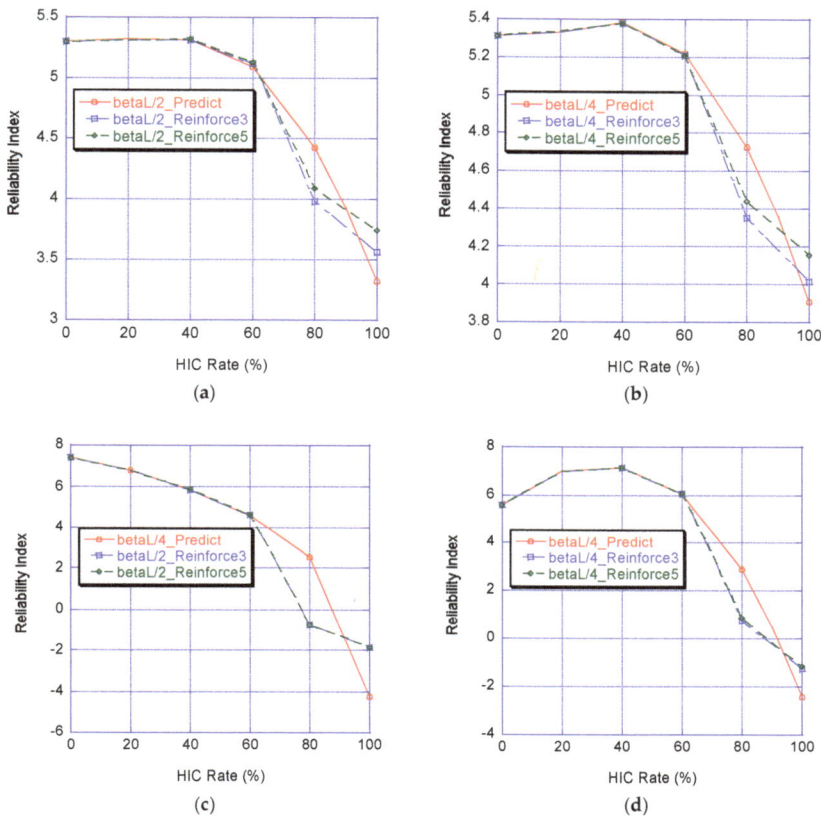

Figure 13. Inverse analysis results for the Bayesian belief network under 80% cracking in wire and reinforced for the cases when the reliability indices increased from: (i) 0.5 to 3.0; and (ii) 0.5 to 5.0. (**a**) Tensile stress at $L/2$. (**b**) Tensile stress at $L/4$. (**c**) Shear stress at $L/2$. (**d**) Shear stress at $L/4$.

It is notable that, compared with the tensile responses, the shear responses indicate higher safety, as the reliability indices increased from 128% and 80%, as increased from -4.25 to -1.86 and from -2.44 to -1.35 in reliability indices, under 100% cracking of the cable wires at $L/2$ and $L/4$, respectively. The tensile responses indicate considerably higher safety levels, as the reliability indices increased from 8.3% and 12.7%, as increased from 3322 to 3600 and from 3322 to 3744 in reliability indices, under 100% cracking of the cable wires at $L/2$, and approximately 5% increased safety at $L/4$. Therefore, it is acknowledged that the shear response is very sensitive to crack propagation in the main cable wires regardless of whether maintenance tasks are carried out.

In Table 3, the numbers of the columns in 25% and 50% reinforced girder represent the mean value of the simulated tensile stresses, shear stresses and cracking. Standard deviations are calculated from the distribution of MCMC results. The importance of the table lies on the comparison of the correlation among response values between the two 25% and 50% reinforced girder cases. Note that even if the responses of tensile stresses have decreased from 4.96 to 3.72 and 2.51, respectively, in terms of the reliability indices, the correlations between tensile and shear stresses are almost unchanged, as summarized in Table 3.

Table 3. Predicted simulation results between tensile and shear responses.

Predicted Variables	Base Variables	25% Reinforced Girder	Standard Deviation (25%)	Correlation (25%)	50% Reinforced Girder	Standard Deviation (50%)	Correlation (50%)
Tensile Stress	Tensile Stress	3.722	0.3722	1	2.48	0.248	1
Tensile Stress	Cracking in cable wires	3.722	0.3722	0.8138	2.48	0.248	0.8138
Tensile Stress	Shear Stress	3.722	0.3722	0.6945	2.48	0.248	0.6945
Cracking in cable wires	Tensile Stress	0.5	0.05	0.8138	0.5	0.05	0.8138
Cracking in cable wires	Cracking in cable wires	0.5	0.05	1	0.5	0.05	1
Cracking in cable wires	Shear Stress	0.5	0.05	0.8993	0.5	0.05	0.8993
Shear Stress	Tensile Stress	2.5052	0.2502	0.6945	2.5052	0.2502	0.6945
Shear Stress	Cracking in cable wires	2.5052	0.2502	0.8993	2.5052	0.2502	0.8993
Shear Stress	Shear Stress	2.5052	0.2502	1	2.5052	0.2502	1

4. Conclusions

A quantitative correlation model was developed and applied for analyzing correlation between hydrogen-induced corrosion cracking in the main cable wires of a cable suspension bridge with degraded stiffening girders, considering the effects of maintenance efforts in time and space. Bayesian inference was applied for predicting the correlation among wires in the main cables owing to HIC of the cable wires of a suspension bridge, by using the improved nonlinear hierarchical Bayesian models developed here, which is based on the Markov Chain Monte Carlo method.

The effects of maintenance interventions for alleviating cracking in the main cable wires due to HIC have been investigated using the developed inverse Bayesian prediction model. While reinforcing the main cable wires, in terms of reliability indices, from negative values to 3.0 and 5.0, the shear responses indicate considerably higher increase in safety than do tensile responses, as the reliability indices increased from 128% and 80% under 100% cracking of the cable wires at $L/2$ and $L/4$, respectively. Therefore, it is acknowledged that shear responses are very sensitive responses to the crack propagation in the main cable wires.

Consequently, the correlations between local corrosion and crack growth in high-strength steel in terms of crack propagation in the main cable wire of a suspension bridge with global responses of the structure are modeled for predicting future degradation, followed by inverse analysis for identifying the effects of maintenance interventions. The proposed inverse Bayesian inference model as the quantified correlation model not only provides highly improved convergence but also shows the possibility of future maintenance control and savings by offering infrastructure managers the opportunity of risk prognosis and mitigation.

Acknowledgments: This research was supported by Human resources Exchange program in Scientific technology through the National Research Foundation of Korea (NRF) funded by the Ministry of Science, ICT and future Planning (No. 2017H1D2A2000600).

Author Contributions: Taejun Cho, David Joaquin Delgado-Hernandez and Tae-Soo Kim conceived and designed the works; Taejun Cho, David Joaquin Delgado-Hernandez and Byung-Jik Son performed the numerical analysis; Kwan-Hyeong Lee and Byung-Jik Son analyzed the data; Taejun Cho, David Joaquin Delgado-Hernandez and Kwan-Hyeong Lee contributed reagents/materials/analysis tools; Taejun Cho and Tae-Soo Kim wrote the paper.

Conflicts of Interest: The authors declare no conflict of interest.

Appendix A

Table A1. Percentile, mean, standard deviation, and error of MCMC, simulation for Figure 11.

Stochastic Variables	Mean	Standard Deviation	Monte-Carlo Error	Lower Limit (2.5 Percent)	Median	Lower Limit (97.5 Percent)
mu[1,1]	24.52	700	19.03	3.221	5.511	7.847
mu[1,2]	16.57	421.6	11.47	3.377	5.109	6.871
mu[1,3]	8.608	143.1	3.902	3.331	4.714	6.091
mu[1,4]	0.6498	135.4	3.662	2.89	4.321	5.664
mu[1,5]	−7.308	413.8	11.23	2.087	3.922	5.618
mu[1,6]	−11.29	553.1	15.01	1.611	3.719	5.678
mu[1,7]	−15.27	692.3	18.79	1.105	3.517	5.775
mu[2,1]	24.31	689.1	18.76	3.233	5.576	7.939
mu[2,2]	16.54	415.2	11.31	3.479	5.243	7.02
mu[2,3]	8.765	141.4	3.862	3.517	4.911	6.296
mu[2,4]	0.9903	132.6	3.586	3.145	4.582	5.944
mu[2,5]	−6.784	406.5	11.03	2.393	4.252	5.971
mu[2,6]	−10.67	543.4	14.76	1.927	4.089	6.067
mu[2,7]	−14.56	680.3	18.48	1.439	3.924	6.207
mu[3,1]	22.41	691.5	18.8	1.353	3.616	6.012
mu[3,2]	13.9	416.7	11.34	0.8279	2.561	4.442
mu[3,3]	5.393	142	3.872	0.1083	1.504	3.071
mu[3,4]	−3.116	132.9	3.595	−0.9538	0.4491	2.024
mu[3,5]	−11.63	407.6	11.06	−2.336	−0.6043	1.293
mu[3,6]	−15.88	545	14.79	−3.107	−1.129	1.006
mu[3,7]	−20.13	682.4	18.53	−3.914	−1.654	0.744
mu[4,1]	27.69	704.9	19.18	5.92	8.566	10.93
mu[4,2]	18.19	424.7	11.57	4.733	6.661	8.414
mu[4,3]	8.69	144.5	3.947	3.351	4.754	6.109
mu[4,4]	−0.8083	135.8	3.671	1.499	2.852	4.244
mu[4,5]	−10.31	415.9	11.29	−0.8075	0.9476	2.832
mu[4,6]	−15.06	556	15.1	−2.063	−0.00627	2.211
mu[4,7]	−19.8	696.2	18.91	−3.336	−0.9572	1.638

References

1. Hopwood, T.; Havens, J.H. Corrosion of Cable Suspension Bridges. Available online: http://uknowledge.uky.edu/cgi/viewcontent.cgi?article=1666&context=ktc_researchreports (accessed on 30 May 2017).
2. Roffey, P. The fracture mechanisms of main cable wires from the forth road suspension. *Eng. Fail. Anal.* **2013**, *31*, 430–441. [CrossRef]
3. Cho, T.; Kim, T.S.; Lee, D.H.; Han, S.H.; Choi, J.H. Reliability analysis of a suspension bridge affected by hydrogen induced cracking based upon response surface method. *ISIJ Int.* **2009**, *49*, 1414–1423. [CrossRef]
4. Cho, T.; Kim, T.S. A prediction model for hydrogen induced cracking in a prestressed wire with a fracture analysis. *ISIJ Int.* **2008**, *48*, 496–505. [CrossRef]
5. Laureys, A.; Depover, T.; Petrov, R.; Verbeken, K. Influence of sample geometry and microstructure on the hydrogen induced cracking characteristics under uniaxial load. *Mater. Sci. Eng. A* **2017**, *690*, 88–95. [CrossRef]
6. Masoumi, M.; Silva, C.C.; de Abreu, H.F.G. Effect of crystallographic orientations on the hydrogen-induced cracking resistance improvement of API 5L X70 pipeline steel under various thermomechanical processing. *Corros. Sci.* **2016**, *111*, 121–131. [CrossRef]
7. Dunne, D.P.; Hejazi, D.; Saleh, A.A.; Haq, A.J.; Calka, A.; Pereloma, E.V. Investigation of the effect of electrolytic hydrogen charging of X70 steel: I. The effect of microstructure on hydrogen-induced cold cracking and blistering. *Int. J. Hydrog. Energy* **2016**, *41*, 12411–12423. [CrossRef]
8. Laureys, A.; Depover, T.; Petrov, R.; Verbeken, K. Characterization of hydrogen induced cracking in TRIP-assisted steels. *Int. J. Hydrog. Energy* **2015**, *40*, 16901–16912. [CrossRef]

9. Shi, X.; Yan, W.; Wang, W.; Zhao, L.; Shan, Y.; Yang, K. Effect of microstructure on hydrogen induced cracking behavior of a high deformability pipeline steel. *J. Iron Steel Res. Int.* **2015**, *22*, 937–942. [CrossRef]

10. Koch, G.H. Fatigue and fracture. *ASM Handb.* **1996**, *19*, 483–506.

11. Tsai, W.T.; Chou, S.L. Environmentally assisted cracking behavior of duplex stainless steel in concentrated sodium chloride solution. *Corros. Sci.* **2000**, *42*, 1741–1762. [CrossRef]

12. Mohtadi-Bonab, M.A.; Szpunar, J.A.; Razavi-Tousi, S.S. A comparative study of hydrogen induced cracking behavior in API 5L X60 and X70 pipeline steels. *Eng. Fail. Anal.* **2013**, *33*, 163–175. [CrossRef]

13. Gjerding-Smith, K.; Johnsen, R.; Lange, H.I.; Leinum, B.H.; Gundersen, G.; Isaksen, B.; Nærum, G. Wire fractures in locked coil cables. *Bridge Struct.* **2006**, *2*, 63–77. [CrossRef]

14. Cho, T.; Kim, T.S. Probabilistic risk assessment for the construction phases of a bridge construction based on finite element analysis. *Finite Elem. Anal. Des.* **2008**, *44*, 383–400. [CrossRef]

15. LaFrance-Linden, D.; Watson, S.; Haines, M. Threat Assessment of Hazardous Materials Transportation in Aircraft Cargo Compartments. *Transp. Res. Rec.* **2001**, *1763*, 130–137. [CrossRef]

16. Sundararajan, C. *Probabilistic Structural Mechanics Handbook*; Chapman & Hall: London, UK, 1994.

17. Nowak, A.S.; Cho, T. Prediction of the combination of failure modes for an arch bridge system. *J. Constr. Steel Res.* **2007**, *63*, 1561–1569. [CrossRef]

18. Cho, T.; Song, M.-K.; Lee, D.H. Reliability analysis for the uncertainties in vehicle and high-speed railway bridge system based on an improved response surface method for nonlinear limit states. *Nonlinear Dyn.* **2010**, *59*, 1. [CrossRef]

19. Bayes, T.; Price, R.; Canton, J. An essay towards solving a problem in the doctrine of chances. *Philos. Trans.* **1763**, *53*, 370–418. [CrossRef]

20. Mayrbaurl, R.M.; Camo, S. Cracking and Fracture of Suspension Bridge Wire. *J. Bridge Eng.* **2001**, *6*, 645–650. [CrossRef]

21. Toribio, J. Hydrogen embrittlement of prestressing steels: the concept of effective stress in design. *Mater. Des.* **1997**, *18*, 81–85. [CrossRef]

22. Forman, R.G.; Shivakuma, V. Growth Behavior of Surface Cracks in the Circumferential Plane of Solid and Hollow Cylinders. In *Fracture Mechanics: Seventeenth Volume*; ASTM International: West Conshohocken, PA, USA, 1986; p. 59.

23. Forman, R.G.; Shivakumar, V.; Newman, J.C., Jr. *Fatigue Crack Growth Computer Program*; NASA/FLAGRO: Houston, TX, USA, 1994.

24. Bucher, C.G.; Bourgund, U. *Efficient Use of Response Surface Methods*; Report No. 9-87; University of Innsbruck: Innsbruck, Austria, 1987.

25. Sagues, A.A.; Wang, H. *Corrosion of Post Tensioning of Strands*; Florida Department of Transportation: Tallahassee, FL, USA, 2005; p. 255.

![metals logo] *metals*

MDPI

Article

Effect of Hot Mill Scale on Hydrogen Embrittlement of High Strength Steels for Pre-Stressed Concrete Structures

Marina Cabrini [1,2,3,*], Sergio Lorenzi [1,2,3], Tommaso Pastore [1,2,3] and Diego Pesenti Bucella [1,2]

[1] Department of Engineering and Applied Sciences, University of Bergamo, Dalmine (BG) 24044, Italy;
    sergio.lorenzi@unibg.it (S.L.); tommaso.pastore@unibg.it (T.P.); diego.pesentibucella@unibg.it (D.P.B)
[2] Consorzio Superfici Grandi Interfase (CSGI), Unit of Research of Bergamo, Dalmine (BG) 24044, Italy
[3] Consorzio Interuniversitario Nazionale per la Scienze e Tecnologia dei Materiali (INSTM),
    Unity of Research of Bergamo, Dalmine (BG) 24044, Italy
* Correspondence: marina.cabrini@unibg.it; Tel.: +39-035-2052-318

Received: 2 February 2018; Accepted: 23 February 2018; Published: 3 March 2018

Abstract: The presence of a conductive layers of hot-formed oxide on the surface of bars for pre or post-compressing structures can promote localized attacks as a function of pH. The aggressive local environment in the occluded cells inside localized attacks has as consequence the possibility of initiation of stress corrosion cracking. In this paper, the stress corrosion cracking behavior of high strength steels proposed for tendons was studied by means of Constant Load (CL) tests and Slow Strain Rate (SSR) tests. Critical ranges of pH for cracking were verified. The promoting role of localized attack was confirmed. Further, electrochemical tests were performed on bars in as received surface conditions, in order to evaluate pitting initiation. The adverse effect of mill scale was recognized.

Keywords: stress corrosion cracking; localized corrosion; pre-stressed concrete; hydrogen embrittlement

---

1. Introduction

Due to the high stress involved, bars made of high strength alloys are preferentially used as short tendons for post-tensioning. In fact, they are very susceptible to small damages, such as surface scratches, corrosion cracks and pits, which usually do not affect the performance of bars for concrete reinforcement [1]. Stress corrosion cracking is one the most probable causes of failure of high strength tendons in pre-stressed or post-stressed concrete structures [2–5]. These corrosion phenomena occur in the form of cracks, which nucleate and propagate due to combined action of tensile stresses and aggressive environment. Failures can take place at stress lower than yield strength of the steel, with catastrophic consequences on the structures stability and for people safety [1,6,7].

Susceptibility to stress corrosion cracking increases with tensile strength of materials and occurs above a threshold stress value. The more load increases above this critical limit, the more time to rupture decreases [8,9]. Since high strength steels, stressed to very high load (up to 80% of tensile yield strength), are utilized for tendons, critical conditions for stress corrosion cracking can be reached.

Susceptibility depends on composition, microstructures, and mechanical properties of the steels [10–12].

The older pre-stressed concrete structures were realized using quenched and tempered steels with high susceptibility to EAC [13,14]. Later, cold drawn eutectoid steels were widely used in pre-stressed concrete structures [15,16]. Moreover, it depends on environmental parameters, such as dissolved salt in water, pH and potential. The presence of conductive layers on the surface can promote localized attacks, with formation of local environment in the occluded cells [17], critic for the initiation of stress corrosion cracking [3,18].

In concrete structures, bars can be exposed to different environments, because these are used as external or internal unbonded tendons or bonded tendons embedded in concrete. The pH of solutions is usually alkaline (concrete pore solutions and bleeding waters), but neutral or acid values can be found, due to humidity condensation and atmospheric water penetration in ungrouted ducts. The composition depends on rainwater, atmospheric pollution (SO_2), distance from the sea (Cl^-) and soluble salts that can be washed away by water percolation on structure.

Literature data explain the SCC of high strength steels through a mechanism based on hydrogen embrittlement. Nürberger [19] reported on a statistical study of failures due to SCC and pointed out the effect of the localized corrosion for achieving critical ranges of potential and pH. However, there is lack of experimental data about the effects of environmental parameters. Furthermore, the SCC susceptibility of steels is normally evaluated by means of standard tests, as FIP thiocyanate test, which is quite different from field conditions.

Aim of the research is to evaluate the behaviour to SCC of commercial bars for pre-stressed concrete structures. The paper reports the results of the tests performed in order to study the critical ranges of pH for the initiation of stress corrosion cracking and the effect on localized attack of surface conditions of the bars.

2. Materials and Methods

The chemical composition and mechanical properties of the steels are reported in Table 1. The bars were produced through an in-line heat treatment by water quenching with self-tempering. Thus, a different microstructure between core and surface layers is observed. The external layer is characterized by tempered martensite, with hardness in the ranges 390–410 $HV_{200,15}$ and 360–380 $HV_{200,15}$ for the two steels respectively (Figure 1). The bars are covered by a mill scale.

Table 1. Chemical composition and mechanical properties of the considered steels.

Steel	Chemical Composition (% Weight)					Mechanical Properties	
	C	Mn	Si	P	S	R_S (MPa)	R (MPa)
A	0.71	0.63	0.22	0.016	0.016	996	1103
B	0.2	1.45	0.45	0.022	0.017	713	818

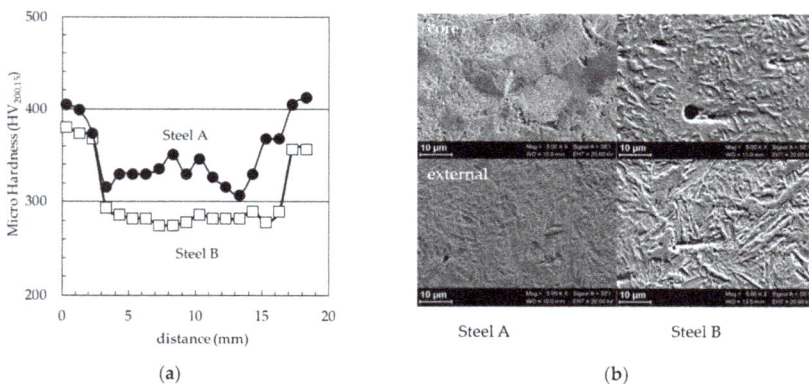

Figure 1. (**a**) Vickers 200g micro-hardness profile ($HV_{200,15}$) and (**b**) microstructure (5000×) of the considered steels.

Constant Load tests and Slow Strain Rate (SSR) tests were performed on 3 mm diameter cylindrical tensile specimens. Since the different microstructures between core and surface of the bars, the specimens were machined from a high strength martensitic steel with microstructure and hardness (400 $HV_{200,15}$) similar to that of the surface layer. The tests were performed in aerated hydrochloric acid solutions with pH from 3 to 6.4, at the free corrosion potential. Some tests were also executed in de-aerated solutions. Constant Load tests were carried out at 90% of yield strength.

SSR tests were performed according to ISO 7539-7 at 10^{-6} s^{-1} strain rate. SCC was evaluated through the reduction in area ratio (i.e., ratio between the value of area reduction after test in solution and the value of area reduction after the test in air) and the presence of brittle areas on fracture surface and secondary cracks.

The electrochemical tests were performed on specimens cut from bars of steel A and steel B, with as received surfaces—covered by the mill scale—and after pickling by inhibited hydrochloric acid. Monitoring of free corrosion potential and polarization resistance of bars during 1 month immersion test, Cathodic and Anodic Polarization Potentiodynamic (PD) and Cyclic Voltammetry tests (CV) were carried out.

Solutions of Portland cement, calcium and sodium hydroxide were used as test environments at pH between 9 and 14. Dilute hydrochloric acid solutions were utilized for more acid solutions (5.5 to 7 pH). The specimens were covered by epoxy resin in order to leave only the lateral surface exposed at the aggressive environments. Corrosion potential was monitored by means of a data recording. Linear Polarization Resistance (LPR) was measured through a potentiodynamic method, in the range of ±10 mV vs E_{corr} at 10 mV/min scan rate.

Potentiodynamic tests were carried out with a scan rate of 0.166 mV/s on rotating electrode at 2500 rpm; cathodic potentiodynamic curves were started from the corrosion potential and finished at −1.5 V vs SCE; anodic cyclic potentiodynamic tests were started from −0.2 V vs E_{corr} with a vertex potential of +1.5 V vs SCE.

Cyclic voltammetry tests were performed after 15 s of equilibration at open circuit potential (OCP) for two consecutive voltammetry cycles, from −1.7 to +0.7 V vs. SCE at 50 mV/s scan rate.

All the tests were twice conducted and a good reproducibility was detected.

3. Results and Discussion

Stress Corrosion Tests

Stress corrosion cracking of high strength steels takes place with a hydrogen embrittlement mechanism. Atomic hydrogen produced from the cathodic reduction of hydrogen ions is adsorbed on the metal surface, diffuses into the metal lattice and can promote the insurgence of cracks in the presence of a tensile stress. The cracks grow under the synergic action of stress and environment until the final mechanical failure is reached [20,21].

Hydrogen ions reduction is the main cathodic reaction in acid environments; in neutral or alkaline media it could take place only under cathodic polarization at very negative potentials, which can be reached only in the presence of cathodic protection systems or galvanic coupling with less noble metals [22,23].

Stress corrosion cracking insurgence increases in acid environments at pH lower than 5. Figure 2 illustrates the environmental effect on the tensile stress-strain curve of the steels during SSR tests. The curves drawn during tests in solution with pH above 4 showed no appreciable differences with respect to the curves obtained in air. Both in aerated and de-aerated solutions with pH below 5, low ductility is found. Such loss in ductility is induced by clear brittle cracking on the fracture surface (Figure 3a). In solutions with pH 3, secondary transgranular cracks, perpendicular to the applied stress, were found (Figure 3b). At higher pH, secondary cracks also grew along slip plane on a 45° direction, in areas very close to the fracture surface of specimens heavily strained, producing the characteristic aspect described (Figure 3b). Thus, the embrittlement phenomena become more pronounced as pH decreases (Figure 4). The loss in ductility due to environment is expressed in Figure 5a as reduction of area ratio (ratio between the value of the reduction of area in environment and the reduction of area in air — R.A.%environment/R.A.%air). Low values of this parameter indicate the insurgence of SCC phenomena.

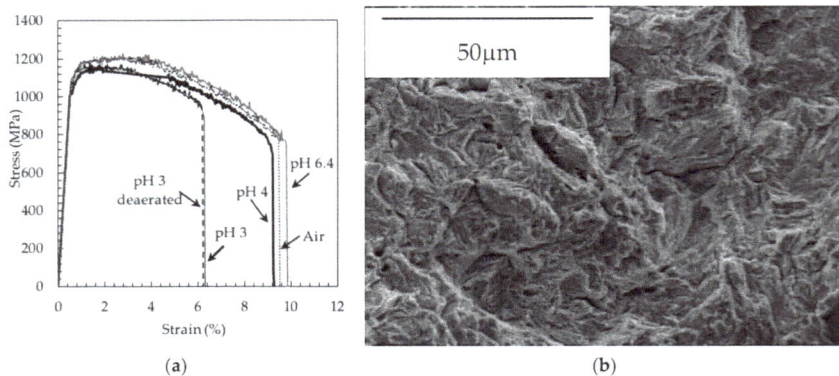

(a)

(b)

Figure 2. (a) Effect of pH solution on the tensile loading curve during SSR tests; (b) Fracture surface of the specimen after the SSR test at pH 3 in de-aerated solution (1000×).

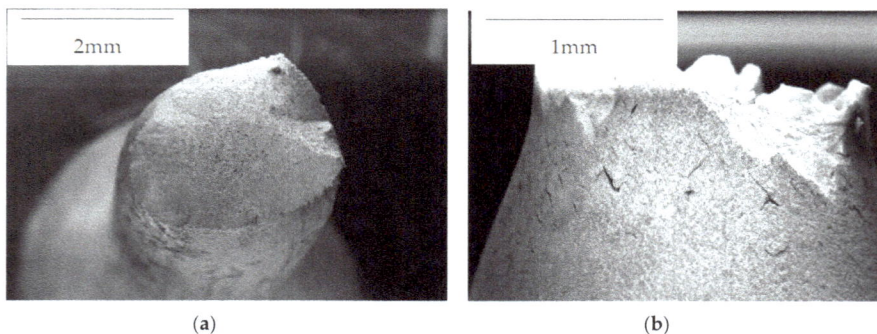

(a)

(b)

Figure 3. (a) Fracture surface of the specimen after the SSR test at pH 3 in aerated solution (50×); (b) Growth along slip plane of secondary cracks on the specimens after the SSR test at pH 4.

Figure 4. Effect of pH on the fracture morphology of the specimens.

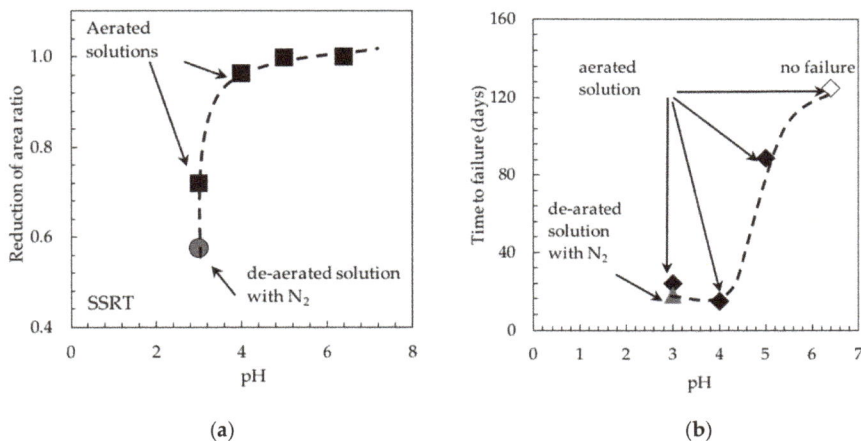

Figure 5. Effect of pH on (**a**) the area ratio (R.A.%environment/R.A.%air); (**b**) the time to failure of the Constant Load test specimens.

The results of Constant Load tests confirm the effect of pH on SCC. Figure 5b shows the time to failure of specimens loaded at 90% of yield strength. The time to failure increases with pH; specimens did not break in the solution with pH 6 after 4 months of testing.

The critical range for SCC is quite lower than the pH of the solutions usually found in the structures. Bonded tendons are usually grouted with alkaline mortar, with pH higher than 12.5. However, the metal surface can be wetted by less alkaline solutions before grouting, or in incompletely grouted ducts and in unbonded tendons without any protection system (such as coating) applied on metal surfaces. Bleeding waters show pH between 9 and 12. Very low pH (below 5) can only be measured in rainwater and condensations in areas with high pollution [24–26].

However, Constant Load tests even showed well detectable SCC phenomena at pH 5 after long time testing. The cracks initiated from localized attack, as shown in Figure 6.

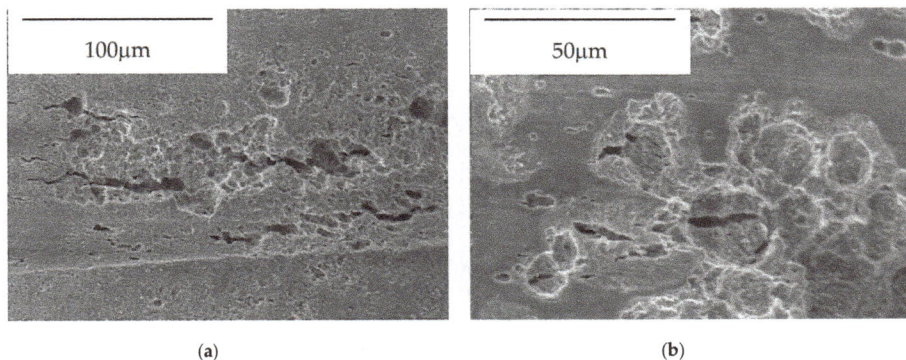

Figure 6. (**a**) Secondary cracks initiated from pit after Constant Load test in solution with pH 4. (**b**) close-up of (**a**) at higher magnification.

Inside the occluded cell, formed by the localized attack, the pH sensibly decreases and can reach values below 4 due to hydrolysis of metal ions. Consequently, the critical range of pH can be reached, promoting stress corrosion initiation. This mechanism was proposed by Nürberger in order to explain the failure of pre-stressing steels observed in neutral and alkaline solutions. The specimen failure takes place after a time needed to form the occluded cell [19,27].

Localized attacks can occur in alkaline environments with pH above 11.5, in the presence of a sufficient concentration of chloride that can break down the passive film on steel. This case may occur in concrete exposed to chloride-containing environment, due to the penetration of these ions until the bar surface. In neutral-alkaline environments, carbon steel is not passive. The main cause of localized attacks is the presence of a mill scale (mainly calamine), formed on the steel surface during hot rolling. The corrosion rate of carbon steel can be enhanced by galvanic coupling with calamine, which is an electronic conductive oxide, inducing localized corrosion on the anodic areas not covered by the scale. The galvanic coupling considerably increases the corrosion potential of the steel (Figure 7) during the initial period. Specimens in as received surface conditions showed a corrosion potential 100–250 mV nobler than the specimens after pickling, when they are just immersed in the test solution. Moreover, the difference tends to increase over time in all the considered solutions. Such an anodic polarization induces very high corrosion rate on the anodic areas.

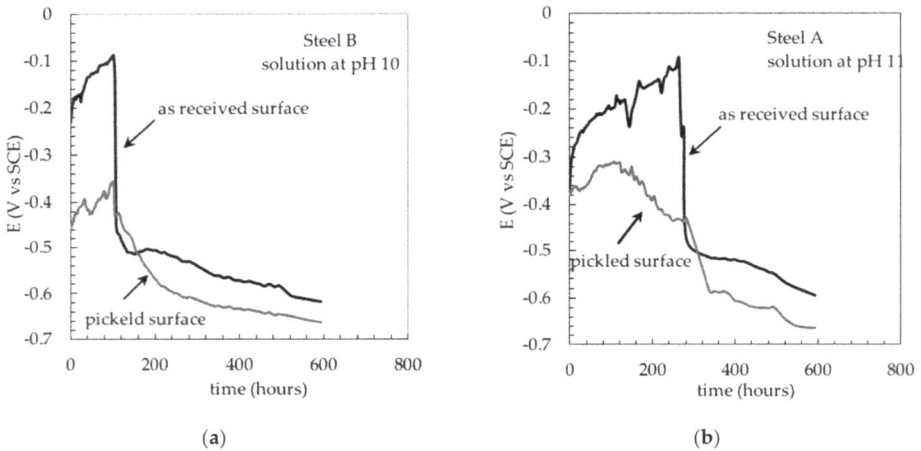

Figure 7. Corrosion potential vs time curves for the tested steels in solution at (**a**) pH 10 and (**b**) pH 11.

After the initial period, depending on the pH of the solution, the anodic polarization due to calamine becomes no more significant. The corrosion potential decreases to active potentials of pickled specimens due to the formation of deep localized attacks and the progressive change of the oxide scale, which is not stable at room temperature (Figure 7). Thus, the effect of calamine is especially evident in the first periods of exposure, which is a function of pH, increasing with it.

Figure 8 compares the polarization resistances measured on the specimens in as received surface conditions with mill scale and after pickling. The average corrosion rate evaluated on the specimens in as received conditions is initially low (high values of the polarization resistance in Figure 8) because of very small anodic areas uncovered by the scale. Afterwards, the corrosion rate of the specimens tends to similar values of pickled specimens.

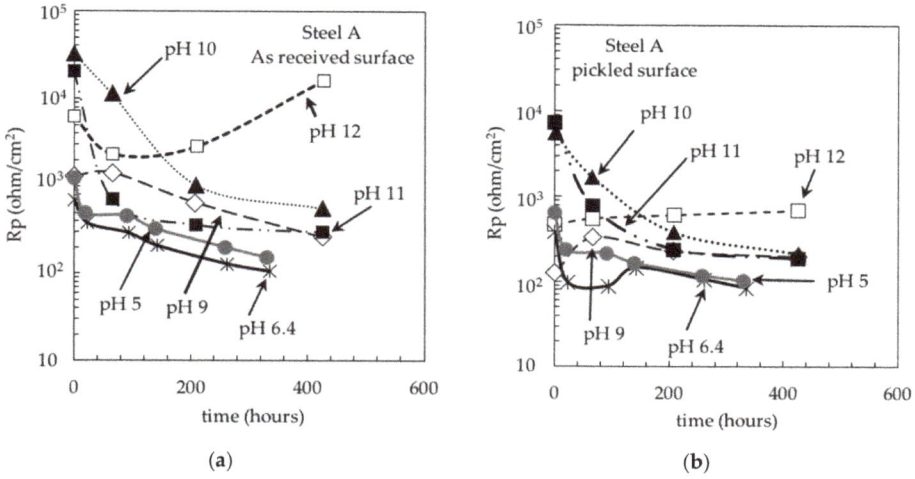

Figure 8. Polarization resistance vs time curves for steel A, as a function of different pH, with (**a**) as received surface and (**b**) pickled surface.

Figure 9 shows the specimens after the immersion tests at different pH. At pH lower than 10 and higher than 12, there are no differences between the specimens with as received or pickled surface. In the first case, a generalized corrosion was observed, whereas at pH higher than 12 all the specimens are in passive conditions, without evidence of corrosion products. In the range of pH from 10 and 11, the pickled specimens showed uniform corrosion; on the contrary, the specimens with as received surface showed un-attacked areas, with the presence of calamine and some corroded regions around the un-attacked areas. No differences between the two considered steels were founded in the electrochemical behavior.

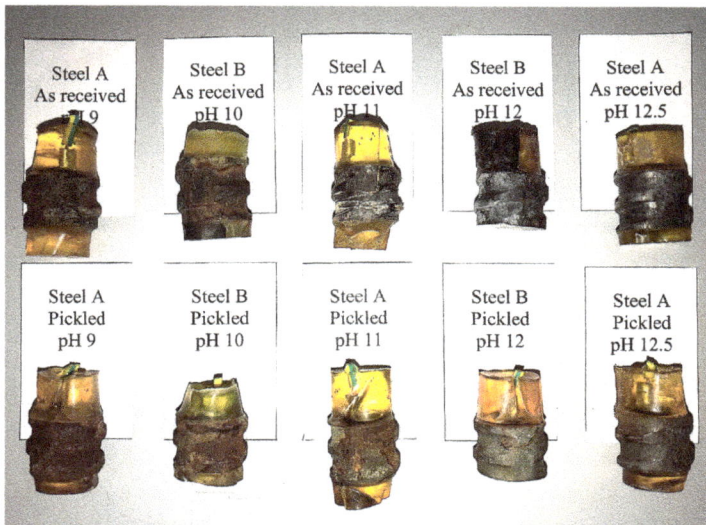

Figure 9. Specimens with as received or pickled surface after the immersion tests at different pH.

Figure 10a shows the potentiodynamic cathodic curves obtained on as received surface and after pickling. The pickled surface was also tested after 48 h of passivation in the solution. It could be noted that the specimens with presence of calamine have low overvoltages at potentials above the oxygen limiting current density zone. The data confirm that calamine can act as a good cathode for the oxygen reduction.

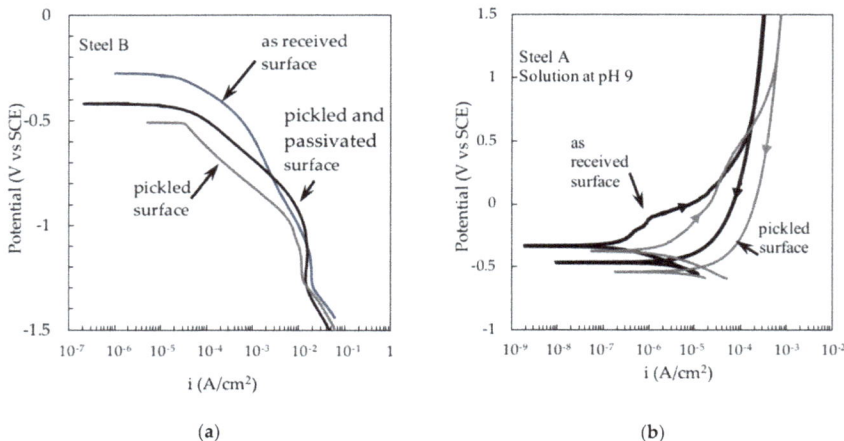

(a)

(b)

Figure 10. Effect of surface conditions on (**a**) cathodic potentiodynamic curves of the steel B in alkaline solution (pH 14) and (**b**) anodic cyclic potentiodynamic curves of the steel A in pH 9 solution.

Figure 10b and Figure 11 show the effect of mill scale on the cyclic potentiodynamic curves as a function of pH. The mill scale modifies the anodic polarization curves of the steels. In solution at pH less than 11, the steels show active dissolution, independently from the surface conditions (Figure 10b). At pH 13, a passive behavior was always observed (Figure 11a). In the pH range between 11 and 13 it is possible to observe noticeable differences between the anodic cyclic potentiodynamic curves.

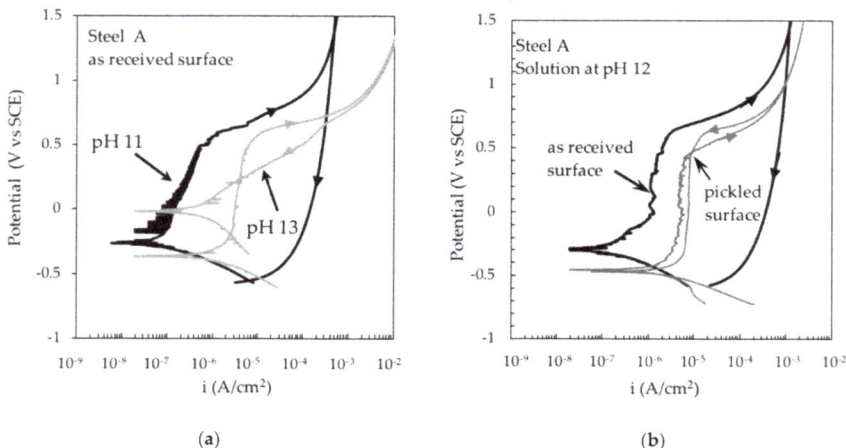

(a)

(b)

Figure 11. (**a**) Effect of pH on anodic cyclic potentiodynamic curves of the steel A in as received conditions, covered by a mill scale; (**b**) Effect of surface conditions on anodic cyclic potentiodynamic curves of the steel A in pH 12 solution.

The anodic current is low for the as received surface specimen at the beginning of the test, due to the presence of the oxide scale that protects the metal surface. Oxygen evolution at high potentials, both on steel and on scale, produces similar current densities on as received specimens and pickled specimens. After the potential scan is inverted, the specimens with as received surface show a great hysteresis, while the specimens with polished surface maintain a passive behavior (Figure 11b).

The calamine oxide scale can stimulate the localized corrosion of the steel in all the situations characterized by the absence of grouting with alkaline mortar or corrosion protection systems that prevent the direct contact between the steel and the solution. In alkaline mortar, the steel becomes passive and shows potentials similar to those of the calamine. Thus, no galvanic couple takes place. In solution with pH below 11.5, which is the limit of pH for the passivity of the steel in solutions without chlorides, the galvanic effect of calamine can be considerable. This behavior is well illustrated by the potentiodynamic curves of Figure 11, in which the pickled surface did not show hysteresis during the reverse of potential; instead, the hysteresis is very evident on the as received surfaces.

In the cyclic voltammetry tests (Figure 12) it is possible to observe the appearance of the passivation peaks of the iron only at pH 12.5. At pH 12, a hint of peaks appears, more pronounced for specimens with pickled surface than those with as received surface. The current values recorded for the pickled specimens are higher than those for specimens with the as received surface, shielded from the calamine scale. Once the passivation of the specimen intervenes, the cathodic reduction of oxygen on as received specimens occurs at a potential lower than the values observed on pickled specimens, confirming the electrocatalytic effect of the calamine. With respect to literature data [28–30], the peak of the Fe/Fe(II) reaction is not present on the cyclic voltammetric curve, for all the surface conditions of the specimens. This could be due to the presence of the hot-formed oxide on the specimens, but it also shows how pickling cannot restore an iron surface completely absent from oxides. On the other hand, the presence for all the specimens of the large oxidation peak from Fe(II) to Fe(III) clearly indicates that the hot-formed oxide film does not isolate the underlying metal from the environment. On the contrary, it allows the irreversible growth of the film by anodic oxidation, as shown by the increase in the amplitude of the peak in the second potential cycle and by the absence of the conjugate reduction peak.

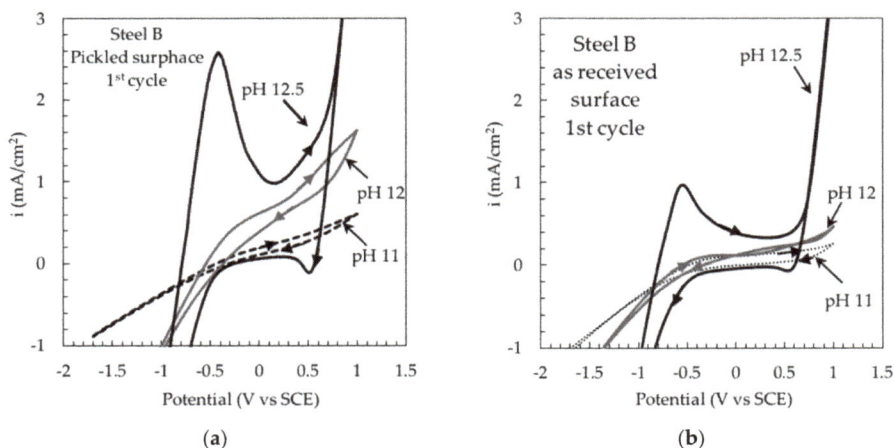

Figure 12. Cyclic voltammetry tests of steel B in solution at different pH; (**a**) specimens with pickled surface; (**b**) specimens with as received surface.

From these tests, it is possible to hypothesize that the film present on the bars with as received surface is not stable in solution, but tends to dissolve with pH greater than 12.5–13, it has a good electrical and ionic conductivity and allows the oxidation of the underlying metal and, finally, it has a low overvoltage for oxygen.

It is therefore conceivable that, in case of partial covering of the surface with this oxide, which occurs as the oxide tends to melt, there may be localized corrosion between the areas still covered with oxide and those with non-coated surfaces. This can be done for all pH investigated up to 12.5: beyond this value, the specimen can passivate. At pH < 12 there is a competition between the location of the attack between shielded areas and free areas, and the dissolution of the oxide film. The lower the pH of the solution, the greater the speed of destruction of the film, so it is conceivable that the maximum localization of the attack is for pH between 11 and 12. These results are confirmed by the observation of the samples recovered after the monitoring of the corrosion potential at different pH (Figure 7).

Finally, the promoting role of calamine on stress corrosion cracking in neutral or slight alkaline solutions can be sketched as proposed in Figure 13.

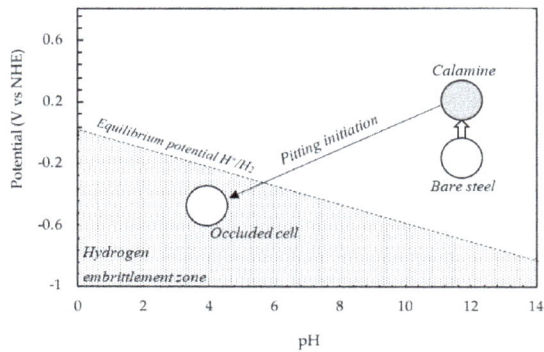

Figure 13. Role of calamine on insurgence of favorable conditions for hydrogen embrittlement in neutral or slight alkaline environment.

When a bar is embedded in alkaline solutions at pH higher than 12.5, this becomes passive, independently from the surface conditions. If the pH of the environment decreases at values lower than 12, the presence of calamine increases the corrosion potential of the bar and, in the area where the oxide scale is not present, some localized attacks may be initiated. Inside the localized attack, the pH decreases and, contemporaneously, the open circuit potential decreases too, so the hydrogen evolution reaction becomes possible. In these conditions, in the presence of susceptible steel and mechanical stress, hydrogen embrittlement can take place. For very high strength steels, with a low critical intensity factor, very small defect generated by hydrogen embrittlement can promote brittle catastrophic fracture [1].

4. Conclusions

The paper studies the condition for initiation of localized attacks on high strength steel for post-tensioning structures as a function of superficial conditions. SCC tests were performed on a martensitic steel with the same hardness compared to that obtained in peripheral zones of hot worked and quenched bars. Stress corrosion can occur in solutions with pH below 5 due to hydrogen embrittlement. The pH inside the localized attacks reaches values lower than 5 owing to the well knows mechanism of the occluded cell. In the absence of chlorides, the initiation of localized attacks occurs due to the presence of oxide scale produced by hot-working and the pH of the environment. Localized attacks initiate in a critical range of pH between 10 and 12. At pH higher than 12.5, the bar is passive independently upon the surface conditions. pH values lower than 10 promote the occurrence of general corrosion. At pH values lower than 12, the presence of calamine increases the corrosion potential of the bar and localized attacks can occur in the zones not protected by the oxide layer.

Hydrogen embrittlement can produce very small defects, which can promote brittle fracture in high strength steels characterized by low fracture toughness. The results pointed out the negative effect played by the mill scale for applications in which the reinforcements are protected by mortars characterized by low alkalinity, mainly below pH values of 12.

Author Contributions: Pastore and Cabrini conceived and designed the experiments; Pesenti Bucella performed the experiments, photographed the specimens, reviewed the English form; Lorenzi analyzed the data. All the authors contributed to write the paper.

Conflicts of Interest: The authors declare no conflict of interest.

References

1. Valient, A.; Elices, M. Premature failure of prestressed steel bars. *Fail. Anal. Case Stud. II* **1998**, *5*, 219–227. [CrossRef]
2. Monfore, G.; Verbeck, C. Corrosion of prestressed wire in concrete. *J. Am. Concr. Inst.* **1960**, *37*, 491–516.
3. Perrin, M.; Gaillet, L.; Tessier, C.; Idrissi, H. Hydrogen embrittlement of prestressing cables. *Corros. Sci.* **2010**, *52*, 1915–1926. [CrossRef]
4. Yuyama, S.; Yokoyama, K.; Niitani, K.; Ohtsu, M.; Uomoto, T. Detection and evaluation of failures in high-strength tendon of prestressed concrete bridges by acoustic emission. *Constr. Build. Mater.* **2007**, *21*, 491–500. [CrossRef]
5. Stoll, F.; Saliba, J.; Casper, L. Experimental study of CFRP-prestressed high-strength concrete bridge beams. *Compos. Struct.* **2000**, *49*, 191–200. [CrossRef]
6. Woodward, R.; Williams, F. Collapse of Ynys-y-Gwas Bridge, west Glamorgan. *Proc. Instn. Civ. Engrs.* **1988**, *84*, 635–669. [CrossRef]
7. Helmerich, R.; Zunkel, A. Partial collapse of the Berlin Congress Hall on May 21st 1980. *Eng. Failure Anal.* **2014**, *43*, 107–119. [CrossRef]
8. Senigaglia, D.; Re, G.; Pedeferri, P. *Cedimento per Fatica e Ambientale dei Materiali Metallici*; CLUP: Milano, Italy, 1979.
9. Pedeferri, P.; Bertolini, L. *La Durabilità del Calcestruzzo Armato*; Mc Graw-Hill: Milano, Italy, 2000; p. 199.
10. Woodtli, J.; Kieselbach, R. Damage due to hydrogen embrittlement and stress corrosion cracking. *Eng. Failure Anal.* **2000**, *7*, 427–450. [CrossRef]
11. Cabrini, M.; Lorenzi, S.; Marcassoli, P.; Pastore, T. Hydrogen embrittlement behavior of HSLA line pipe steel under cathodic protection. *Corros. Rev.* **2011**, *29*, 261–270. [CrossRef]
12. Cabrini, M.; Lorenzi, S.; Pellegrini, S.; Pastore, T. Environmentally assisted cracking and hydrogen diffusion in traditional and high-strength pipeline steels. *Corros. Rev.* **2015**, *33*, 529–545. [CrossRef]
13. Mietz, J. Investigation on hydrogen-induced embrittlement of quenched and tempered prestressing steels. *Mater. Corros.* **2000**, *51*, 2–80. [CrossRef]
14. Darmawan, M.; Stewart, M. Effect of pitting corrosion on capacity of prestressing wires. *Mag. Concr. Res.* **2007**, *59*, 131–139. [CrossRef]
15. Toribio, J.; Ovejero, E. Effect of Cold Drawing on Microstructure and Corrosion Performance of High-Strength Steel. *Mech. Time-Depend. Mater.* **1998**, *1*, 307–319. [CrossRef]
16. Enos, D.; Scully, J. A Critical-Strain Criterion for Hydrogen Embrittlement of Cold-Drawn, Ultrafine Pearlitic Steel. *Metall. Mater. Trans. A* **2002**, *33A*, 1151–1166. [CrossRef]
17. Zitrou, E.; Nikolaou, J.; Tsakiridis, P.; Papadimitriou, G. Atmospheric corrosion of steel reinforcing bars produced by various manufacturing processes. *Constr. Build. Mater.* **2007**, *21*, 1161–1169. [CrossRef]
18. Vehovar, L.; Kuhar, K.; Vehovar, A. Hydrogen-assisted stress corrosion of prestressing wires in a motorway viaduct. *Eng. Failure Anal.* **1998**, *5*, 21–27. [CrossRef]
19. Nürnberger, U. *Korrosion und Korrosionsschutz im Bauwesen*; Grundlagen, B., Metallbau, K., Eds.; Bauverlag BV GmbH: Gütersloh, Germany, 1995.
20. Hirth, J. Effects of Hydrogen on the Properties of Iron and Steel. *Metall. Trans. A* **1980**, *11A*, 861–890. [CrossRef]

21. Lynch, S. Hydrogen embrittlement phenomena and mechanisms. *Corros. Rev.* **2012**, *30*, 105–123. [CrossRef]

22. Bockris, J.O.; McBreen, J.; Nanis, L. The Hydrogen Evolution Kinetics and Hydrogen Entry into a-Iron. *J. Appl. Electrochem.* **1965**, *112*, 1025–1031. [CrossRef]

23. Cabrini, M.; Lorenzi, S. Pipeline Steels: Hydrogen Diffusion and Environmentally Assisted Cracking. In *Encyclopedia of Iron, Steel, and Their Alloys*; George, E., Totten, R.C., Eds.; CRC Press Taylor and Francis Group: Boca Raton, FL, USA, 2016; pp. 2547–2599.

24. Camuffo, D.; Bernardi, A.; Zanetti, M. Analysis of the Real-Time Measurement of the oH of Rainfall at Padova, Italy: Seasonal Variatino and Meteorological Aspects. *Sci. Total. Environ.* **1988**, *71*, 187–200. [CrossRef]

25. Mrose, H. Measurements of pH, and chemical analyses of rain-, snow-, and fog-water. *Tellus* **1966**, *18*, 266–270. [CrossRef]

26. Singh, A.; Agrawal, M. Acid rain and its ecological consequences. *J. Environ. Biol.* **2008**, 15–24.

27. Nürnberger, U. Corrosion induced failure mechanisms of prestressing steel. *Mater. Corros.* **2002**, *53*, 591–601. [CrossRef]

28. Schrebler Guzmán, R.S.; Vilche, J.R.; Arvia, A.J. The potentiodynamic behaviour of iron in alkaline solutions. *Electrochim. Acta* **1979**, *24*, 395–403. [CrossRef]

29. Cabrini, M.; Lorenzi, S.; Pastore, T. Cyclic voltammetry evaluation of inhibitors for localised corrosion in alkaline solutions. *Electrochim. Acta* **2014**, *124*, 156–164. [CrossRef]

30. Schrebler Guzman, R.S.; Vilche, J.R.; Arvia, A.J. The voltammetric detection of intermediate electrochemical processes related to iron in alkaline aqueous solutions. *J. Appl. Electrochem.* **1981**, *11*, 5. [CrossRef]

metals

MDPI

Article

Brittle Fracture Behaviors of Large Die Holders Used in Hot Die Forging

Weifang Zhang, Hongxun Wang, Jingyu Zhang, Wei Dai * and Yuanxing Huang

School of Reliability and Systems Engineering, Beihang University, Beijing 100191, China;
zhangweifang@buaa.edu.cn (W.Z.), wanghongxun@buaa.edu.cn (H.W.), jingyuzhang@buaa.edu.cn (J.Z.),
linhaihyx@126.com (Y.H.)
* Correspondence: dw@buaa.edu.cn; Tel.: +86-10-8233-8673

Academic Editors: Ricardo Branco, Filippo Berto and Daolun Chen
Received: 4 March 2017; Accepted: 25 May 2017; Published: 30 May 2017

Abstract: Brittle fracture of large forging equipment usually leads to catastrophic consequences. To avoid this kind of accident, the brittle fracture behaviors of a large die holder were studied by simulating the practical application. The die holder is used on the large die forging press, and it is made of 55NiCrMoV7 hot-work tool steel. Detailed investigations including mechanical properties analysis, metallographic observation, fractography, transmission electron microscope (TEM) analysis and selected area electron diffraction (SAED) were conducted. The results reveal that the material generated a large quantity of large size polyhedral $M_{23}C_6$ (M: Fe and Cr mainly) and elongated M_3C (M: Fe mainly) carbides along the martensitic lath boundaries when the die holder was recurrently tempered and water-cooled at 250 °C during the service. The large size carbides lead to the material embrittlement and impact toughness degradation, and further resulted in the brittle fracture of the die holder. Therefore, the operation specification must be emphasized to avoid the die holder being cooled by using water, which is aimed at accelerating the cooling.

Keywords: brittle fracture; property degradation; mechanical properties; failure analysis; 55NiCrMoV7

1. Introduction

Forging is one of the major production processes in the mechanical industry. It is also an advanced manufacturing technique that provides important engine components. Forging technology is widely used to produce high-performance mechanical parts in both military and civil fields, such as engine blades that are manufactured by hot die forging. A major disadvantage of hot die forging is the poor process reliability, as well as the low durability of dies and die holders. In addition, the expense of hot forging equipment is quite high, which usually takes a proportion of 8–15% of all the productive task costs. If considering the failures of hot forging equipment and manufacturing loss during the production, the expense will increase to 30–50% [1,2]. Therefore, avoiding the failures of hot forging equipment is one of the most significant tasks for the forging industry.

The failures of dies and die holders are usually caused by wear, deformation, corrosion and fatigue during the production process [3,4]. Especially, the dies and die holders have to suffer long-term cyclic mechanical and thermal loads [5,6]. The cyclic mechanical loads can lead to fatigues, and the cyclic thermal loads can cause mechanical property degradation of dies and die holders. When the property degradation occurs, the die and die holder will easily fracture under the effect of the cyclic mechanical loads. The brittle fractures caused by property degradation usually lead to catastrophic consequences, and the preventive actions are difficult to find comparing with other failure modes [7–10]. The large die holder studied in this paper is made of 55NiCrMoV7 hot-work tool steel, which is similar to 5CrNiMo in chemical composition except with higher content of Cr and V. Compared with 5CrNiMo, the 55NiCrMoV7 has better hardenability and abrasion resistance. Furthermore, it has been widely used in aviation and motor industries [11]. Zhang et al. [12] studied the low cycle fatigue behaviors of 55NiCrMoV7 under different working temperatures. The results show that the cyclic load and working temperature can cause the cyclic softening and microstructure variation, which can lead to the property degradation of 55NiCrMoV7. Therefore, it is important to learn the brittle fracture mechanism of the large die holder and to exclude the factors resulting in material embrittlement. Moreover, it is necessary to take preventive actions to avoid brittle fracture, reduce economic losses and personnel casualties.

In this paper, the brittle fracture behaviors of the large press die holder was studied. Firstly, the failure background and the failure mode of the die holder are described. Then, the analysis techniques and methods, such as chemical composition analysis, metallographic observation, mechanical properties testing, tempering and water-cooling treatment, microstructure analysis and selected area electron diffraction (SAED), are scheduled. Thirdly, the experiments are conducted, and the results are comprehensively discussed. Finally, some conclusions and suggestions are made.

2. Failure Background of the Die Holder

The die holder studied in this paper is used on a large screw press for hot die forging. The profile of the die holder is shown in Figure 1. The outline size of the die holder is 3640 mm (length) × 2430 mm (width) × 1150 mm (height) and the inner cavity size is 2200 mm (length) × 850 mm (width). The weight of the die holder is 62.26 t. The production processes of the die holder include vacuum smelting, forging, heat treatment and machining. According to the technical requirement, the material of the die holder is set to be 55NiCrMoV7 hot-work tool steel. The initial heat treatment of the die holder is as follows. Firstly, the die holder was roughly forged. Then, it was handled by using high temperature diffusion annealing. After that, the die holder was heated up from 20 °C to 850 °C at the speed of 50 °C/h. Then, it was austenitized for 15 h at 850 °C. Thirdly, the die holder was processed by using step quenching (water-oil quenching) one time, which is first water-quenched for 30 min and then oil-quenched to room temperature. Finally, it was submitted to a double tempering treatment, first at 620 °C and then at 560 °C. After the heat treatment, the microstructure is tempered martensite.

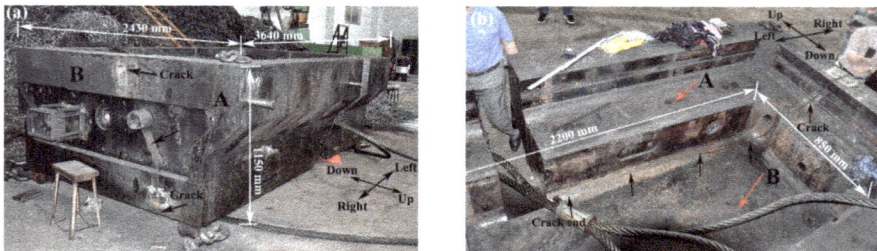

Figure 1. The profile of the die holder: (**a**) the outline; (**b**) the inner cavity.

The die holder generated a large crack during the process of forging aero-engine blades. The crack penetrates the bottom and the right side of the die holder whose length extends as long as 2316.7 mm as shown in Figures 1 and 2, where the crack orientation is marked by using the arrows. Figure 2a is the crack morphology, where A represents the up side of the crack and B represents the down side, as the coordinates show in figures. Figure 2b is the numbered fracture of the A side that was cut from the die holder along the crack. On the inner bottom of the die holder, which is the top edge of the fracture, there is an obvious semicircular fatigue area (region 1). Furthermore, along the inner bottom surface, there are three surface cracking zones where the crack propagation directions are different from that of the fatigue area. The fracture characteristics of the die holder have been studied in our previous investigation [13], and the information about the fracture characteristics are only briefly described in this section. The failure analysis of the die holder [13] has revealed that the crack initiated from the semicircular fatigue area (region 1) and the surface cracking zones (the top edges of region 2, region 4 and region 9, which are on the surface of the inner bottom as shown in Figure 2), where the fatigue is point source and the surface cracking zones are linear sources. The crack propagation region (the whole fracture except for region 1 and region 5) is rapid brittle fracture with large radial ridges. The microscopic morphology is quasi-cleavage, which is characterized by river patterns and tear ridges as shown in Figure 3.

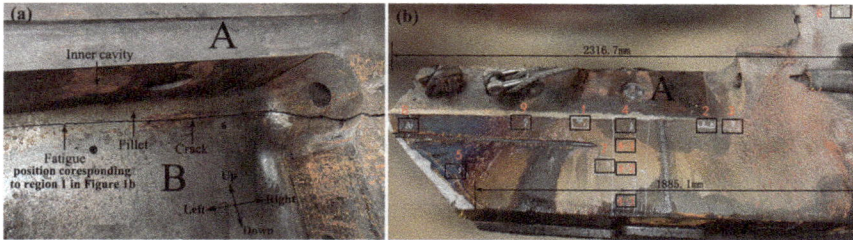

Figure 2. The macroscopic crack and fracture morphology of the die holder: (**a**) macroscopic crack morphology; (**b**) macroscopic fracture morphology.

Figure 3. The low and high magnification morphology of brittle fracture region on the bottom of the die holder: (**a**) the low magnification morphology; (**b**) the high magnification morphology.

The semicircular fatigue region and surface cracking zones of the die holder are shown in Figure 4. The semicircular fatigue region is about 40 mm in diameter, and the region accounts for 0.02% of the whole fracture. The thickness of the surface cracking zones are 5–10 mm, as the red arrows show in Figure 4, which take up 0.04% of the whole fracture. The die holder had been working for two years before failure. During the working process, the die holder had forged 113,025 products made of high strength stainless steel, titanium alloy and high temperature alloy at 150 °C to 350 °C, and it had endured the impact loads 351,089 times. The impact loads are usually between 180 MN and 280 MN. The maximum and minimum loads are 350 MN and 83 MN, respectively. The bearing capacity of the die holder is 350 MN, so the service stress of the die holder does not exceed the ultimate strength of the material. Considering that the fracture sources only account for 0.06% of the total fracture, the service stress is not large enough to explain such a large and rapid brittle fracture. Therefore, it must be other reasons that led to the brittle fracture. Thus, further study was made in this paper.

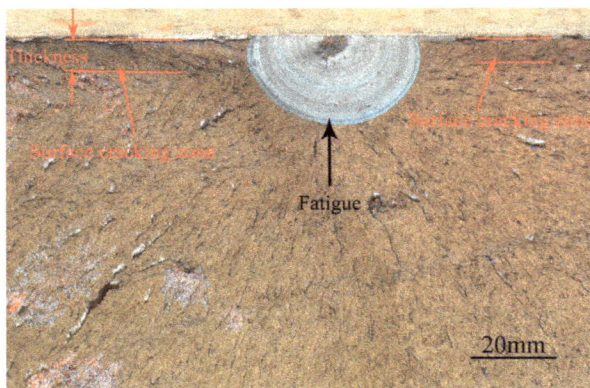

Figure 4. The semicircular fatigue and surface cracking zones.

3. Methods

In order to determine the reasons why brittle fracture of the die holder happens, extensive samples are cut from the broken die holder, and the research scheme of the paper is designed as follows. In this paper, region 4 contains three parts of the bottom, which is the main brittle fracture region and suffered the most severe environment stress. The regions 4-1, 4-2 and 4-3 refer to the upper, middle and lower regions of the die holder bottom, respectively. Samples cut from region 4, which can represent the mechanical properties of the whole bottom, were tested by using the chemical composition and metallographic analysis, hardness testing, tensile experiment and impact property testing. Meanwhile, the impact fracture features were investigated. Additionally, there is no original material of the die holder, so region 6, which suffered the least temperature and mechanical load effects, are used for comparison with region 4. Furthermore, the tempering and water-cooling experiment was designed to simulate the practical environment stress of the die holder, and the effect of cyclic tempering and water cooling on the material impact property was studied as well [14]. That is, the samples of region 6 were tempered for 40 min at the temperature of 150 °C, 250 °C, 350 °C and 450 °C, respectively, and then water-cooled. After tempering and water-cooling treatment, the impact property of region 6 was tested and the impact fractures were observed by using a scanning electron microscope (SEM) (TESCAN CHINA, LTD., Shanghai, China). Finally, the transmission electron microscope (TEM) (JEOL (BEIJING) CO., LTD., Beijing, China) and selected area electron diffraction (SAED) were applied to analyze the microstructure and precipitates of the regions 4 and 6 in detail to investigate the reasons for brittle fracture.

4. Results

4.1. The Chemical Composition and Metallography

The chemical composition analysis shows that the chemical composition meets the technical requirement of the die holder as shown in Table 1. The metallographic samples removed from the broken die holder were eroded by using the 1% HNO_3 + C_2H_5OH solution. The grain size was measured by using the comparison method, which is to compare the grain size under 100× magnification with the standard grain size chart. The results show that the die holder has a fine grain size of level 7 to 8. The microstructure is homogeneous with tempered martensite and there is no oxidation and decarbonization as shown in Figure 5a. In addition, the material has dot shaped nonmetallic inclusions. There is no large inclusion and obvious nonmetallic inclusion segregation zone as shown in Figure 5b. In conclusion, the material fits the technical requirements and there are no metallurgical defects.

Table 1. The chemical composition of the die holder (wt %).

Element	C	Cr	Mo	Ni	V	Si	Mn	P	S
Standard	0.5–0.6	1.0–1.2	0.45–0.55	1.5–1.8	0.07–0.1	0.1–0.4	0.65–0.95	≤0.02	≤0.02
Tested	0.56	1.07	0.52	1.61	0.096	0.26	0.79	0.007	0.003

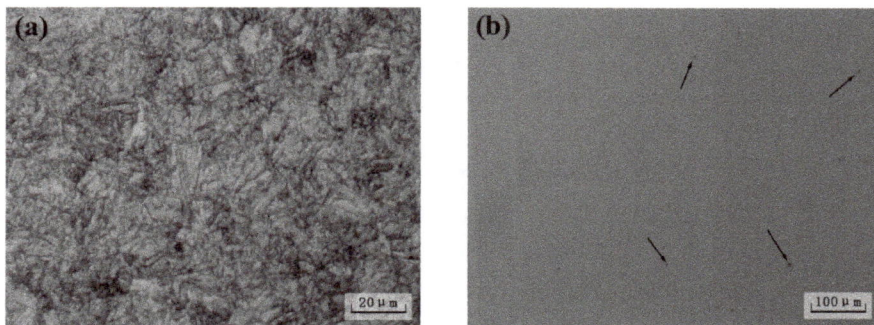

Figure 5. Metallographic structure and nonmetallic inclusions of the die holder: (**a**) metallographic phase; (**b**) nonmetallic inclusions.

4.2. Hardness Analysis

The Rockwell hardness of samples cut from region 4 and region 6 were measured as shown in Table 2. The results show that the hardness of region 4 is distributed between HRC 32.9 and HRC 33.1, which is lower than the technical requirement (HRC 36–HRC 40). As a whole, the hardness of region 4 is relatively homogeneous. However, the hardness of region 6 is HRC 37.2, which conforms to the technical requirements. Compared with region 6, the hardness of region 4 decreased by 11.3%, which indicates that the property of region 4 has degraded. According to Zhang's study, the service temperature of 55NiCrMoV7 can cause the tempering effect (aging effect), which results in the hardness decreasing, even though the service temperature is lower than the initial tempering temperature. Furthermore, the cyclic loads also can lead to the hardness diminution. As the hardness of region 6 has no obvious degradation, it can be concluded that region 6 had less cyclic mechanical and thermal loads than region 4, which is consistent with the practical service situation of the die holder. The hardness of region 6 is the most similar with the original material, so the samples of region 6 are used to compare with region 4 in the following.

Table 2. The Rockwell harness (HRC) of region 4 and region 6.

Number	HRC	HRC	HRC	HRC	HRC	Average
Region 4-1	32.9	32.9	32.8	32.9	32.9	32.9
Region 4-2	32.9	33.0	33.0	33.2	33.2	33.1
Region 4-3	33.1	33.3	32.6	32.9	32.9	33.0
Region 6	37.7	37.2	37.3	36.6	37.4	37.2
Standard	-	-	-	-	-	HRC 36–40

4.3. Tensile Properties

The tensile properties of region 4 and region 6 were tested. The results are shown in Table 3. It turns out that the tensile properties of region 4 increase in turn from the upper region to the lower region. Moreover, the tensile properties of the upper region and the middle region are almost the same, while the lower region is higher than those of the other two regions. The tensile strengths of region 4 are between 1017 MPa and 1033 MPa. It meets the technical requirement, which is 1000–1250 MPa. The yield strength of region 4 is between 819 MPa and 833 MPa, which is in accordance with the technical requirements ($\geq$650 MPa). The elongation (A) of the lower region is 16.5%, which meets the technical requirements. However, the elongations of the upper region and the middle region are lower than 15%. As a whole, the tensile properties of the lower region are better than those of the upper and middle region. Additionally, the tensile properties of region 6, which fit the technical requirement well, are superior to those of region 4, which indicates that the properties of region 6 are the most similar with the original material. It has to clarify that the tensile properties of region 4 and region 6 just meet the minimum value of the technical requirement, which indicates that the properties of both regions have degraded. Furthermore the degradation degree of region 6 is lower than region 4. Similarly, the different properties of region 4-1, 4-2 and 4-3 are also related to the degradation degree, which will be discussed in Section 5.

Table 3. The tensile test results of region 4 and region 6.

Sample Locations	Rm (MPa)	$Rp0.2$ (MPa)	A (%)	Z (%)
Region 4-1 (the upper region)	1017	819	13.0	39
Region 4-2 (the middle region)	1018	821	14.0	47
Region 4-3 (the lower region)	1033	833	16.5	50
Region 6	1046	850	16.5	51
Standard	1000–1250	$\geq$650	$\geq$15.0	–

Rm is tensile strength; $Rp0.2$ is yield strength; A is elongation; and Z is reduction in cross section.

4.4. Impact Property and Fracture Morphology

Impact toughness is one of the most important parameters to evaluate the brittle fracture resistance of material. Therefore, the impact property of v-notched samples removed from region 4 and region 6 were measured. The results are shown in Table 4. The impact toughness of region 4 is between 9.0 J/cm^2 and 9.8 J/cm^2, which are much lower than the technical requirements ($\geq$25 J/cm^2). Moreover, the impact toughness of region 4-1, 4-2 and 4-3 decreases slowly in turn. However, the impact toughness of region 6 is 40.0 J/cm^2, which fits the technical requirements well. Compared with region 6, the impact toughness of region 4 declines by 76.5%, which reveals that the impact toughness of region 4 has seriously degraded. The low impact toughness of region 4 also indicates that region 4 has pretty low impact resistance, which is easy to fracture under impact loads.

Table 4. The impact property test results of region 4 and region 6.

Sample Locations	ff_{kv} (J/cm^2)			Average
Region 4-1	7.0	11.3	11.1	9.8
Region 4-2	8.2	7.3	12.8	9.4
Region 4-3	6.4	11.1	9.4	9.0
Region 6	43.6	42.4	34.0	40.0
Standard	-	-	-	$\geq$25

To compare the impact fracture morphology of region 4 with region 6 as well as the brittle fracture, SEM observation was carried out. The impact fractures of region 4 and region 6 are shown in Figure 6. It can be seen that there is little fiber region and shear lip on the fracture surfaces of region 4, and most of the fracture surface is the radial region with a large quantity of radial ridges, which indicates that the impact toughness of region 4 is very low. However, the impact fracture of region 6 has obvious shear lip, which is much larger than that of region 4. It indicates that the impact toughness of region 6 is much higher than that of region 4, which is consistent with the tested impact toughness.

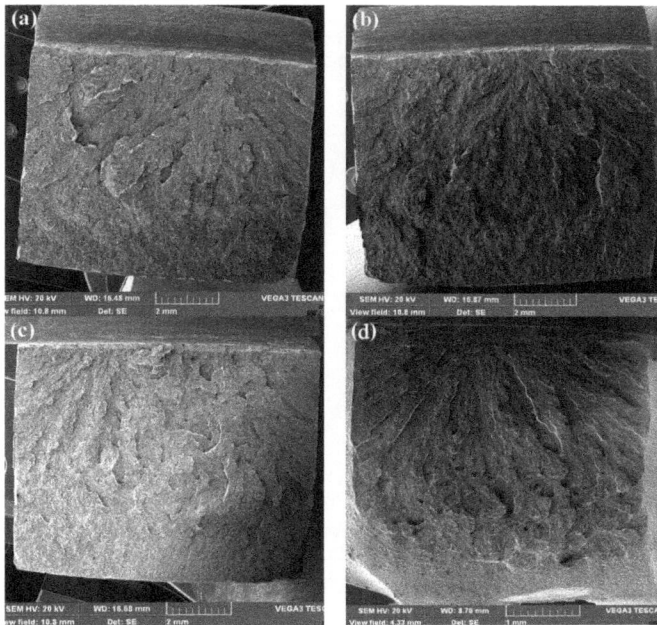

Figure 6. The impact fracture of region 4 and region 6: (**a**) region 4-1; (**b**) region 4-2; (**c**) region 4-3; (**d**) region 6.

To further study the property differences of region 4 and region 6, the high magnification morphology of radial regions are shown in Figure 7. It is obvious that the microscopic morphology of the radial region on region 4 is classical quasi-cleavage with large cleavage step patterns and tear ridges. The microscopic morphology of region 6 is also quasi-cleavage, but the surface is more flat than that of region 4, and there are no obvious cleavage step patterns, which also indicates that the impact toughness of region 4 is much lower than that of region 6.

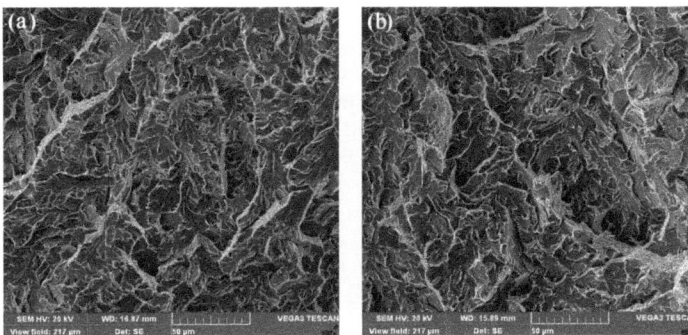

Figure 7. *Cont.*

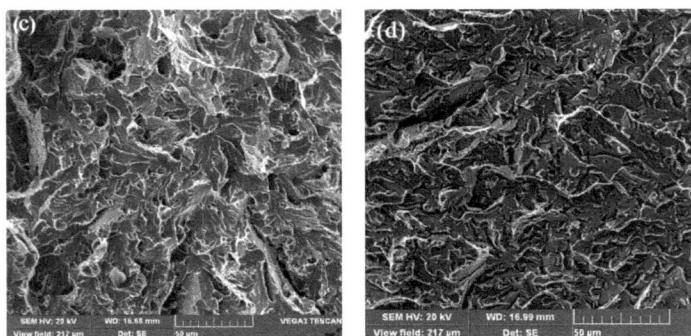

Figure 7. The radial region morphology with high magnification of region 4 and region 6 at room temperature: (**a**) region 4-1; (**b**) region 4-2; (**c**) region 4-3; (**d**) region 6.

4.5. The Tempering and Water-Cooling Effect on Impact Property

The above analysis has demonstrated that the mechanical properties of region 6 are much superior to those of region 4. Furthermore, all the mechanical properties of region 6 meet the technical requirements, while those of region 4 do not meet them. It reveals that the properties of region 4 have degraded, especially the impact toughness which represents the impact resistance of the material. Furthermore, the mechanical properties of region 6 are the most similar to the original material of the die holder. Therefore, samples cut from region 6 are employed for tempering and water-cooling experiments.

To investigate the reason for impact toughness degradation, the tempering and water-cooling experiment was used on the samples of region 6 to simulate the practical environment stress of the die holder. Each experiment has three parallel samples, and the results given are the average values. The results are shown in Figure 8. The impact toughness of region 6 is 40 J/cm^2 before tempering and water-cooling treatment, which is much higher than the minimum value of technical requirements (25 J/cm^2). The results demonstrate that region 6 is of excellent impact resistance. However, the impact toughness decreases to 33.8 J/cm^2 after tempering and water-cooling treatment at 150 °C. The impact toughness increases with the tempering temperature rising. When the sample is tempered and water-cooled at 450 °C, the impact toughness reaches 41.1 J/cm^2, which is slightly above the impact toughness before tempering and water-cooling treatment. In a word, the impact toughness degradation is caused by tempering and water-cooling treatment at the temperatures between 150 °C and 350 °C. During the service process of the die holder, the temperature which region 4 endured is right between 150 °C and 350 °C. It means that the impact toughness degradation of region 4 is related to the tempering and water-cooling effects. It should be noticed that the minimum impact toughness after tempering and water-cooling treatment is 33.8 J/cm^2, but the impact toughness of region 4 is much lower than 33.8 J/cm^2. The reason is that the impact toughness is measured after tempering and water-cooling treatment once. However, the die holder has endured the tempering and water-cooling effects at 150 °C to 350 °C for 113,025 times as well as the cyclic loads for 351,089 times, which would result in much lower impact toughness as measured in region 4.

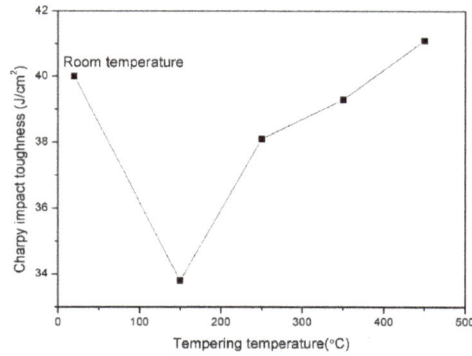

Figure 8. The impact toughness of region 6 after tempering and water-cooling experiment at different temperatures.

The impact fractures of region 6 after tempering and water-cooling treatment at different temperatures are shown in Figure 9. It can be seen that the fracture morphology with low magnification is similar, which includes three regions. The upper part of the fracture is the fiber region. The middle is the radial region. The lower, left and right sides of the fracture are shear lip. Moreover, the size of different regions on each fracture changes with the tempering temperature. The proportion of radial region and shear lip on two sides increase with the temperature rising, but the lower shear lip decreases. The radial region after being tempered was at 250 °C, which is the same as the fracture of region 4, as shown in Figure 3a.

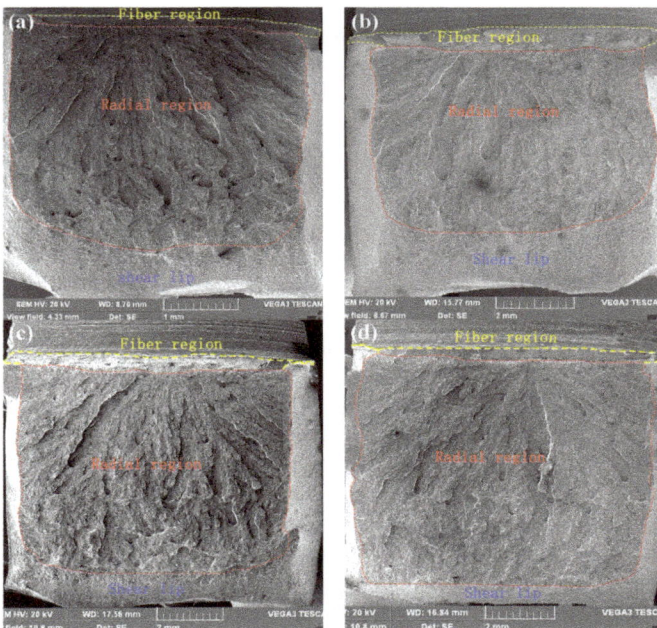

Figure 9. The impact fracture of region 6 after tempering and water-cooling experiment at different temperatures: (**a**) room temperature; (**b**) 150 °C; (**c**) 250 °C; (**d**) 350 °C.

The brittle fracture morphology of the die holder is the same as the radial region of the impact fracture at low magnification. Therefore, the radial region morphology of region 6 was observed by using SEM at high magnification as shown in Figure 10. Though the samples were tempered and water-cooled at different temperatures, the microscopic morphology of radial regions are all quasi-cleavage. However, the quasi-cleavage patterns and quasi-cleavage steps are various at different tempering temperatures. The microscopic fracture of region 4 has the same quasi-cleavage morphology as that tempered at 250 °C, both of which have obvious quasi-cleavage steps as shown in Figures 3b and 10c.

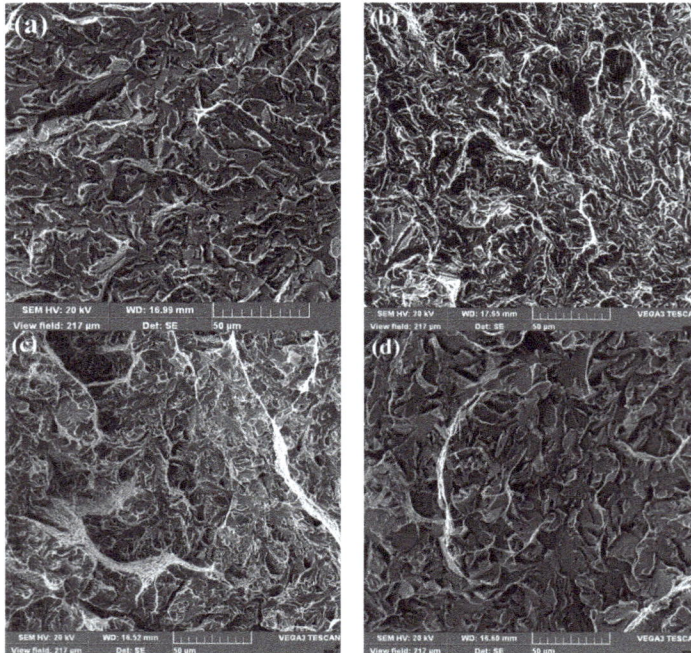

Figure 10. The high magnification morphology of impact fracture in region 6 after tempering and water-cooling experiments at different temperatures: (**a**) room temperature; (**b**) 150 °C; (**c**) 250 °C; (**d**) 350 °C.

4.6. Transmission Electron Microscope Analysis

To investigate the mechanism of impact toughness degradation of region 4, the microstructures of region 4 and region 6 before tempering and water-cooling treatment were compared by using the TEM. Figure 11 is the bright-field images of region 4 and region 6. The microstructure of the two regions are both lath martensite and carbides [11].

The observation of precipitate morphology results indicate that there are many precipitates with different shapes in region 4 and region 6. There are three kinds of precipitates in region 4: the hexagonal precipitates whose sizes are 200 nm to 300 nm as shown in Figure 12a. The tetragonal precipitates whose sizes are 400 nm to 640 nm as shown in Figure 12b. The elongated precipitates, most of which are between 50 nm and 1.5 μm, and a few of them are less than 50 nm as shown in Figure 12c.

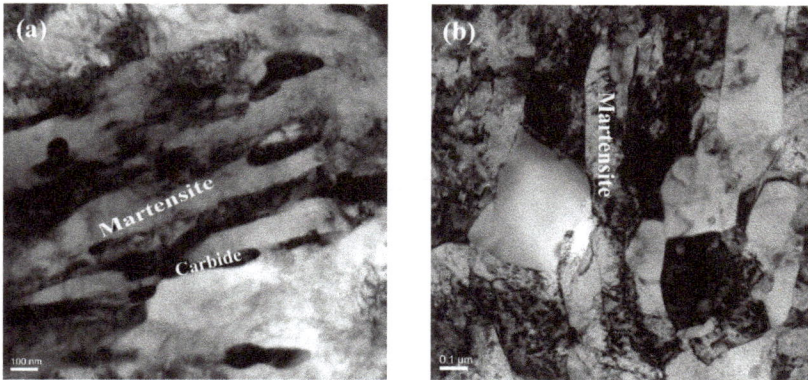

Figure 11. Bright-field TEM images of region 4 and region 6: (**a**) region 4; (**b**) region 6.

In order to determine the phase of the precipitates, the SAED was utilized on precipitates. The results reveal that the SAED patterns of hexagonal precipitates and tetragonal precipitates belong to the different crystal zone axes of the same face-centered lattice as shown in Figure 12. Furthermore, they have identical chemical composition. Therefore, the hexagonal precipitates and the tetragonal precipitates are the same substance, which are collectively called the polyhedron precipitates. Compared with the standard SAED patterns data of precipitates in steel, the SAED pattern of elongated precipitates is the same as the pattern of the [001] crystal zone axis of orthorhombic lattice.

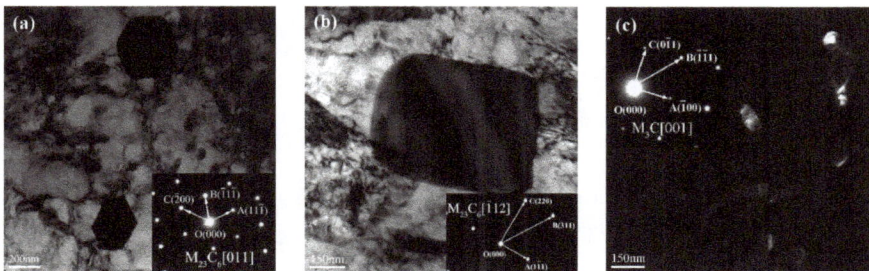

Figure 12. TEM images of the precipitates and corresponding selected area electron diffraction (SAED) patterns in region 4: (**a**) the bright-field image of hexagonal precipitates; (**b**) the bright-field image of tetragonal precipitates; (**c**) the dark-field image of elongated precipitates.

The precipitates were affirmed by comparing the lattice parameter of indexed patterns with the standard PDF cards, as well as by analyzing the chemical composition of precipitates (Table 5). The polyhedron precipitates are $M_{23}C_6$ carbides, where M are mainly Fe and Cr. The elongated precipitates are M_3C carbides, where M are mostly Fe. In region 4, the small size M_3C carbides are distributed mainly in lath martensite, while the large size M_3C carbides are distributed mainly along the boundaries of lath martensite as shown in Figure 12a.

Table 5. The chemical composition of precipitates from region 4 (wt %).

Precipitate Elements	NiK	FeK	MnK	CrK	MoK	VK
Polyhedron precipitate	0.59	77.46	-	13.80	7.58	0.57
Elongated precipitate	0.88	93.41	4.31	1.40	-	-

As shown in Figure 13, the precipitates of region 6 are mainly elongated precipitates whose sizes are mainly within 100 nm. Most of the precipitates are almost globular, except for a few that are elongated. Therefore, the precipitates' size in region 6 is much smaller than those of region 4. In addition, region 6 has no polyhedron precipitates. The SAED patterns of elongated precipitates were indexed, and the results are the same as that of region 4, which is shown in Figure 13. The elongated precipitates are M_3C carbides, where M are mainly Fe [15,16]. However, the M_3C carbides of region 6 are distributed mainly in lath martensite, and only a few of them are located at the martensitic lath boundaries.

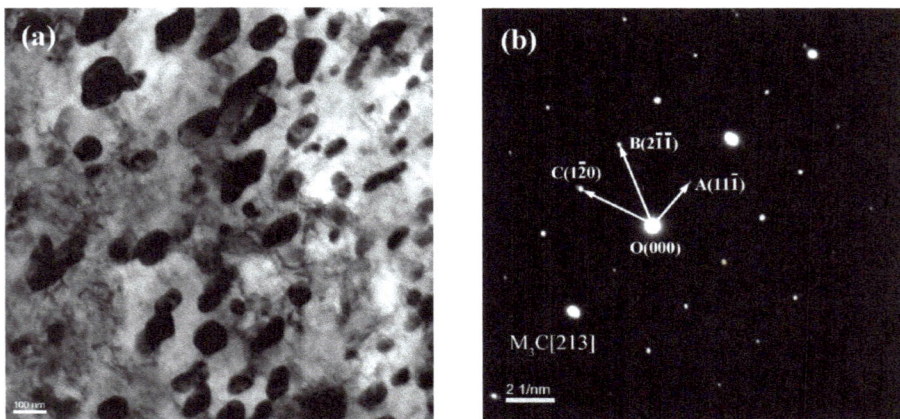

Figure 13. The bright-field TEM image of region 6 and corresponding indexed SAED pattern: (**a**) the elongated precipitates; (**b**) indexed SAED pattern of elongated precipitates.

The $M_{23}C_6$ and M_3C carbides are both hard and brittle phases in tempered steel. $M_{23}C_6$ carbides are mainly formed in the process of heat treatment, which are transformed by undissolved MC carbides during the tempering process. However, the carbides in tempered 55NiCrMoV7 are not always changeless. The existing small carbides can coalesce and grow up under the interaction of cyclic mechanical and thermal loads during the service process [12]. Especially, the fatigue loads can accelerate the growth of $M_{23}C_6$ carbides [17–19]. In region 4, plenty of large size $M_{23}C_6$ carbides that are formed along the martensitic lath boundaries can decrease the impact toughness. Moreover, the large size M_3C carbides can seriously decrease the impact toughness and increase the material brittleness, especially the elongated M_3C carbides that are distributed in the boundaries of lath martensite. Finally, the interaction of $M_{23}C_6$ and M_3C carbides result in the impact toughness degradation [20,21].

5. Discussion

During the service process, the die holder needs to be preheated to 240 °C by using heating rods built in the bottom of the die holder before working. The die holder is used to install the forging die. Furthermore, the inner bottom (region 4) of the die holder would directly contact with the die where the temperature is about 250 °C. Therefore, the inner bottom temperature of the die holder remains at 250 °C during the forging process, while the temperature of the undersurface and region 6 may slowly decrease because they are exposed in air. According to the operation rule, each forging product uses one exclusive forging die, and the die and die holder must be air-cooled when replacing forging product. However, in order to accelerate cooling, operators usually cool the die holder by using water, which results in the cyclic tempering and a water-cooling effect on the bottom of the die holder. The die holder has endured the tempering and water-cooling effect at 150 °C to 350 °C 113,025 times as well as the cyclic mechanical loads 351,089 times. In this paper, the tempering and water-cooling experiment successfully simulates the effect. In addition, the results demonstrate that the long-term combined action of tempering and water-cooling effect at 150–350 °C, as well as the cyclic mechanical loads, caused the growth and coalescence of carbides in the microstructure of 55NiCrMoV7 steel. Then, the large size $M_{23}C_6$ and M_3C carbides distributed at the boundaries of lath martensite lead to the degradation of impact toughness and the material embrittlement, which can easily result in brittle fracture. Additionally, the forging forces act mainly on the inner bottom of the die holder, so region 4 suffers more severe cyclic mechanical loads, as well as tempering and a water-cooling effect, than other regions. However, region 6 only suffers the cyclic tempering without water-cooling, as it is far away from the water-cooling region. Therefore, it was the long-term cyclic water-cooling as well as the cyclic mechanical loads during the service process that result in the serious impact toughness degradation in region 4. The temperature of the inner bottom is higher than the undersurface, and the water cooling also occurs on the inner bottom, so the tempering and water-cooling effect on the inner bottom is more obvious than that of the undersurface. Therefore, the mechanical properties of region 4-1, 4-2 and 4-3 are different. Because region 6 is far away from the regions of the tempering and water-cooling treatment as well as the cyclic mechanical loads, the large carbides of region 6 have not been formed, and, therefore, it does not present the brittleness. Finally, because the impact resistance of region 4 has seriously degraded, the brittle fracture whose microscopic feature is quasi-cleavage occurred on the whole bottom of the die holder under impact load during forging.

6. Conclusions

- The serious material embrittlement and brittle fracture of the die holder are caused by the long-term tempering and water-cooling effect as well as the cyclic mechanical loads. In addition, the operation specification must be emphasized to avoid the die holder being cooled by using water during the working process.

- The tempering and water-cooling treatment at 150 °C to 350 °C could lead to the impact toughness degradation and material embrittlement. The effect of material embrittlement declines with the tempering temperature rising.

- The material embrittlement are related to the large size $M_{23}C_6$ (M: mainly Fe and Cr) and M_3C carbides (M: mainly Fe) distributed in the martensitic lath boundaries.

Acknowledgments: The research was supported by the National Technology Foundation of China (No. JSZL2014601B004). The authors are grateful to Yaozhong Zhang and Hongxiang Jing from the Wuxi Turbine Blade Co., Ltd., Wuxi, China for the technology and equipment support. The guidance and help of Fanchang Zeng from the Confederation of Chinese Metalforming Industry and Peidao Zhong from Beijing Institute Aeronautical Materials are greatly acknowledged.

Author Contributions: Hongxun Wang and Weifang Zhang conceived and designed the experiments; Jingyu Zhang and Yuanxing Huang performed the experiments; Hongxun Wang and Weifang Zhang analyzed the data; Wei Dai contributed reagents/materials/analysis tools; Hongxun Wang and Wei Dai wrote the paper.

Conflicts of Interest: The authors declare no conflict of interest.

References

1. Gronostajski, Z.; Kaszuba, M.; Hawryluk, M.; Zwierzchowski, M. A review of the degradation mechanisms of the hot forging tools. *Arch. Civ. Mech. Eng.* **2014**, *14*, 528–539. [CrossRef]
2. Kchaou, M.; Elleuch, R.; Desplanques, Y.; Boidin, X.; Degallaix, G. Failure mechanisms of H13 die on relation to the forging process—A case study of brass gas valves. *Eng. Fail. Anal.* **2010**, *17*, 403–415. [CrossRef]
3. Gronostajski, Z.; Kaszuba, M.; Polak, S.; Zwierzchowski, M.; Niechajowicz, A.; Hawryluk, M. The failure mechanisms of hot forging dies. *Mater. Sci. Eng. A* **2016**, *657*, 147–160. [CrossRef]
4. Okazaki, Y. Comparison of fatigue properties and fatigue crack growth rates of various implantable metals. *Materials* **2012**, *5*, 2981–3005. [CrossRef]
5. Kim, T.H.; Kim, B.M.; Choi, J.C. Prediction of die wear in the wire-drawing process. *J. Mater. Process. Technol.* **1997**, *65*, 11–17. [CrossRef]

6. Alimi, A.; Fajoui, J.; Kchaou, M.; Branchu, S.; Elleuch, R.; Jacquemin, F. Multi-scale hot working tool damage (X40CrMoV5-1) analysis in relation to the forging process. *Eng. Fail. Anal.* **2016**, *62*, 142–155. [CrossRef]
7. Jhavar, S.; Paul, C.P.; Jain, N.K. Causes of failure and repairing options for dies and molds: A review. *Eng. Fail. Anal.* **2013**, *34*, 519–535. [CrossRef]
8. Chen, C.; Wang, Y.; Ou, H.; He, Y.; Tang, X. A review on remanufacture of dies and moulds. *J. Clean. Prod.* **2014**, *64*, 13–23. [CrossRef]
9. Brnic, J.; Turkalj, G.; Lanc, D.; Canadija, M.; Brcic, M.; Vukelic, G. Comparison of material properties: Steel 20MnCr5 and similar steels. *J. Constr. Steel. Res.* **2014**, *95*, 81–89. [CrossRef]
10. Li, J.; Huang, Q.; Ren, X. Dynamic initiation and propagation of multiple cracks in brittle materials. *Materials* **2013**, *6*, 3241–3253. [CrossRef]
11. Zhang, Z.; Delagnes, D.; Bernhart, G. Microstructure evolution of hot-work tool steels during tempering and definition of a kinetic law based on hardness measurements. *Mater. Sci. Eng. A* **2004**, *380*, 222–230. [CrossRef]
12. Zhang, Z.; Qi, Y.; Delagnes, D.; Bernhart, G. Microstructure variation and hardness diminution during low cycle fatigue of 55NiCrMoV7 steel. *J. Iron Steel Res. Int.* **2007**, *14*, 68–73. [CrossRef]
13. Wang, H.; Jiang, P.; Zhang, W.; Zhang, Y.; Song, T. Failure analysis of large press die holder. *Eng. Fail. Anal.* **2015**, *64*, 13–25. [CrossRef]
14. Roberti, R.; Faccoli, M. On the step cooling treatment for the assessment of temper embrittlement susceptibility of heavy forgings in superclean steels. *Metals* **2016**, *6*, 239. [CrossRef]
15. Song, Y.Y.; Ping, D.H.; Yin, F.X.; Li, X.Y.; Li, Y.Y. Microstructural evolution and low temperature impact toughness of a Fe-13%Cr-4%Ni-Mo martensitic stainless steel. *Mater. Sci. Eng. A* **2010**, *527*, 614–618. [CrossRef]
16. Ning, A.; Mao, W.; Chen, X.; Guo, H.; Guo, J. Precipitation behavior of carbides in H13 hot work die steel and its strengthening during tempering. *Metals* **2017**, *7*, 70. [CrossRef]
17. Flora, M.G.D.; Pellizzari, M. Behavior at elevated temperature of 55NiCrMoV7 tool steel. *Mater. Manuf. Process.* **2009**, *24*, 791–795. [CrossRef]
18. Li, Z.; Xiao, N.; Li, D.; Zhang, J.; Luo, Y.; Zhang, R. Effect of microstructure evolution on strength and impact toughness of G18CrMo2-6 heat-resistant steel during tempering. *Mater. Sci. Eng. A* **2014**, *604*, 103–110. [CrossRef]
19. Garcia-Mateo, C.; Morales-Rivas, L.; Caballero, F.G.; Milbourn, D.; Sourmail, T. Vanadium effect on a medium carbon forging steel. *Metals* **2016**, *6*, 130. [CrossRef]
20. Paul, V.T.; Saroja, S.; Vijayalakshmi, M. Microstructural stability of modified 9Cr-1Mo steel during long term exposures at elevated temperatures. *J. Nucl. Mater.* **2008**, *378*, 273–281. [CrossRef]
21. Lee, T.-H.; Lee, Y.-J.; Joo, S.-H.; Nersisyan, H.H.; Park, K.-T.; Lee, J.-H. Intergranular M23C6 carbide precipitation behavior and its effect on mechanical properties of Inconel 690 tubes. *Metall. Mater. Trans. A* **2015**, *46*, 4020–4026. [CrossRef]

metals

MDPI

Article

AISI 304 Welding Fracture Resistance by a Charpy Impact Test with a High Speed Sampling Rate

Bambang Riyanta [1,2,*], I. N. G. Wardana [2], Yudy Surya Irawan [2] and Moch. Agus Choiron [2]

[1] Department of Mechanical Engineering, Muhammadiyah University of Yogyakarta,
   Bantul Yogyakarta 55183, Indonesia
[2] Department of Mechanical Engineering, Brawijaya University, Malang 65145, Indonesia;
   wardana@ub.ac.id (I.N.G.W.); yudysir@ub.ac.id (Y.S.I.); agus_choiron@ub.ac.id (M.A.C.)
* Correspondence: bambangriyanta@umy.ac.id; Tel.: +62-0274-375798

Received: 24 October 2017; Accepted: 29 November 2017; Published: 5 December 2017

Abstract: The purpose of this study was to evaluate fracture resistance in AISI 304. The J-R curve was constructed from data, which resulted from an impact test by Charpy Impact machine equipped with high-speed sampling rate data acquisition equipment. The critical values of fracture resistance in fusion zones (FZ), high temperature heat affected zones (HTHAZ), low temperature heat affected zones (LTHAZ) and unaffected base metals (UBM) were obtained by calculation methods using some formulas and by graphical methods. Laboratory experiments demonstrated the relationships among the values of energy absorption along the impact test with the obstruction of dislocation movement due to the presence of chromium interstitial solute in all zones and chromium rich carbide precipitates in fusion zones and heat affected zones.

Keywords: fracture resistance; Charpy impact test; J-R curve; dislocation; precipitate

1. Introduction

The critical values of such fracture parameters are known as fracture resistance. Fracture resistance is the ability of a ductile material to accept load or deform plastically and resist fractures in the presence of cracks based on the Elastic Plastic Fracture Mechanics (EPFM) approach [1–3]. J-R curves have been widely used to measure the fracture resistance, and their determination has been standardized in ASTM E 1820, which combines the testing and analysis procedures of E 813 and E 1152. The recommended specimen in the standard were fatigue pre-crack [C(T)] and [SE(B)] specimen. Load versus load line displacement were recorded by digital instruments or autographically by x-y plotter. The record of load versus displacement was used to determine crack length. Since Load P, displacement v and crack length have been estimated at each data point, referring to ASTM E 1820, J could be evaluated to construct J-R curves.

The procedure needs precise specimen preparation and high accuracy instruments to run. The overall J-R curve construction procedure was time consuming and costly. It was also not easy to implement. On the other hand, impact toughness tests evidently could be done at reasonable costs and have been developed for metals and non-metal testing with various purposes easily [4–7]. Some studies have successfully determined J-R curves by applying the normalization method on Charpy Impact test data [2,8–10]. The normalization method, which uses the principle of load separation, has been introduced as the convenient way for J-R curve determination. The method related three variables, load P, displacement v, and crack length during the fracture process, which provides a prediction of any one among three variables from the other two. Using the normalization method, there were many studies that developed further techniques for J-R curve determination in order to overcome some difficult conditions such as: unavailability of crack length, displacement or load measurement [1,2]. Other studies investigate the possibility to apply a Charpy V-notched (CVN) test to determine fracture toughness K_{IC} and critical crack resistance values, J_{IC} [11–13]. Like the previous method, the basic concept of the crack length was proportional to the square of absorbed energy.

Stainless steel is a metallic alloy consisting of at least 10.5% chromium (Cr) and 50% iron (Fe). There are various grades of stainless steel such as: Austenitic Stainless Steel, Ferritic Stain less steel, Martensitic stainless steel, duplex stainless steel and precipitation hardening stainless steel. Austenitic Stainless Steel is the most widely used due to its corrosion resistance, mechanical properties and economic value [14]. As, in general, AISI 304 welding potentially faced crack problems, which were considered to be the most serious problems among a variety of physical defects. When there were some cracks in structural components, it could fail at a lower stress level than the ultimate strength of the material [1,2]. Many failures have occurred as a result of fracture, even though the yield stress was not reached.

The main purpose of this study was to apply the previous findings on fracture resistance determination by the Charpy Impact Test to investigate fracture resistance profiles and fracture resistance critical values of American Iron and Steel Institute Standard 304 (AISI 304) welding. Charpy Impact machine equipped with high-speed sampling rate data acquisition equipment is expected to exhibit precisely the relation among the values of energy absorption along the impact test with the presence of chromium interstitial solute and chromium rich carbide precipitation. The relationships were suspected as important keywords in describing fracture resistance profile in each AISI 304 welding zone. When energy absorption along the impact test was increased, it very likely has a strong relationship with obstruction of dislocation movement by the presence of Cr as a dominant particle in AISI 304, which has a high hardness and yield strength even at high temperature. The formation of chromium carbides and its properties on AISI 304 welding also hamper the movement of dislocation. Both of these obstruction agents were suspected to be the cause of increased energy absorption capability of post welding AISI 304 through the length of Charpy Impact tests, and, finally, influence the fracture resistance profile.

2. Experimental Method

AISI 304 strip plate with the dimension of 1000 mm × 100 mm × 10 mm as shown in Figure 1 was used as a basic material for all necessary testing to investigate the values of AISI 304 welding fracture resistance. Composition of the basic material is presented in Table 1.

Figure 1. AISI 304 Stainless Steel Strip plate with the dimensions of 1000 mm × 100 mm × 10 mm.

Table 1. Composition of AISI 304.

Element	C	Cr	Mn	Mo	Ni	P	Si	S	O	N
Weight %	0.041	18.4	0.19	0.19	8.5	0.031	0.54	0.014	0.025	0.062

The material was then welded by TIG with the welding specification as performed in Table 2.

Table 2. Welding specification.

Welding Process	141/Gas Tungsten Arc Welding (GTAW)
Welding position	PA/1G-Down hand
Joint Type	Butt Joint
Parents Material Specification	AISI 304
Preparation Cleaning	Mechanical
Welding Sequence	(1) Clean the surface (brush & grinding); (2) Dry 50 °C, tack welding & welding; (3) 1 Layer filled weld only in straight layer; (4) Visual control of welding seam; (5) Next layer in straight layer
Filler Metal	High-Performance Stainless Steel Welding Electro Rod-ER 308 L, Post drying: dry store, Slow cooling, argon shielding gas
Other Information	Preheat temp: plate dray > 25 °C, welder's test: welding position European Standard EN 287-1.PF

In order to place the notch precisely in the fusion zones, high temperature heat affected zones, low temperature heat affected zones and unaffected base metals, microstructure observation by optical microscope with a magnification of $100\times$ was conducted for each zone. The microstructure images of each zone and the distance of notches from the center of welding were shown in Figure 2a–d, respectively.

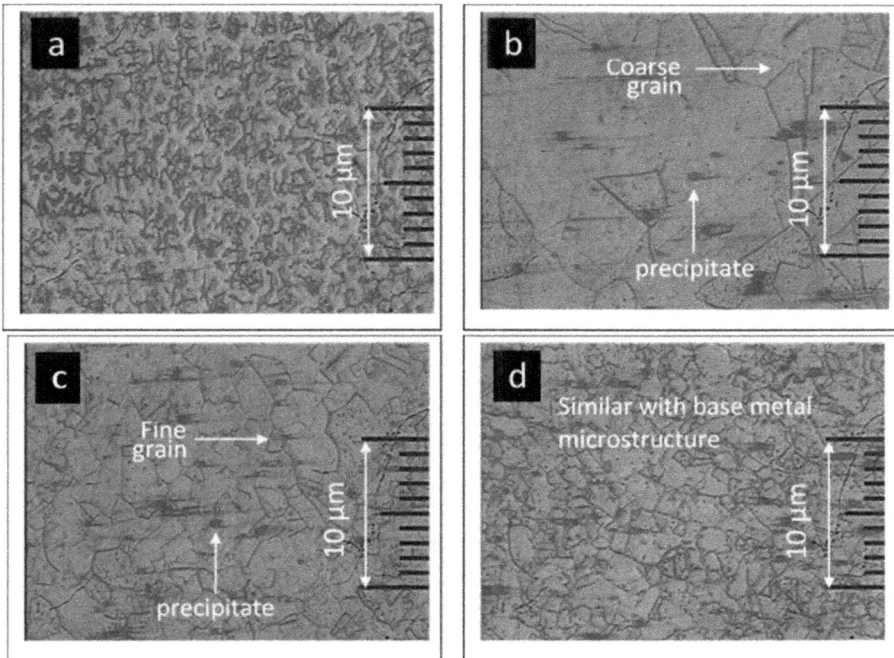

Figure 2. Microstructure of AISI 304 welding. (a) fusion zone, notch position at the center of welding; (b) high temperature heat afected zone, notch position: 4 mm from the center of welding; (c) low temperature heat affected zone, notch position: 5.5 mm from the center of welding; (d) unaffected base metal zone notch position: 7 mm from the center of welding.

Seven AISI 304 three points bending welded impact specimens for each zone with the dimensions of 55 mm × 10 mm × 10 mm were prepared using Electrical Discharge Machining (EDM) referring to the ASTM E 23-02a standard [13]. The relative position notch from the center weld bead was presented in Figure 3.

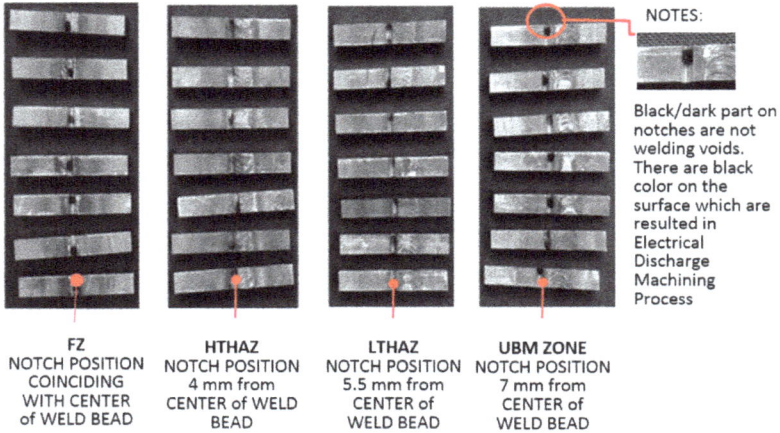

Figure 3. Notch position in an AISI 304 welding Charpy Impact specimen.

Based on ASTM E 23-02a, Charpy Impact tests were conducted with a 300 Joules × 2 Charpy Impact machine equipped with a high-speed sampling rate data acquisition instrument (see Figure 4) in order to examine the energy absorption profile in each zone. The data acquisition instrument (manufactured by Advantech Corporation, Taipei, Taiwan) that consists of a S-type Load Cell, ADAM 3016 signal conditioner, and USB 4702-AE analog to digital converter (ADC) was set to record 45,000 samples per second.

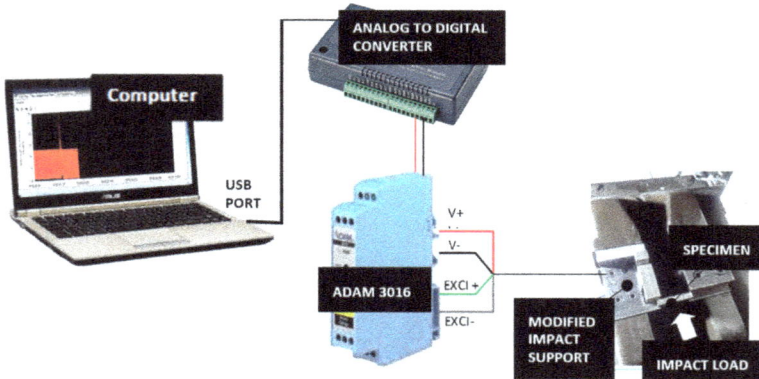

Figure 4. High-speed sampling rate data acquisition instrument.

Fracture surface was observed by a Hitachi SU 3500 Scanning Electron Microscope (Tokyo, Japan) and Energy Dispersive X-ray Spectroscopy, which has the ability to magnify 10–300,000× with a depth of field of 4–0.4 mm and 3 nm resolution.

3. Study Results

The chart presented in Figure 5 is one of the Charpy Impact test results, representing six other similar results for figures of Charpy Impact tests on seven specimens for each zone.

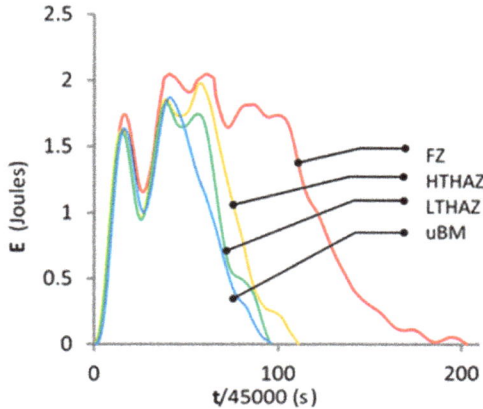

Figure 5. Energy versus unit of time chart for Charpy Impact tests of AISI 304 welding.

The reproducibility of the test was high, considering the resemblance between the test results with Charpy Impact tests conducted with impact velocity of 3.4 m/s, which was obtained in the previous study by Janssen et al. [15]. Load versus time chart as performed in Figure 5 visually shows that area under the load line, which represents the energy absorbed by the zone during impact loading, is narrower as the distance from the center of welding gets farther. The fusion zone has the ability to absorb the most energy followed by HTHAZ and two other zones, respectively. The load line of LTHAZ and Unaffected Base Metal Zone coincide with each other, which means that they are not significantly different. Total impact energy of each zone could be obtained by estimating the areas under the load line on the chart of each zone and then multiplying by 2 (2 support). Impact energy value of each zone was obtained by calculating the area under load line, and impact energy values from the indicator on the impact machine are displayed in Table 3.

Table 3. Impact energy.

Methods	Total Impact Energy (Joules)			
	Fusion Zone	HTHAZ	LTHAZ	Unaffected Base Metal
Area under load line	252.9	130.9	109.9	100.36
Indicator on impact machine	214.8	114.5	99.5	90

Some interesting findings were performed in the load versus time chart as shown in Figure 5 and in the load versus displacement chart in Figure 6a–d. The charts in all of the zones perform a negative overshoot "a", which was occurred typically in the time range of 20 to 30 ms.

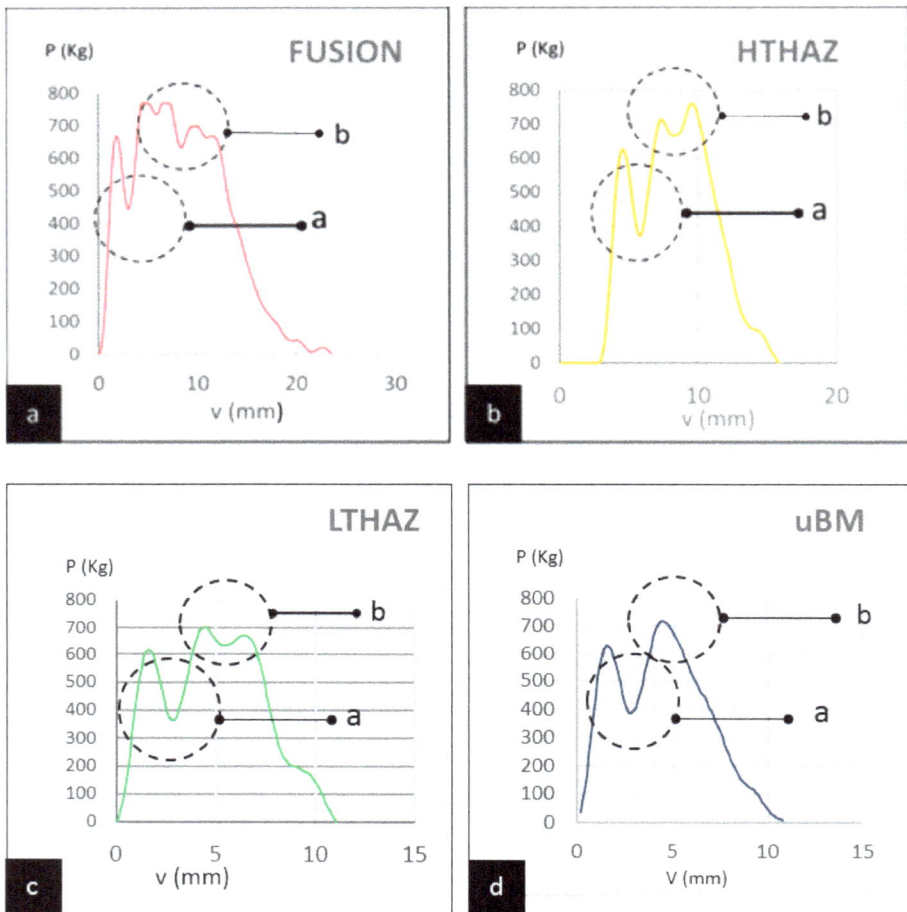

Figure 6. Findings: negative overshoots "a" and number of peaks "b". (**a**) a negative overshoot "a" and lots of peak "b" in fusion zone (FZ). (**b**) a negative overshoot "a" and moderate number of peak "b" in high temperature heat affected zone (HTHAZ). (**c**) a negative overshoot "a" and moderate number of peak "b" in (LTHAZ). (**d**) a negative overshoot "a" and a peak "b" in unaffected Base Metal (UBM) zone.

The negative overshoot "a" indicates a plastic deformation. Previous research reported that, during plastic deformation, stainless steel 304 transformed into α' (bcc) martensite from γ (fcc) austenitic. The other study also reported the cold working process, and most of the 300 series stainless steels transform into ε (hcp) martensite and α' (bcc) martensite [16].

The presence of Cr in this alloy was suspected as a plastic deformation obstruction agent. The dislocation theory could be explained by the Cr atoms' presence dominantly in γ-Fe-Ni, they would play a role as an interstitial solute, which becomes an obstacle that hampers the motion of dislocation. Once a dislocation has stopped, an extra force was required to make the dislocation move, producing an observed upper loading in a load versus displacement graph. The presence of Cr atoms can be seen in the image by a Scanning Electron Microscope (SEM) and Energy-dispersive X-ray spectroscopy (EDS) investigation, as shown in Figure 7.

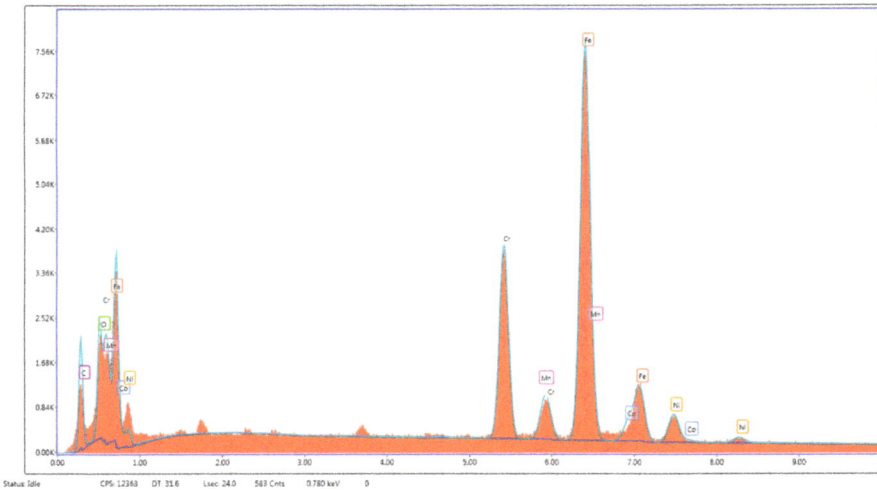

Figure 7. Scanning electron microscope (SEM)/Energy-dispersive X-ray spectroscopy (EDS) investigation indicates the presence of Cr.

The combination of the formation of martensite structure after plastic deformation and the presence of chromium interstisial solute, which has been described above, can be expected to be the main cause of a negative overshoot "a" occurrence.

The numbers of peak "b" variation, which was shown in the previous charts, was another interesting finding. The number of peak decreases as the increase of distance from the center of welding. The number of peaks considered has a relationship with void growth mechanism and the presence of precipitates in austenitic steel welding. Simple austenitic steels contain between 0.03% and 0.1% carbon. The solubility limit of carbon is about 0.05% at 800 °C up to 0.5 wt % at 1100 °C. The treatment at the temperature of 1050 °C to 1150 °C followed by rapid cooling would produce a solid solution saturated austenite at room temperature. This would lead to rejection of the carbon solid solution at slow cooling or reheating in the range 550–800 °C, even with the carbon content of steel being less than 0.05%. At room temperature, the structure in equilibrium condition contains austenite γ, α ferrite and carbides [17–19]. Precipitate phase occurred at temperatures below 900 °C. When heated to 1100–1150 °C, carbide went into solution and, on cooling, a precipitate-free austenite was obtained.

When saturated austenite is heated to high temperatures, further precipitation will take place at the austenite grain boundaries. In Stainless steel welding, especially in Heat affected zones, when Stainless steel gained a heating process at more than 300 °C, the process above also takes place. The precipitation in Stainless steel welding, usually called sensitization, was observed by an optical microscope and a Scanning Electron Microscope (SEM). The presence of precipitates with dimensions of about 0.5 μm are performed in Figure 8.

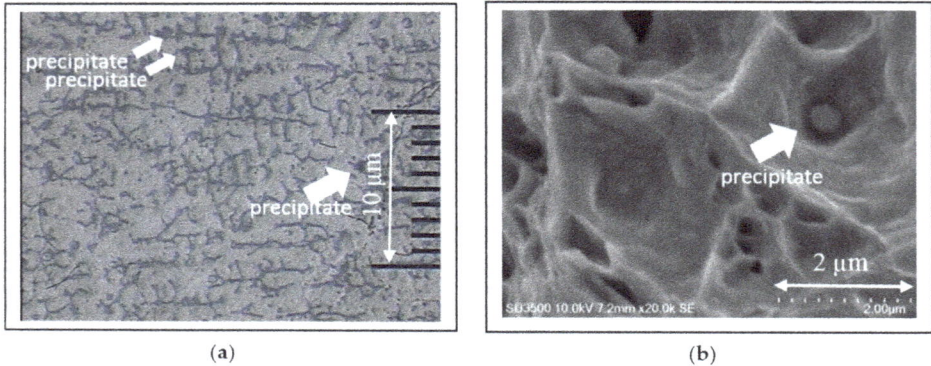

(a)   (b)

Figure 8. The presence of precipitates. (**a**) precipitate observation by optical microscope magnification = 100; (**b**) precipitate observation by SEM accelerating voltage = 10,000 volts; magnification = 20,000; working distance = 7220 μm.

The sensitization process results in reduction of chromium content to the weight level of 12%, as displayed in Figure 7, with respect to the origin chromium content at the weight level of 18.4%, as shown in Table 1. This is because the segregated carbon takes up the chromium forming chromium carbides. The dimension of the chromium carbide precipitate corresponded with previous literature, which reported that the precipitate size was varying in the approximated range of 0.5–1.5 μm [20]. Energy dispersive X-ray spectroscopy (EDS) was used for compositional point analysis of the precipitates. The EDS results are provided in Figure 9 and Table 4.

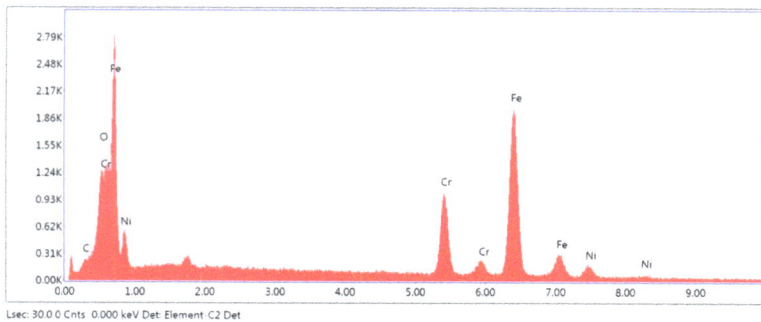

Figure 9. Energy dispersive X-ray spectroscopy (EDS) spectrum of precipitate.

Table 4. Composition of precipitates.

Element	Weight %	Atomic %	Net Int.	Error %	Kratio	Z	R	A	F
C K	5.72	20.77	117.24	10.85	0.0202	1.3170	0.8525	0.2677	1.0000
O K	2.53	6.90	148.64	11.45	0.0168	1.2605	0.8775	0.5259	1.0000
Cr K	17.37	14.57	421.19	4.46	0.2045	0.9726	1.0013	0.9964	1.2145
Fe K	66.70	52.07	915.48	3.39	0.6566	0.9667	1.0094	0.9841	1.0348
Ni K	7.67	5.69	67.14	12.08	0.0738	0.9751	1.0145	0.9501	1.0383

The fracture resistance could not be obtained from the load versus time chart directly. It must be converted to the load versus displacement chart as seen in Figure 6. The conversion could be done by applying a series of calculation methods based on the theory of conventional dynamics [1,2].

Load versus displacement chart could then be analyzed to construct J-R curves, as shown in Figure 10, by predicting the crack length using normalization methods [2].

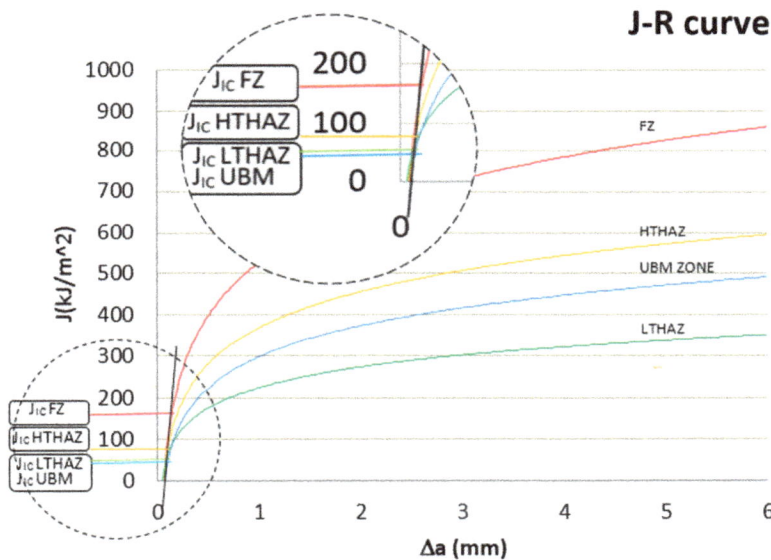

Figure 10. J-R curve of AISI 304 welding.

The fracture resistance profiles were determined by analyzing the fracture resistance curve (R-curve) of each zone. The R-curve in elastic-plastic materials acknowledges the fact that the resistance to fracture increases with growing crack size. The R-curve is a plot of the total energy dissipation rate as a function of the crack size. Figure 10 shows the fracture resistance curve (R-curve) of the Fusion Zone was the highest one, and then followed by HTHAZ, UBM zone and LTHAZ, respectively. The order of R-curve on Figure 10 indicates that fusion zone has the ability to absorb the most energy while LTHAZ has the ability to absorb the least one. The fact was not clearly seen on the energy versus unit of time chart in Charpy Impact test as performed in Figure 5, which made the energy absorption curve on LTHAZ coincide with the energy absorption curve in the UBM zone. The matter is probably due to insignificant precipitate content in LTHAZ. When the temperature is not enough to meet the needs of the sufficient precipitate formation, then precipitate hardening is not evoked adequately in LTHAZ. Unfortunately, the rapid loading on impact test contributes to making the differences of the energy absorption curve as a function of time between LTHAZ and UBM zone less visible. A large amount of energy that is absorbed in the impact process at a medium loading speed will be converted into various material responses such as plastic deformation, hysteresis effects, and inertia. The Charpy Impact Specimens receive high-speed loading/rapid loading on impact tests, resulting in a high strain rate. Due to the very high strain rate on the impact test, there is not enough time for the dislocation to move to a grain boundary so that no plastic deformation occurs. The material will suffer transgranular fracture. High strain rate also causes the material to have no chance to maintain its shape. When plastic deformation and inertia effect do not occur due to high strain rate, the material doesn't have an ability to absorb more energy before failure. J-R curve, as performed in Figure 10, displays a more obvious difference regarding the capabilities of LTHAZ and UBM zones in absorbing energy as a function of Δa before failure. The ability of the fusion zone to absorb the most energy indicates that the zone has a better ductility then three other zones. Impact fracture macro image in Figure 11 and scanning electron microscope observation as seen in Figure 8b show ductile fracture in fusion zones clearly. The failures by ductile fracture were performed by large and deep dimple rupture. The brittle failure especially in LTHAZ was performed by a smooth surface fracture as seen in Figure 11.

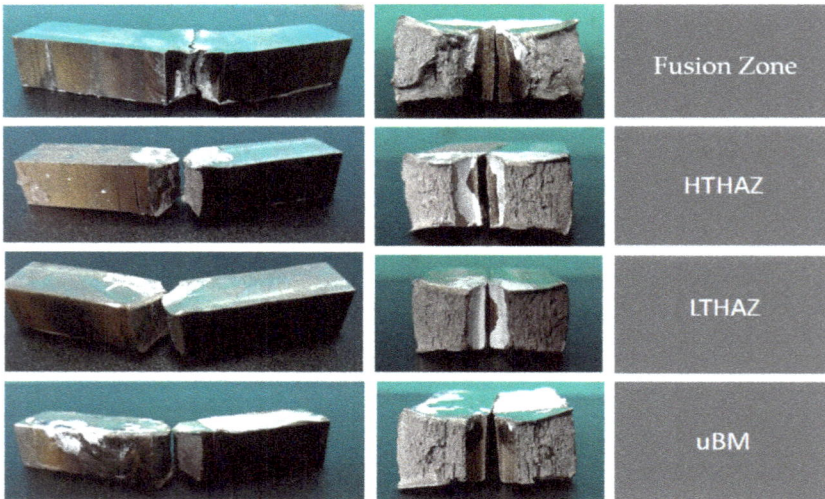

Figure 11. Macro image of impact fracture.

The sequence of R-curves differs from the sequence of critical fracture resistance values (J_{IC}). The J_{IC} values were obtained by graphical projection of intercepts of the R-curve with 0.2 mm offset line horizontally to the ordinate axis. Critical fracture resistance value (J_{IC}) of the UBM zone was the lowest one, which is not significantly different from the Critical fracture resistance value of LTHAZ.

When the values of fracture toughness K_{IC} could be obtained by using the relation with the Charpy V-Notch test result (CVN), as mentioned in the introduction and shown in Table 5 adapted from reference [11], then critical values of fracture resistance J_{IC} is determined by [3]:

$$K_{IC} = \sqrt{EJ_{IC}},\qquad(1)$$

where K_{IC} is critical fracture toughness value ; E is Young modulus for plane stress condition and J_{IC} is a critical fracture resistance value.

Table 5. Relationship between K_{IC}–CVN (Source: Terán et al., [11]).

Formula	Units	Equation Number
Rolfe–Novak–Barson [$(\frac{K_{IC}}{\sigma_{ys}})^2 = 0.64 (\frac{CVN}{\sigma_{ys}} - 0.01)$]	Mpa $\sqrt{m}$, Mpa, J	(2)
Welding Research Council (WRC) 265 [$(\frac{K_{IC}}{\sigma_{ys}})^2 = 0.54 (\frac{CVN}{\sigma_{ys}} - 0.02)$]	Mpa $\sqrt{m}$, Mpa, J	(3)
Robert and Newton [$K_{IC} = 0.804\sigma_{ys} (\frac{CVN}{\sigma_{ys}} - 0.0098)^{0.5}$]	Mpa $\sqrt{m}$, Mpa, J	(4)
Sailors and Corten [$K_{IC} = 14.6 (CVN)^{0.50}$]	Mpa $\sqrt{m}$, Mpa, J	(5)
Marandet and Sanz [$K_{IC} = 19 (CVN)^{0.50}$]	Mpa $\sqrt{m}$, Mpa, J	(6)
Materials Standard Institution, Stockholm INSTA [$K_{IC} = 12 \sqrt{CVN}$]	Mpa $\sqrt{m}$, J	(7)

The values of fracture toughness K_{IC} resulting from the calculation using the relationships shown in Table 5 and calculated J_{IC} using Equation (1) are performed in Table 6.

Table 6. Calculated K_{IC} by various formulas.

Formula	FZ		HTHAZ		LTHAZ		UBM	
	K_{IC}	J_{IC}	K_{IC}	J_{IC}	K_{IC}	J_{IC}	K_{IC}	J_{IC}
Rolfe–Novak–Barson	185.750	172.516	133.101	88.58	121.763	74.132	116.248	67.568
Welding Research Council (WRC) 265	169.889	144.312	121.236	73.491	110.725	61.301	105.605	55.763
Robert and Newton	174.275	151.859	89.497	40.049	74.905	28.054	68.275	23.308
Sailors and Corten	269.540	363.261	139.513	97.319	117.131	68.598	106.963	57.206
Marandet and Sanz	456.484	1041.89	236.274	279.128	198.369	196.752	181.149	164.076
INSTA	182.088	165.780	94.248	44.413	79.128	31.306	72.259	26.107

The critical values of fracture resistance (J_{IC}) of the fusion zone estimated from the graphical procedure is 170 kJ/m^2 followed by HTHAZ with critical fracture resistance value of 85 kJ/m^2 and two other zones, LTHAZ and UBM, with similar critical fracture resistance values of approximately 60 kJ/m^2 and 50 kJ/m^2, respectively.

The most corresponding critical fracture resistance values on the comparison list in all zones by both methods is provided in Table 7 as follows:

Table 7. The critical fracture resistance (J_{IC}) comparison list.

Methods	FZ (kJ/m^2)	HTHAZ (kJ/m^2)	LTHAZ (kJ/m^2)	UBM (kJ/m^2)
Graphical methods	170	85	60	50
The most corresponding calculated values	172.516 Rolve–Novak–Barson	88.58 Rolve–Novak–Barson	61.301 WRC-265	55.763 WRC-265

The critical fracture resistance values (J_{IC}) of all zones were obtained by the graphical methods are looked similar with the J_{IC} values which calculated by the formula of Rolve–Novak–Barson. The similar critical fracture resistance values relates to the fit between AISI 304 yield stress of 215 MPa with Rolve–Novak–Barson material yield stress usage restrictions in the range of 170–760.

4. Discussion

The critical values of fracture resistance by both methods were strongly associated with the previous findings of chromium interstitial solute (see Figure 7), which was allegedly responsible for strengthening the material. Strength of the material depends on the material's ability to resist the movement of the dislocation. Dislocations generate a stress that will interact with local stresses formed by the presence of chromium as solute atoms. When dislocation movement was inhibited, the magnitude of shear stress required moving dislocations in a material is greater, causing an increase in the yield stress of the material, which also means an increase in strength of the material. The solid solution strengthening depends on a concentration of the solute atoms, modulus of the solute atoms, size of the solute atoms, valence of the solute atoms, and the symmetry of the solute field [21,22]. In the present study, the critical values of fracture resistance are also related with chromium rich carbide precipitation, which was already outlined in the previous section. The energy versus unit of time chart in Figure 5 shows that the fusion zone as a zone with the highest number of peak "b" has an ability to absorb energy more than other zones. As mentioned in the previous section, the number of peaks on the graph considered have a relationship with the presence of precipitates in the zone. In most binary systems, the second phase will be formed when the concentration of alloy is above the concentration given by the phase diagram. The alloy in the the form of the second phase in the solid solution at elevated temperatures becomes precipitate upon quenching and aging at lower temperature The precipitate has an important role in strengthening mechanism of alloy, which is called precipitation hardening [22,23]. The precipitate particle acts as a barrier to dislocation in several ways. If the precipitate atoms' radii are small, the dislocations will cut through the precipitate. As a result, new surfaces get exposed to the matrix and lead the increasing of the particle-matrix interfacial energy. As the size of the second phase particle increases, it becomes increasingly difficult for the particles to cut through the material. Hence, dislocations tend to loop/to bow around the particle by Orowan Looping. At a critical diameter of about 10 nm–60 nm, dislocations will preferably cut across the obstacle, while, for a diameter more than 60 nm, the dislocations will readily loop or bow to overcome the obstacle. The precipitate with the small size (up to 0.05 μm) and intermediate size (0.05–0.5 μm) makes a contribution to strengthening the stainless steel by improving hardness and yield strength [3]. When the precipitate in austenite boundaries is getting coarse, then it can easily become a void nucleation site or crack [23,24]. In the present study, a few and less dominant large precipitates appeared as shown in Figure 12.

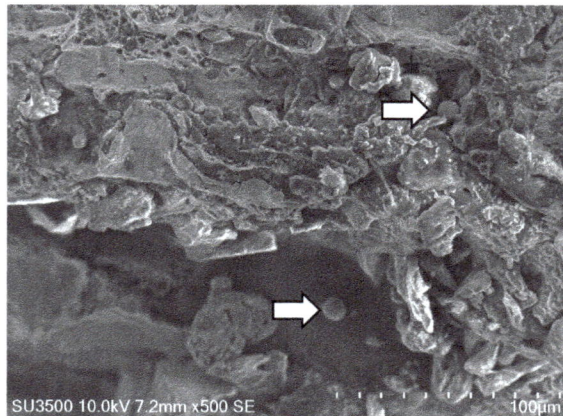

Figure 12. The presence of less dominant large precipitate.

Impact fracture surface images are performed in Figure 13. Void growth mechanism on the fusion zone is dominant in the stable crack extension region as indicated with very large dimples with size of about 2 μm. The dimples are less and shallower in other zones, as shown in Figure 13b–d, indicating a more brittle material.

High stress triaxiality (plane-strain state) is suspected as triggering fast void growth, leading to formation of large dimples [25].

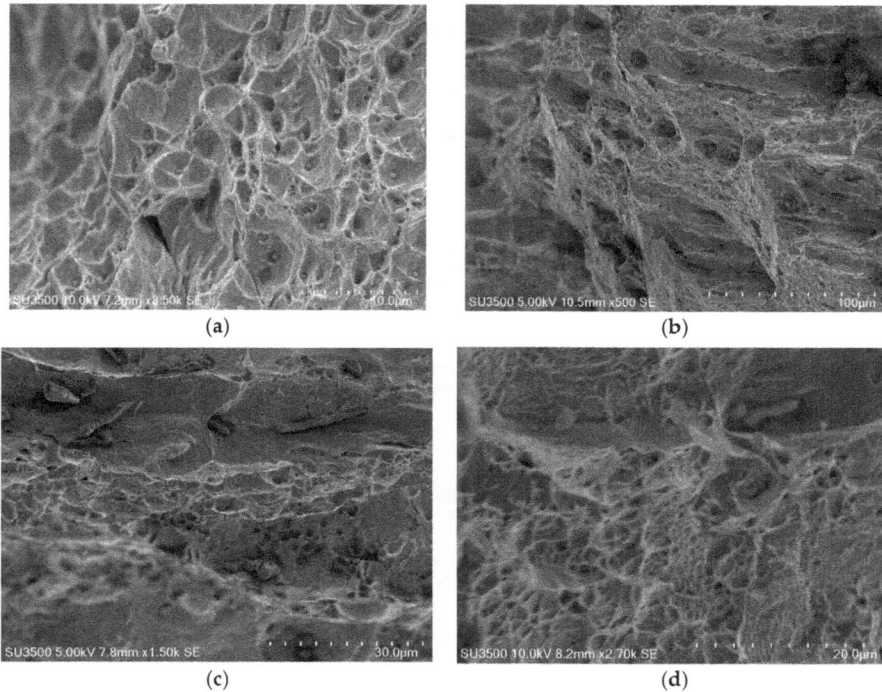

Figure 13. Scanning electron microscope of impact fracture surface. (**a**) fusion zone accelerating voltage = 10,000 Volt; magnification = 3500; working distance = 7210 μm; (**b**) high temperature heat affected zone accelerating voltage = 5000 volts; magnification = 500; working distance = 1030 μm; (**c**) low temperature heat affected zone accelerating voltage = 5000 volts; magnification = 1500; working distance = 7820 μm; (**d**) unaffected base metal accelerating voltage = 10,000 volts; magnification = 2700; working distance = 8190 μm.

5. Conclusions

The study succesfully performed the fracture resistance chart (J-R curve) of AISI 304 welding by a high speed sampling rate Charpy Impact Test. The R-curve of the Fusion Zone was the highest one, then followed by HTHAZ, the UBM zone and LTHAZ, respectively. The sequence of R curves differs from the sequence of critical fracture resistance values (J_{IC}), and critical fracture resistance value of the Fusion Zone was highest followed by HTHAZ, LTHAZ and UBM, respectively. Compared with energy vs. unit time graph, the J-R curve chart has advantages in displaying a more obvious difference in the capabilities of the zones in absorbing energy before failure, especially between LTHAZ and UBM zones.

The present study also found the presence of Cr in all zones, which act as a plastic deformation obstruction agent, showed on the impact graph as a negative overshoot followed by increasing energy absorption. The increase of energy absorption capability, due to dislocation obstruction by the presence of a hard and strong Cr atom as a dominant particle, increases the fracture resistance profile of AISI 304. The observation throughout the impact test also showed a fine chromium carbide precipitate particle in an elevated temperature zone during the welding act as a barrier to dislocation, strengthening the AISI 304 and eventually influencing the fracture resistance profile, especially on fusion zones.

Acknowledgments: The authors gratefully acknowledge support from the Government of the Republic Indonesia for providing the fellowships and financial support, which made the program possible. Any kind of support from Universitas Muhammadiyah Yogyakarta are also acknowledged.

Author Contributions: Bambang Riyanta, I. N. G. Wardana, Yudy Surya Irawan and Moch. Agus Choiron conceived and designed the experiments; Bambang Riyanta performed the experiments; Bambang Riyanta analyzed the data; Bambang Riyanta contributed reagents/materials/analysis tools; Bambang Riyanta, I. N. G. Wardana, Yudy Surya Irawan and Moch. Agus Choiron wrote the paper.

Conflicts of Interest: The authors declare that there is no conflict of interest in the manuscript.

References

1. Brandon, P.P. Fracture Toughness: Evaluation of Testing Procedure to Simplify Jic Calculations. Master's Thesis, University of Tennessee, Knoxville, TN, USA, May 2005.
2. Lee, K. Elastic-Plastic Fracture Toughness Determination under Some Difficult Conditions. Ph.D. Thesis, University of Tennessee, Knoxville, TN, USA, August 1995.
3. Broek, D. *Elementary Engineering Fracture Mechanics*, 4th ed.; Kluwer Academic Publishers: Dordrecht, The Netherlands, 1986.
4. Zhang, X.; Gao, H. *A Study of Impact Toughness of Intercritically Reheated Coarsed-Grain of Heat Effected Zone of Two Type X80 Grade Pipeline Steel*; Spesial Issue on WSE2011; Transaction of JWRI: Osaka, Japan, 2011.
5. Purnomo, P.; Soenoko, R.; Suprapto, A.; Irawan, Y.S. Impact Fracture Toughness Evaluation by Essential Work of Fracture Method in High Density Polyethylene Filled with Zeolite. *FME Trans.* **2016**, *44*, 180–186. [CrossRef]
6. Kim, H.; Kang, M.; Jung, H.J.; Kim, H.S.; Bae, C.M.; Lee, S. Mechanisms of toughness improvement in Charpy impact and fracture toughness tests of non-heat-treating cold-drawn steelbar. *Mater. Sci. Eng.* **2013**, *571*, 38–48. [CrossRef]
7. Pamnania, R.; Jayakumarb, T.; Vasudevanb, M.; Sakthivelb, T. Investigations on the impact toughness of HSLA steel arc welded joints. *J. Manuf. Process.* **2016**, *21*, 75–86. [CrossRef]
8. Landes, J.D.; Zhou, Z.; Lee, K.; Herrera, R. Normalization Method for Developing J-R Curves with the LMN function. *J. Test. Eval.* **1991**, *19*, 305–311.
9. Lee, K.; Landes, J.D. Developing J-R Curves without Displacement Measurement Using Normalization. In *Fracture Mechanics: Twenty-Third Symposium*; Chona, R., Ed.; ASTM International: Philadelphia, PA, USA, 1993; pp. 133–167.
10. Chaouadi, R.; Puzzolante, J.L. Crack resistance determination from the charpy impact test. In Proceedings of the 16th European Conference of Fracture, Alexandroupolis, Greece, 3–7 July 2006; pp. 515–516.
11. Terán, G.; Colindres, S.C.; Herrera, D.A.; Velázquez, J.C.; Cueto, F.M. Estimation of fracture toughness KIC from Charpy impact test data in T-welded connections repaired by grinding and wet welding. *Eng. Fract. Mech.* **2016**, *153*, 351–359. [CrossRef]
12. Kapp, J.A.; Underwood, J.H. *Correlation between Fracture Toughness, Charpy V-Notch Impact Energy, and Yield Strength for ASTM A723 Steel*; Technical Report, ARCCB-MR-92008; U.S. Army Ardec Benet Laboratories: Watervliet, NY, USA, 1992.
13. ASTM E23-02. *Standard Test Methods for Notched Bar Impact Testing of Metallic Materials*; ASTM Committee E28: West Conshohocken, PA, USA, May 2005.
14. McGuire, M.F. *Stainless Steels for Design Engineers*, 1st ed.; ASM International: Geauga County, OH, USA, 2008.
15. Janssen, M.; van Leeuwen, M.B.; de Leon Mendes, M.F. Fracture Toughness of High-Chromium White Cast Iron in Relation to the Primary Carbide Morphology. In Proceedings of the 13th European Conference on Fracture, San Sebastian, Spain, 6–9 September 2000.
16. Ding, Y. Effects of Elevated Temperature Exposure on the Microstructural Evolution of Ni(Cr)-Cr3C2 Coated 304 Stainless Steel. Ph.D. Thesis, University of Nottingham, Nottingham, UK, June 2009.
17. Young, S.Y. Thermodynamic Study on B and Fe Substituted Cr23C6 Using First-Principles Calculations. Master's Thesis, Pohang University of Science and Technology, Pohang, Korea, 15 April 2010.
18. Fujita, N. Modelling Carbide Precipitation in Alloy Steels. Ph.D. Thesis, University of Cambridge, Cambridge, UK, February 2000.
19. Bhadeshia, H.K. Design of Ferritic Creep-resistant Steels. *ISIJ Int.* **2001**, *41*, 626–640. [CrossRef]

20. Gharehbaghi, A. Precipitation Study in a High Temperature Austenitic Stainless Steel Using Low Voltage Energy Dispersive X-ray Spectroscopy. Master's Thesis, Royal Institute of Technology, Stockholm, Sweden, March 2012.
21. Joshua, P. *Mechanical Properties of Materials*; Springer: New York, NY, USA, 2013; pp. 236–239, ISBN 978-94-007-4341-0.
22. Dieter, G.E. *Mechanical Metallurgy*; McGraw-Hill Publishing Company: New York, NY, USA, 1986.
23. Callister, W.D. *Fundamentals of Materials Science and Engineering*, 2nd ed.; Wiley & Sons: Hoboken, NJ, USA, 2004.
24. Thosmas, H.C. *Mechanical Behavior of Materials*, 2nd ed.; Waveland Press, Inc.: Long Grove, IL, USA, 2005.
25. Chen, J.; Verreman, Y.; Lanteigne, J. On fracture toughness JIC testing of martensitic Stainless Steels. In Proceedings of the13th International Conference on Fracture, Beijing, China, 16–21 June 2013.

metals

MDPI

Review

Distinct Fracture Patterns in Construction Steels for Reinforced Concrete under Quasistatic Loading— A Review

Fernando Suárez [1,†], Jaime C. Gálvez [2,*], David A. Cendón [3] and José M. Atienza [3]

[1] Departamento de Ingeniería Mecánica y Minera, Universidad de Jaén, 23071 Jaén, Spain; fsuarez@ujaen.es
[2] Departamento de Ingeniería Civil-Construcción, Universidad Politécnica de Madrid, E.T.S.I. Caminos, Canales y Puertos, 28040 Madrid, Spain
[3] Departamento de Ciencia de Materiales, Universidad Politécnica de Madrid, E.T.S.I. Caminos, Canales y Puertos, 28040 Madrid, Spain; david.cendon.franco@upm.es (D.A.C.); josemiguel.atienza@upm.es (J.M.A.)
* Correspondence: jaime.galvez@upm.es; Tel.: +034-913-365-350
† Current address: Departamento de Ingeniería Mecánica y Minera, Universidad de Jaén, Campus Científico-Tecnológico de Linares, Cinturón Sur, 23700 Linares (Jaén), Spain.

Received: 6 February 2018; Accepted: 4 March 2018; Published: 9 March 2018

Abstract: Steel is one of the most widely used materials in construction. Nucleation growth and coalescence theory is usually employed to explain the fracture process in ductile materials, such as many metals. The typical cup–cone fracture pattern has been extensively studied in the past, giving rise to numerical models able to reproduce this pattern. Nevertheless, some steels, such as the eutectoid steel used for manufacturing prestressing wires, does not show this specific shape but a flat surface with a dark region in the centre of the fracture area. Recent studies have deepened the knowledge on these distinct fracture patterns, shedding light on some aspects that help to understand how damage begins and propagates in each case. The numerical modelling of both fracture patterns have also been discussed and reproduced with different approaches. This work reviews the main recent advances in the knowledge on this subject, particularly focusing on the experimental work carried out by the authors.

Keywords: steel; fracture mechanics; tensile test; cohesive zone model; internal damage; XRCT

---

1. Introduction

Steel is, with concrete, the most extended material in construction and civil engineering works. Its strength and ductility make it of special interest when addressing structural safety issues, since it enables stress distribution with adjacent elements, allowing a higher amount of energy to dissipate before failure. However, some aspects related with its failure behaviour remain unclear, especially in those steels that do not present the classical cup–cone fracture surface, which has been extensively studied in the past.

The mechanical characterization of these types of materials is usually reduced to obtaining their elastic parameters, elastic modulus E and Poisson's ratio ν. These values are generally obtained by means of a tensile test, which is standardised by EN ISO 6892 [1], and allows obtaining with precision the stress–strain diagram up to the maximum loading point. Nevertheless, difficulties arise when the behaviour after the maximum load point needs to be defined, which usually leads to neglecting that information from the test. This final part of the stress–strain diagram is, however, of great interest since it is directly related to the maximum energy that can be absorbed by a structural element before collapsing, which goes together with the structural safety. This may help, for instance, to distinguish between accidental damage and induced damage.

Regarding the fracture mechanisms that can be found in distinct construction steels, the cup–cone fracture pattern, typical of very ductile steels and shown in Figure 1b, is very well known and has been extensively studied in the past [2–9]. The mechanisms involved in it are clearly identified: the central zone corresponds to a process of nucleation, growth and coalescence of microvoids while the surrounding inclined lips, usually referred to as shear lips [2,10], develop due to a combination of normal and shear separation [5]. Nevertheless, not all steels show the same fracture behaviour, for instance, the pearlitic steel used for manufacturing prestressing wires presents a flat fracture surface, perpendicular to the loading direction and with a circular dark area in the center (see Figure 1a). For the sake of clarity, in the rest of the text, the flat pattern will be referred to as type 1 and the cup–cone pattern as type 2.

Other authors have studied the steel used in prestressing steel wires, which shows the flat fracture pattern called type 1, focusing their interest on the effect of cold-drawing on the developed fracture mechanism [11–13].

(a) (b)

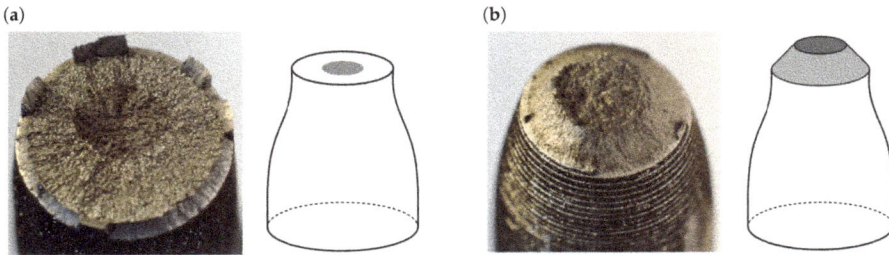

Figure 1. Fracture surfaces on 9 mm-diameter specimens of two steels with different fracture patterns after testing under tension: (**a**) fracture pattern type 1; (**b**) fracture pattern type 2. Reproduced from [14], with permission from Elsevier, 2016.

Despite the characteristics of steels being studied for a long time, there are still some issues that remain unclear, especially regarding the plastic behaviour after maximum loading and, more precisely, the mechanisms that unchain final failure in steels exhibiting different fracture patterns. The objective of this work is to review recent experimental and numerical advances dealing with fracture mechanics in steels, with a special emphasis on the type 1 pattern, flat fracture surface, since the type 2, cup–cone pattern, is better known and has been widely studied in the past. Firstly, some relevant experimental results that analyse both fracture patterns are presented and, secondly, the main numerical models used with metals are briefly overviewed, also describing a lately approach based on Linear Elastic Fracture Mechanics (LEFM) considerations.

2. Experimental Results on Steel Specimens under Tensile Loading

The following results correspond to two steels that are representative of both mentioned fracture patterns. Table 1 shows their chemical composition, with Material 1 being an eutectoid steel used for manufacturing prestressing wires before cold-drawing, which exhibits fracture pattern type 1 and Material 2 a low-carbon steel (<0.24% C), used as reinforcement in concrete structures, which shows the fracture pattern denoted as type 2.

Table 1. Chemical composition of both materials in %.

Mat.	C	Si	Mn	P	S	Cr	Mo	Ni	Cu	Al	Ti	Nb	V	N
1	0.83	0.25	0.72	0.012	0.004	0.24	<0.01	0.02	0.01	<0.003	<0.005	<0.005	<0.01	0.0097
2	0.22	0.18	1.00	0.024	0.042	0.08	0.03	0.14	0.46	<0.003	<0.005	<0.0r 05	<0.01	0.0113

2.1. Stress–Strain Diagrams

When a steel bar is tested under tension, there are several issues that may affect the results, such as the specimen length, the specimen radius, the initial gage length used or even the technique employed for measuring strain. For a complete description of the following results, the reader is referred to [14,15].

2.1.1. Strain Measuring Technique

As already mentioned, the last part of the stress–strain diagram is usually discarded because when the specimen is loaded beyond its maximum bearing capacity, strain is localised in a necking region, which makes the analysis difficult. In fact, this is true if the classical strain measuring devices are used, such as conventional extensometers as that shown in Figure 2a, which only provide information of elongation between two especific points in the specimen and their results are only valid up to the maximum load point, since necking may take place out of the gauge length or, at least, not centered with the reference points used to measure strain. Nevertheless, in the last decades, digital image correlation systems (DIC) have become widely extended, which overcome this limitation [16–21]. These systems use one or two high-definition video cameras (depending on whether results are needed in 2D or 3D), which record a marked specimen throughout the test; Figure 2b shows the experimental setup for a 2D case. The obtained pictures are post-processed with a specific software that can keep track of the speckles marked on the specimen surface, providing relative displacements between any desired pair of points or even providing a strain mapping of the specimen surface. This has a major advantage for analysing the last part of the stress–strain diagram in a tensile test: the deformation can always be tracked between points that are equally spaced from the eventual fracture plane.

(a)

(b)

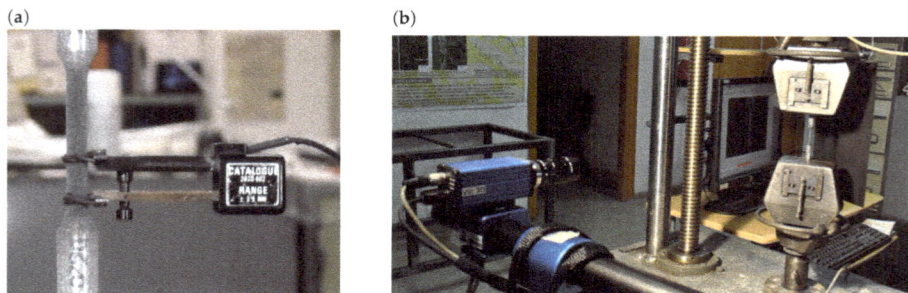

Figure 2. (**a**) a conventional extensometer on a cylindrical specimen; (**b**) setup for measuring strains with a digital image correlation system [15].

Figure 3 shows the fracture diagram obtained for a tensile test where strain was monitored with both systems; DIC proves its capability for providing data up to the eventual fracture instant, while the conventional extensometer only provides valid information up to the maximum load point. Therefore, if DIC systems are used for monitoring strain, there is in principle no reason why the information obtained after peak should be neglected.

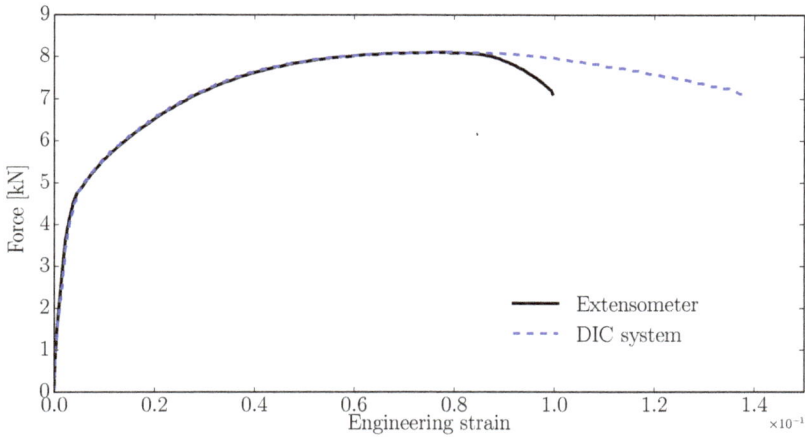

Figure 3. Force-engineering strain curves of a tensile test simultaneaously obtained with a conventional extensometer and a digital image correlation (DIC) system [15].

2.1.2. Influence of the Specimen Radius and the Initial Gauge Length

When testing a cylindrical steel bar under tension, several parameters must be defined, such as the initial gauge length and the diameter and length of the specimen. Figures 4 and 5 show several engineering stress–strain diagrams for Materials 1 and 2, respectively; all tests were monitored with DIC, so the strain was measured with a gauge length centered with the fracture cross-section and obtained up to the eventual failure of the specimen. Three specimen diameters were considered, 3 mm, 6 mm and 9 mm, and three gauge lengths compared, one of them of a fixed length equal to 12.5 mm and two proportional to the specimen diameter, once and twice the specimen diameter.

In both materials, when proportional-to-the-diameter gauge lenths are considered, the results obtained are always similar, no matter the specimen diameter, but when a fixed gauge length is employed, clear differences are found. When necking develops, the strain field is not uniform anymore along the specimen and concentrates on very localised region of the bar; therefore, strains measured with a fixed gauge length on specimens of different diameters provide a certain average strain that cannot be compared among them. See Section 2.2 for further details.

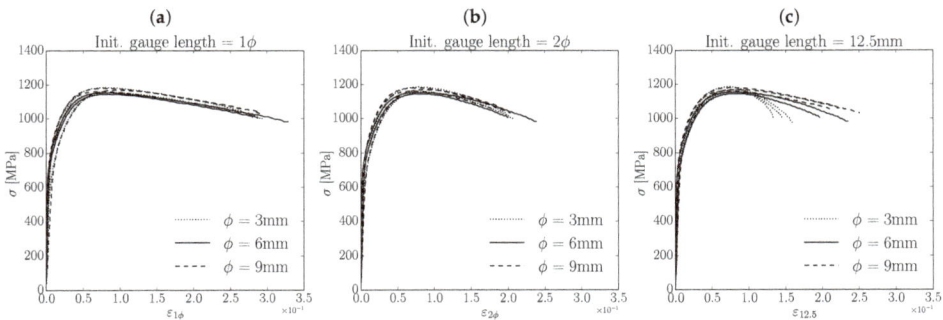

Figure 4. σ-ε curves obtained with Material 1 specimens of distinct radii and using several initial gauge lengths: (**a**) initial gauge length = 1ϕ; (**b**) initial gauge length = 2ϕ; (**c**) initial gauge length = 12.5 mm . Adapted from [22], with permission from Elsevier, 2016.

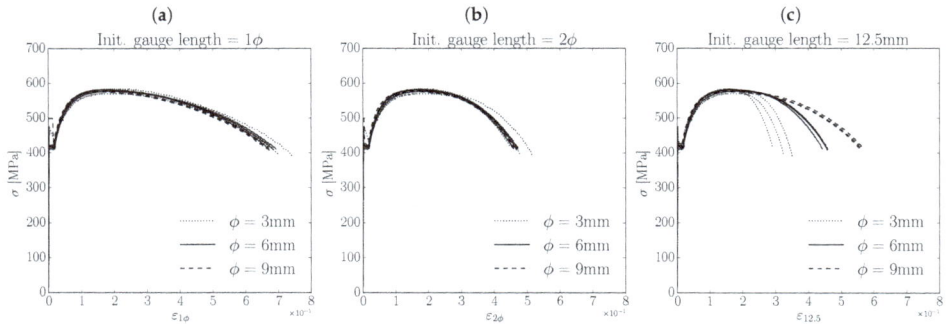

Figure 5. σ-ε curves obtained with Material 2 specimens of distinct radii and using several initial gauge lengths: (**a**) initial gauge length = 1φ; (**b**) initial gauge length = 2φ; (**c**) initial gauge length = 12.5 mm . Adapted from [22], with permission from Elsevier, 2016.

2.1.3. Influence of the Specimen Length

The length of the constant cross-section length in a specimen (L_1 in Figure 6a) is limited by the standards to avoid too short specimens that cannot develop the whole necking process. In principle, once the specimen dimensions meet the requirements defined in [1], this can be designed with any desired length. In [15], this issue was studied by testing 6 mm-diameter specimens with increasing values of L_1. In both materials, 1 and 2, the specimen length did not seem to affect the ultimate strain as can be observed in Figure 6, which shows the results for distinct specimen lengths (L_1) using a fixed initial gauge length of 12.5 mm.

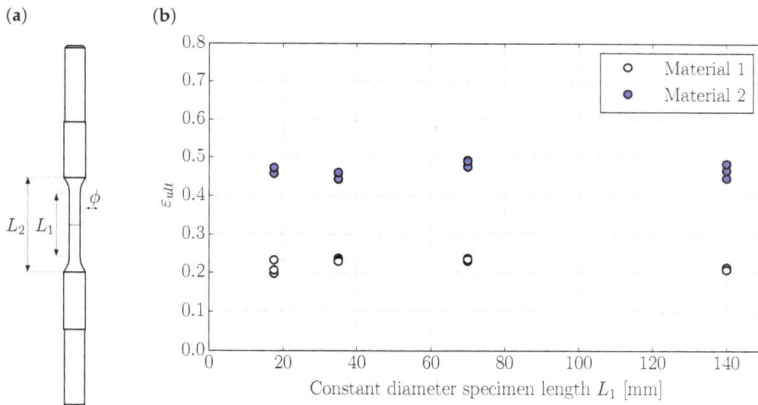

Figure 6. (**a**) specimen parameters for tensile tests; (**b**) influence of the specimen length on the ultimate strain under tensile loading for an initial gauge length of 12.5 mm [15].

2.2. Strain Fields Maps

One of the reasons that make DIC a powerful technique is that, differently from the conventional extensometers, which provide elongation values between only two points of the specimen, it can be used to obtain the strain maps at the surface of the specimen. Figures 7 and 8 show the longitudinal strain evolution on specimens of Materials 1 and 2, respectively; in both cases, sub-figure (a) represents the strain map at the maximum load instant and (d) at the very last instant before failure, with (b,c) being intermediate images between (a,d).

The strain maps help to understand the differences observed when a fixed gauge length is used with distinct diameter specimens. In Figures 4c and 5c, the curves for all diameters were coincident up to the maximum load point and bifurcated later, showing smaller values of strain for smaller diameters. The reason is that, as Figures 7 and 8 show, the strain gradient becomes greater as the tensile test progresses; therefore, if a proportional-to-the-diameter gauge length is used, the strain measured is comparable among distinct diameter specimens, proving that the necking phenomenon is proportional to the diameter, which explains why the curves in these cases are coincident beyond the maximum load point. Nevertheless, when a fixed gauge length is used, the gauge provides an average strain value in the part of the specimen that is between the selected points; therefore, in the case of smaller diameter specimens, the gauge length includes parts of the specimen relatively distant from the necking region, which results in smaller values of ε, while in the case of larger diameter specimens the gauge length includes regions that present comparatively higher strain gradients and, therefore, the average values of ε are larger. The strain gradient is higher in the case of Material 2 compared with the strain maps of Material 1, which explains why the bifurcation is more evident in Figure 5 than in Figure 4.

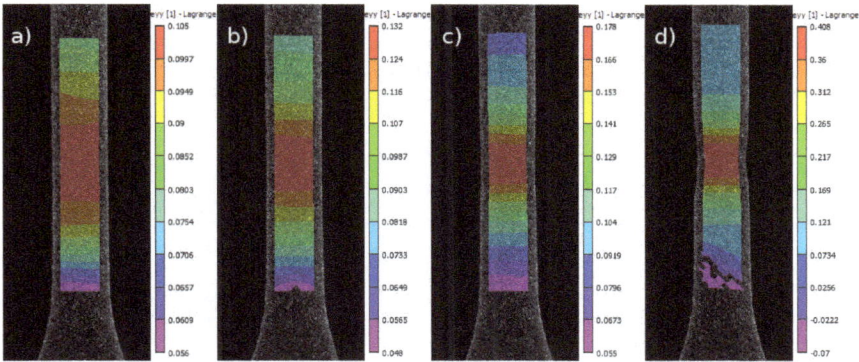

Figure 7. Evolution of the vertical strain field from the maximum load point (**a**) to the instant just before failure (**d**) for a 3 mm-diameter specimen of Material 1; (**b**,**c**) are intermediate images between (**a**,**d**). Reproduced from [22], with permission from Elsevier, 2016.

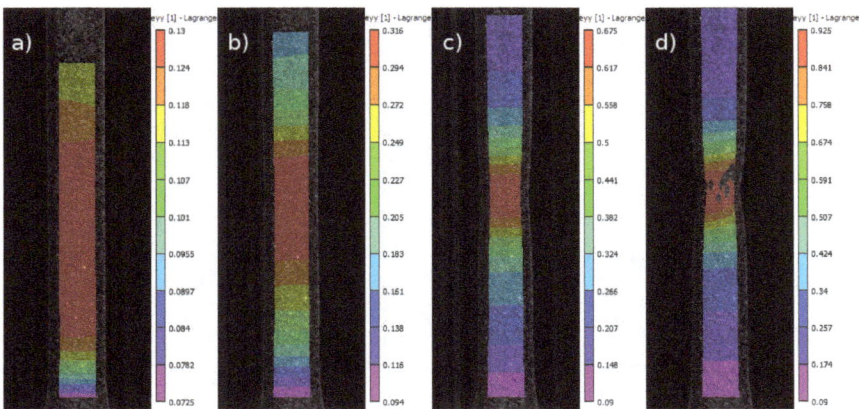

Figure 8. Evolution of the vertical strain field from the maximum load point (**a**) to the instant just before failure (**d**) for a 3 mm-diameter specimen of Material 2; (**b**,**c**) are intermediate images between (**a**,**d**). Reproduced from [22], with permission from Elsevier, 2016.

2.3. Analysis of the Fracture Surface

In 1966, Bluhm and Morrisey carried out an ambitious and pioneering study of the damage evolution on cylindrical bars made of three ductile metals under tensile loading [2]. To do it, they used a device (see Figure 9a) that could apply loading in a progressive manner, allowing a gradual crack propagation, and ultrasonic and metallographic techniques for identifying internal damage. In this study, they already identified several phenomena that helped to understand better the load-elongation curve of a tensile test with these types of materials (see Figure 9b):

- After necking localises, small microvoids develop, interfacial non-connected cracks at inclusions appear, as well as short intercrystalline cracks.
- Later, as strain increases, the volume of voids and cracks also increase, weakening the material matrix in the center of the necking region. Around this weakened zone, an esentially non-fractured region remains under a low hydrostatic stress state, thus being under high shear stresses that eventually lead to cracking out of the initial fracture plane, which are the so-called shear lips.

This explains the typical cup–cone fracture pattern, which would be deeply studied later by several researchers.

In this section, the fracture surface on both analysed patterns, type 1 (flat fracture pattern) and type 2 (cup–cone pattern), are studied by means of a scanning electronic microscope (SEM) and by analysing the geometry of both fracture surfaces for specimens of distinct diameters.

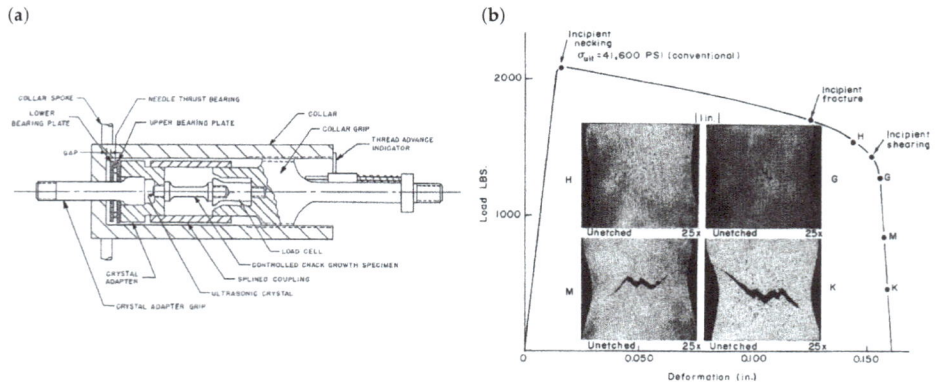

Figure 9. (**a**) Device used by Bluhm and Morrisey to study fracture on cylindrical bars made of different metals; (**b**) evolution of the damage process in the necking region of a copper cylindrical specimen under tension. Reproduced from [2], with permission from the Army Materials Research Agency, 1966.

2.3.1. Fractographs

Figure 10 shows a 9 mm-diameter specimen of Material 1 observed with a SEM; the dashed circumference represents the dark area that can be observed on the fracture surface with a mere visual inspection (see Figure 1a). The central region, which corresponds to the characteristic dark area of this pattern, presents dimples, usually related with a nucleation-growth-coalescence mechanism, while the surrounding region shows the so-called river marking related with a cleavage mechanism. Looking at the general view of the fracture surface, it is also interesting to observe radial marks that are related to the crack opening that initiates at the center of the cross-section and propagates from inside to outside.

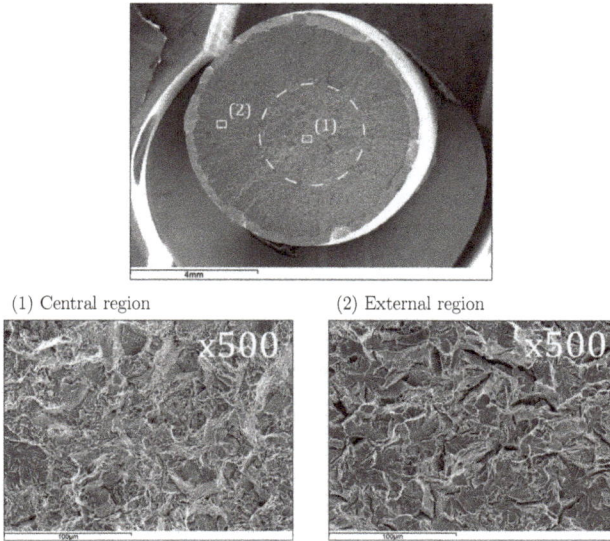

Figure 10. SEM analysis on the fracture surface of a 9 mm-diameter specimen of Material 1 [15].

In Figure 11, the analogous SEM analysis of a 9 mm-diameter specimen of Material 2 can be observed; the dashed circumference identifies the circular flat plane of the cup–cone pattern. These images show that the central region is the result of a nucleation-growth-coalescence process while the surface observed in the shear lips show elongated dimples in the radial direction, typical of this shape, which are the result of a combination of normal and shear crack separation.

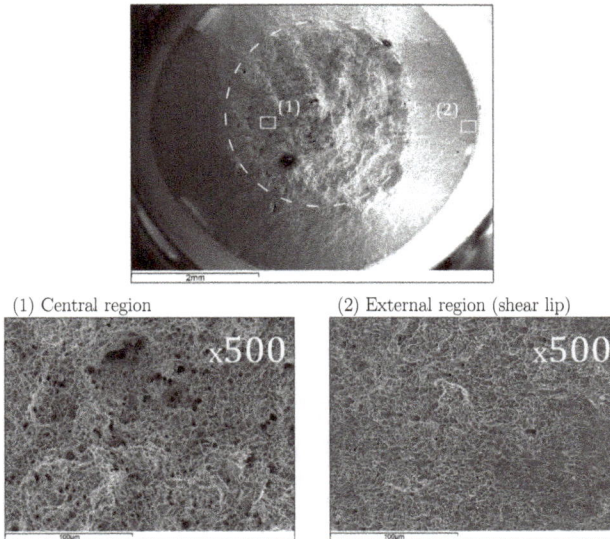

Figure 11. SEM analysis on the fracture surface of a 9 mm-diameter specimen of Material 2 [15].

2.3.2. Size of the Internal Damage

The dark region observed in the fracture pattern of type 1 can be seen as an internal notch induced by a nucleation-growth-coalescence process developed during the loading process. Following this idea, the material behaviour would be ductile, although the final failure mechanism would be triggered by a quasi-brittle behaviour, and such quasi-brittle behaviour is explained by the presence of the dark region, which acts as an internal crack that, once large enough, concentrates stresses around a very specific region of the damage cross-section and leads to a brittle fracture process at the very end of the test.

The area of the dark region observed in fracture surfaces of Material 1 does not seem to be proportional to the specimen cross-section at first sight. If the dark region diameter ($2r$) and the minimum diameter of the specimen at necking ($2R_{min}$) are measured, the evolution of r/R_{min} is observed to decrease with the specimen size, as shown in Figure 12a. When the same analysis was carried out on specimens of Material 2, considering $2r$ as the diameter of the flat circular region of the cup–cone surface, the proportion r/R_{min} remained constant for any specimen size (see Figure 12b).

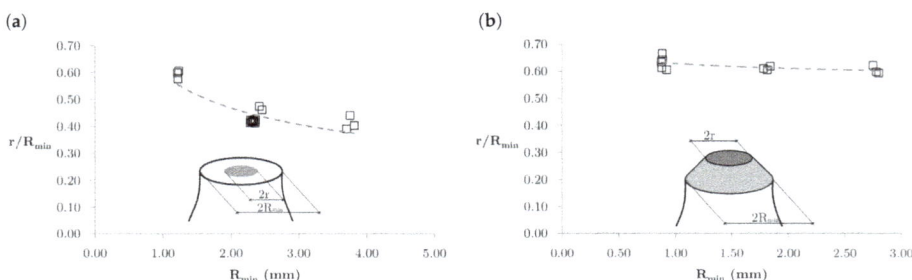

Figure 12. r/R_{min} relations experimentally obtained on specimens with diameters of 3, 6 and 9 mm for (**a**) Material 1; (**b**) Material 2. Adapted from [22], with permission from Elsevier, 2016.

Under the aforementioned approach, the final fracture process can be approached by using expressions based on LEFM for the case of Material 1, which are valid for brittle materials. Thus, considering that the fracture process can be studied as the fracture of a cylindrical specimen with an internal circular crack, the formula provided by Guinea, Rojo and Elices [23,24] can be used:

$$\frac{K_I}{K_0} = 1 + \sum_{i=1}^{5} C_{i0}\left(\frac{r}{R}\right)^{\frac{2i+1}{2}} + \sum_{i=1}^{3}\left\{\ln\left[1+\left(\frac{r}{R}\right)^{2i}\right]\cdot\left[C_{i1}\ln^2\left(\frac{b}{R}\frac{r}{R}\right)+\frac{C_{i2}}{\sqrt{\frac{b}{R}\frac{r}{R}}}\right]\right\}, \tag{1}$$

where the reference stress intensity factor K_0 is:

$$K_0 = \frac{2}{\pi}\cdot\sigma\sqrt{\pi r},$$

where (see Figure 13a):

- K_I is the stress intensity factor.
- b stands for the smaller distance between the crack boundary and the specimen boundary.
- r is the internal crack radius.
- R is the specimen radius.
- σ is the tensile stress away from the fracture zone.
- C_i are the coefficients defined in Table 2.

Table 2. C_i coefficients used in the Guinea–Rojo–Elices expression for the computation of K_{Ic} in cylindrical fibers under tension with eccentric internal crack perpendicular to the specimen axis. Reproduced from [23], with permission from Elsevier, 2004.

	C_{i0}	C_{i1}	C_{i2}
$i = 1$	0.01242	−0.3097	1.185
$i = 2$	−6.388	1.547	−3.723
$i = 3$	16.89	−0.8769	2.628
$i = 4$	−9.838	-	-
$i = 5$	−1.228	-	-

The value of the fracture toughness K_{Ic} can be obtained by means of a three-point bending test, described by the ASTM. E 399-90 [25]. Therefore, once K_{Ic} is known, for each value of R_{min}, the value of r that satisfies (1) can be easily obtained. Figure 13 compares the values of proportion r/R_{min} obtained experimentally and those obtained with (1), thus based on the LEFM. These results show that the experimental values decrease with the specimen size, which agrees with the tendency predicted by the LEFM, which supports the hypothesis of an eventual fracture driven by brittle mechanisms. In contrast to these results, the almost constant r/R_{min} ratio exhibited by Material 2 specimens suggest a final failure not ruled by LEFM considerations.

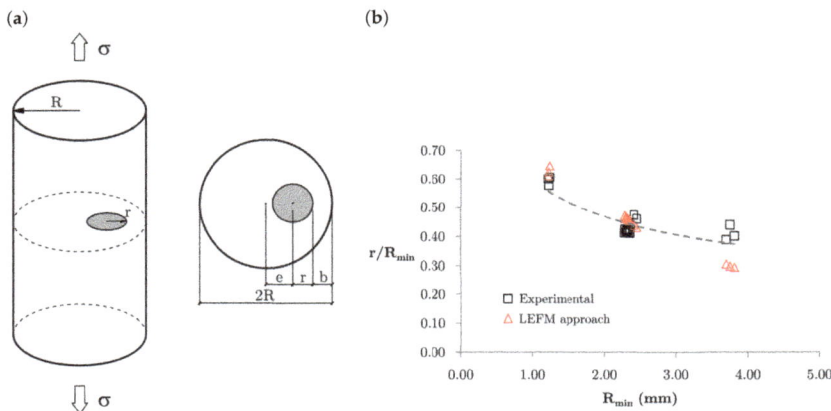

Figure 13. (**a**) parameters used in the expression of Guinea, Rojo and Elices ; (**b**) r/R_{min} values obtained for Material 1 with the expression based on the Linear Elastic Fracture Mechanics (LEFM) compared with the experimental results . Reproduced from [22,23], with permission from Elsevier, 2004 and 2016.

2.4. Evolution of Damage

All the works mentioned before seem to confirm that, in the case of fracture pattern 1, damage initiates inside the specimen producing something similar to an internal defect that eventually leads to a brittle fracture process, but there is no evidence of when this damage begins to develop. In this section, two approaches are described that have helped to identify that the damage develops in the very late stages of the tensile test.

2.4.1. Tests on Embrittled Specimens by Means of Liquid Nitrogen

The stress–strain diagram of Material 1 is very repetitive, so a tensile test can be carried out to a certain desired strain level with notable precision. In [14], four 6 mm specimens of this material were tested up to increasing strain levels that ranged from the maximum load instant up to fracture, with the highest of them being very close to failure. Each specimen was then unloaded and tested again until failure at extremely low temperatures in order to induce an embrittled fracture that would allow identification of the previously developed internal damage. Figure 14a shows the experimental setup; the specimens were submerged into liquid nitrogen and the test was carried out with a rapid displacement of the clamping jaws once the temperature, measured with a thermo-couple, was lower than -100 °C.

This procedure proved to be efficient and allowed for identifying internal damage only in the specimen tested at the more advanced strain level. In addition, this specimen was also analysed by X-ray computed tomography (XRCT) before the embrittle test. The tomographic image obtained also identified the internal damage that could be observed after carrying out the embrittle test. Figure 14b shows both results compared, which allows considering this region as the beginning of the dark region that can be usually observed in this fracture pattern, which can be considered as an internal decohesion process.

(a)   (b)

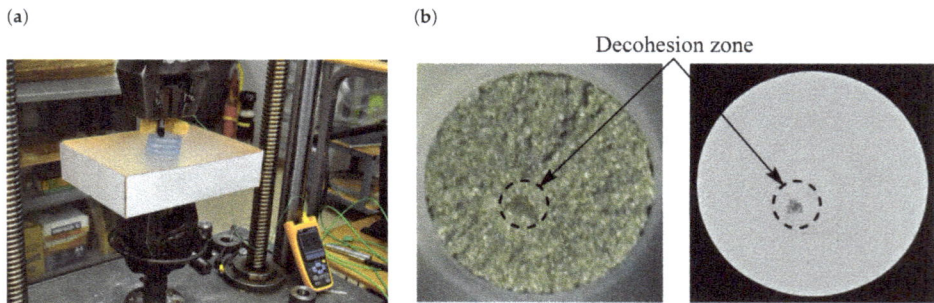

Figure 14. (a) experimental setup for the test on specimens embrittled with liquid nitrogen; (b) comparison between the visual inspection after failure and the internal damage observed with X-ray computed tomography before failure [15].

2.4.2. Tests Analysed with X-ray Computed Tomography

As seen before, XRCT allows identifying internal damage in the specimens, so a more detailed study of the damage evolution was done using this technique on both materials. For each of them, a 3 mm-diameter sample was tested up to a certain strain level, then unloaded and analysed with XRCT. Later on, the same sample was retested up to a higher strain level, unloaded and analysed with XRCT again. This process was repeated increasing the strain level until the sample was broken. Figures 15 and 16 show the tomographic images obtained for both materials at each of the analysed strain levels.

In the case of Material 1, the matrix looks compact and homogeneous in the intact material and, as strain increases after maximum loading, the formation of small voids can be observed. Interestingly, some of these voids are formed following a longitudinal direction, which could be due to the manufacturing process. In this case, voids only coalesce at the very last stage of the test, corresponding to step 4 in the figure, in agreement with the behaviour observed in the 6 mm samples analysed with liquid nitrogen. Here, again, the decohesion process is identified, which will result in the dark region observed on the fracture surface of this material.

In the case of Material 2, the initial tomography, obtained with the intact specimen before testing, already shows a very heterogeneous matrix, with inclusions that are also lined up in the longitudinal direction, probably as a result of the manufacturing process. As strain increases, damage seems to develop all around the necking region.

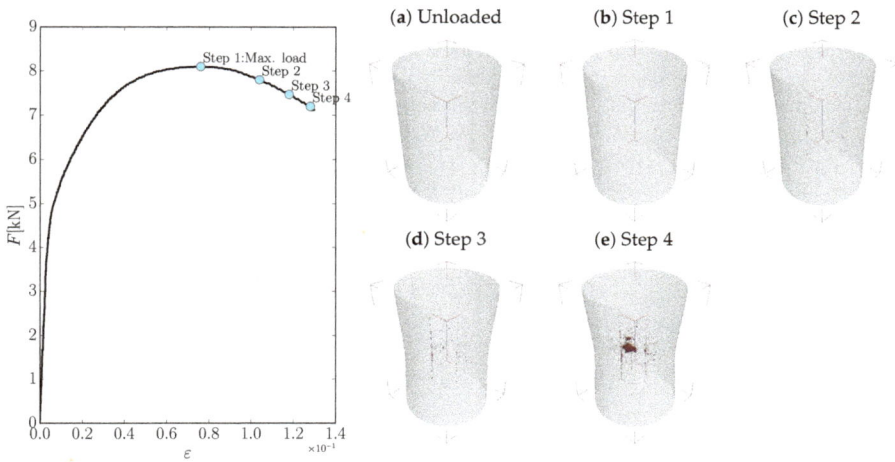

Figure 15. Evolution of internal damage on a 3-mm diameter specimen of Material 1. Images obtained by means of X-ray computed tomography [15].

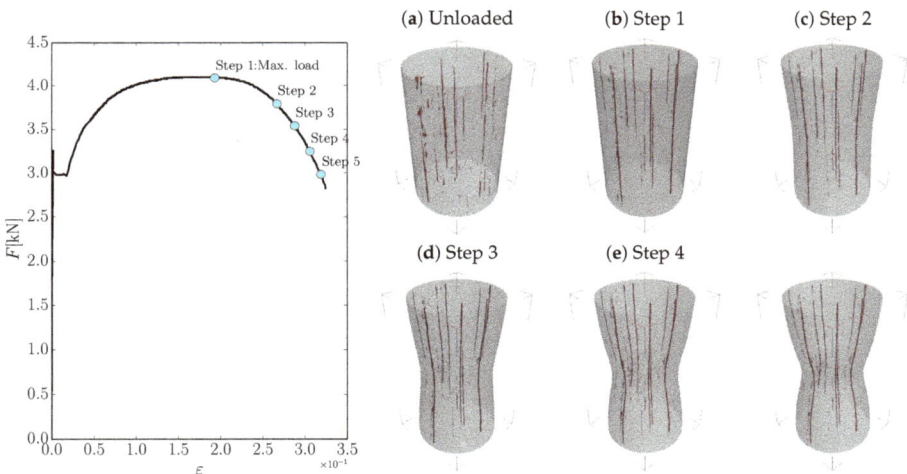

Figure 16. Evolution of internal damage on a 3-mm diameter specimen of Material 2. Images obtained by means of X-ray computed tomography [15].

2.5. Influence of Stress Triaxiality on Ductile Fracture

Stress triaxiality is defined as the ratio $\sigma_H / \bar{\sigma}$, with σ_H being the hydrostatic pressure and $\bar{\sigma}$ the von Mises equivalent stress. This value provides information of how balanced or unbalanced the principal stresses σ_1, σ_2 and σ_3 are.

Mirza et al. [26] studied the influence of the stress triaxiality using copper specimens; they performed tensile tests stopped before failure in order to, once unloaded, polish them to observe the matrix inside the specimen and identify internal damage. They concluded that the stress triaxiality plays a paramount role in how the nucleation-growth-coalescence mechanism develops. In this sense, Toribio et al. [27–29] also studied the influence of triaxiality using notched specimens to induce distinct triaxiality states in the fracture process.

The main material models developed until the mid-nineties included the stress triaxiality in their formulation in an implicit way, for example, by means of parameters p and q in the Gurson model [30] and by means of the Bridgman equations [31] in the Johnson–Cook [32] model.

The works by Bao and Wierzbicki [33–35] help to understand how this issue affects fracture on metallic specimens. They conducted a large experimental programme with alluminum specimens covering a wide range of triaxiality states, from compression to multiaxial tension. Figure 17 presents one of their most relevant results, a diagram that shows how the ultimate strain is dependent on the stress triaxiality. One of their main conclusions is that the influence of the stress triaxiality cannot be modelled by a monotonic function, as other authors had done in the past [32,36], but three distinct regions must be considered; in each region the governing fracture mechanisms are different:

- Zone I: low triaxialities, where fracture is mainly due to shearing.
- Zone II: medium triaxialities, where fracture is the result of a combination of shearing and the nucleation-growth-coalescence mechanism.
- Zone III: high triaxialities, where the nucleation-growth-coalescence mechanism drives fracture.

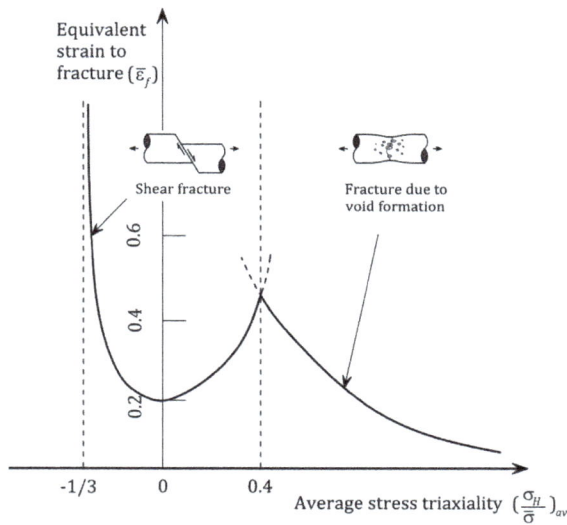

Figure 17. Influence of stress triaxiality on the equivalent strain to fracture. Adapted from [34], with permission from Elsevier, 2004.

The Lode Angle

In the last years, several models and approaches have been based on the so-called Lode angle [37]. If working with the principal stresses coordinate system (σ_1, σ_2, σ_3), any stress state can be represented by means of the Haigh–Westergaard coordinates ξ, ρ and θ:

$$\xi = \frac{I_1}{\sqrt{3}} = \sqrt{3}p,$$

$$\rho = \sqrt{2J_2},$$

$$\theta = \frac{1}{3}\cos^{-1}\left(\frac{3\sqrt{3}}{2}\frac{J_3}{J_2^{3/2}}\right),$$

where I_1 is the first invariant of the Cauchy stress tensor, J_2 and J_3 stand for the invariants of the stress deviator tensor and p for the hydrostatic stress. Parameter θ is usually referred to as the Lode angle.

In [38], a new approach to ductile fracture was proposed by using the Lode parameter, which in combination with the stress triaxiality is considered as fundamental to identify the fracture direction in ductile materials. In that work, Zhang et al. confirmed that the stress triaxiality is not enough to define the fracture behaviour of a ductile material and another value, in this case the Lode parameter, must be used. Since then, many authors have used the Lode parameter in ductile fracture models. For instance, Wue and Wierzbicki [39] proposed a model where the damage criterion was based on the stress triaxiality and a parameter related to the Lode angle, Bai and Wierzbicki [40] proposed later a ductile fracture model based on the stress triaxiality and the Lode parameter, and Erice and Gálvez [41,42] developed a coupled elastoplastic damage model with a failure criterion dependent on the Lode angle. The importance of the Lode angle in ductile fracture processes was confirmed by Mirone and Corallo [43], who studied its influence on fracture of different metals, and Barsoum and Faleskog [6,7,44], who based their work on the mechanisms of growth and coalescence of microvoids, concluded that the Lode parameter has a high influence on the nucleation and growth of voids but not much if the triaxiality is high, when it is the stress triaxiality, which drives the fracture process.

3. Numerical Models

3.1. Models Usually Employed with Metals

There are a number of material models that are currently used for simulating fracture in metals. The election of the model depends on the user's needs, for example, if the parameters that want to be considered in the fracture process involve strain rate, temperature or pressure, the Johnson–Cook model that will be briefly described later would be a good choice.

If we consider the coupling of the elastoplastic behaviour and the fracture behaviour, we can divide these models in coupled and uncoupled. In the following lines, a brief overview of the main models commonly used for reproducing fracture in metals is provided. Some of these formulations are the result of an analytical treatment of the problem and others are based on experimental adjustments of the problem. In any case, the reader will find that all of them include triaxiality as a key factor in the ductile fracture process and that some of them also include the Lode parameter.

After this overview, a model developed by the authors is described. This model is based on the cohesive crack concept and its formulation is affected by triaxiality, showing to reproduce reasonably well the fracture behaviour of Material 1.

3.1.1. Uncoupled Models

In these models, damage process does not affect the elastoplastic constitutive equations. A damage parameter must be defined, usually dependent on the plastic strain accumulation.

These models are usually phenomenological and, since the elastoplastic and failure criteria can be observed independently, are also usually easier to calibrate. Stress is not affected by a progressive deterioration process, which lead to rather an abrupt failure behaviour. Some fracture patterns, such as the cup–cone shape, cannot usually be modelled by these types of formulations.

Johnson–Cook Model

The Johnson–Cook model [32,45] is very extended when strain rate and thermal softening or pressure are important factors to be considered in the fracture process, i.e., ballistic applications [46–50] and blast loadings [51–54]. This model defines a damage parameter D expressed by Equation (2), where $\Delta\bar{\varepsilon}_p$ is the equivalent plastic strain rate and $\bar{\varepsilon}_p^R$ the equivalent plastic strain to failure, which is dependent on the strain rate, temperature and pressure, as Equation (3) shows. In this expression, D_1, D_2, D_3, D_4 and D_5 are material constants that must be calibrated, σ^* stands for the stress triaxiality, $\dot{\bar{\varepsilon}}_p^*$

for a dimensionless plastic strain rate ($\dot{\bar{\varepsilon}}_p^* = \dot{\bar{\varepsilon}}_p / \dot{\varepsilon}_0$, with $\dot{\varepsilon}_0$ being a reference plastic strain rate) and T^* takes into account the effect of temperature:

$$D = \sum \frac{\Delta \bar{\varepsilon}_p}{\bar{\varepsilon}_p^R}, \tag{2}$$

$$\bar{\varepsilon}_p^R = [D_1 + D_2 \exp{(D_3 \sigma^*)}] \left[1 + D_4 \ln \dot{\bar{\varepsilon}}_p^*\right] [1 + D_5 T^*]. \tag{3}$$

Wilkins et al. Model.

Wilkins et al. material model [55] includes a scalar parameter A that plays a similar role as the Lode parameter in later models and depends on the principal deviatoric stresses (s_1, s_2, s_3) by means of Equation (4):

$$A = \max\left(\frac{s_2}{s_3}, \frac{s_2}{s_1}\right). \tag{4}$$

Then, the hardening function reads:

$$Y = Y_T(\bar{\varepsilon}_p) A^\lambda + Y_S(\bar{\varepsilon}_p)(1 - A^\lambda), \tag{5}$$

where Y_T is the equivalent strain hardening function for uniaxial tension/compression, Y_S the equivalent strain hardending function for pure shear/torsion and $\bar{\varepsilon}_p$ the equivalent plastic strain. The parameter λ is a material constant and must be experimentally adjusted.

The damage is dependent on the history of the plastic strain and its evolution depends on the plastic strain rate and two separate weighting functions (see Equation (6)), one depending on the hydrostatic pressure (p) and another depending on the parameter A:

$$\dot{D} = w_1(p) w_2(A) \dot{\bar{\varepsilon}}_p. \tag{6}$$

Bai–Wierzbicki Model

The model proposed by Bai and Wierzbicki [40] explicitly introduces triaxiality dependence in the hardening function, expressed by Equation (7), where $\bar{\sigma}(\bar{\varepsilon}_p)$ represents the equivalent strain hardening function for a reference test, σ_0^* the reference value of the stress triaxiality in the reference test, $\bar{\varepsilon}_p$ the equivalent plastic strain, σ^* the stress triaxiality, c_{σ^*}, c_θ^s and m are material constants and γ and c_θ^{ax} parameters dependent on the Lode angle θ:

$$\phi = \bar{\sigma}(\bar{\varepsilon}_p) \left[1 - c_{\sigma^*}(\sigma^* - \sigma_0^*)\right] \left[c_\theta^s + (c_\theta^{ax} - c_\theta^s)\left(\gamma - \frac{\gamma^{m+1}}{m+1}\right)\right]. \tag{7}$$

The damage accumulation is the same as in the Johnson–Cook model and thus depends on five material constants that must be calibrated.

3.1.2. Coupled Models

These models are usually based on micro-mechanical observations and the elastoplastic behaviour is affected by the degradation process that leads to eventual failure. Because of their coupled nature, their calibration is usually more tedious than in the case of uncoupled models, but are quite often selected by the scientific community due to their solid framework usually based on micromechanics.

Lemaitre's Model

In this model [56,57], strain is obtained by Equation (8), where D is a scalar variable that ranges from 0 to 1 and describes how damaged the material is, and C is the elastic fourth-order tensor and ($\varepsilon - \varepsilon_p$) the elastic strain:

$$\sigma = (1 - D)\boldsymbol{C} : (\boldsymbol{\varepsilon} - \boldsymbol{\varepsilon}_p). \tag{8}$$

The yield function is defined by Equation (9), where $\bar{\sigma}$ is von Mises equivalent stress and σ_y is the isotropic hardening rule, dependent on a hardening variable r:

$$\phi = \frac{\bar{\sigma}}{(1 - D)} - \sigma_y(r). \tag{9}$$

Damage evolution is ruled by an expression that depends on the energy release rate, defined by Equation (10):

$$Y = \frac{\bar{\sigma}^2}{2E(1 - D)^2} \left[\frac{2}{3}(1 + \nu) + 3(1 - 2\nu)(\sigma^*)^2 \right], \tag{10}$$

where E and ν are the material elastic parameters and σ^* the stress triaxiality.

Xue–Wierzbicki Model

This model [58,59] couples damage and plasticity by means of Equation (11), which resembles the expression (8), from Lemaitre's model. β is a material constant that must be calibrated using the matrix material stress–strain curve:

$$\sigma = (1 - D^\beta)\boldsymbol{C} : (\boldsymbol{\varepsilon} - \boldsymbol{\varepsilon}_p). \tag{11}$$

The yield condition is defined by Equation (12), where σ_M represents the yield strength of the matrix material and is a function of the plastic strain ε_p:

$$\phi = \bar{\sigma}^2 - \left[(1 - D^\beta)\sigma_M \right]^2. \tag{12}$$

Damage evolves according to Equation (13), where m is a material constant and $\bar{\varepsilon}_p^f$ is the fracture strain from monotonic loading, which depends on the hydrostatic pressure p and the Lode angle θ:

$$\dot{D} = m \left[\frac{\bar{\varepsilon}_p}{\bar{\varepsilon}_p^f(p, \theta)} \right]^{(m-1)} \frac{1}{\bar{\varepsilon}_p^f(p, \theta)} \dot{\bar{\varepsilon}}_p. \tag{13}$$

Modified Johnson–Cook Model

This model was proposed by Borvik et al. [60] and, based on the uncoupled Johnson–Cook model, reformulated as a coupled version. In this formulation, the strain rate is composed as the sum of three strain rates: elastic, plastic and thermally induced:

$$\dot{\varepsilon} = \dot{\varepsilon}^e + \dot{\varepsilon}^p + \dot{\varepsilon}^t.$$

The stress tensor rate is defined by Equation (14):

$$\dot{\sigma} = (1 - D)\boldsymbol{C} : \dot{\varepsilon}^e - \frac{\dot{D}}{(1 - D)}\sigma, \tag{14}$$

where $\boldsymbol{C}$ is the fourth-order elastic tensor and $\dot{D}$ accounts for the damage evolution.

As in the original Johnson–Cook model, the damage evolution is defined as:

$$\dot{D} = \frac{\dot{\bar{\varepsilon}}_p}{\bar{\varepsilon}_p^R}. \tag{15}$$

The proposed expression for the equivalent plastic strain to fracture $\bar{\varepsilon}_p^R$, Equation (16), is similar to that in the original Johnson–Cook model, although a different influence of the strain rate is considered:

$$\bar{\varepsilon}_p^R = [D_1 + D_2 \exp{(D_3 \sigma^*)}] \left[1 + \dot{\bar{\varepsilon}}_p^*\right]^{D_4} [1 + D_5 T^*]. \tag{16}$$

Gurson-Like Models

The Gurson–Tvergaard–Needleman (GTN) model is one of the most successful when fracture in metals is to be modelled. The seminal model was proposed by Gurson in 1977 [30], who based its formulation on the growth of a spherical void inside a material matrix, studied earlier by Rice and Tracey [61]. The model developed by Gurson could predict the loss of strength due to the nucleation and growth mechanisms, but not the total material loss of strength. Tvergaard and Needleman [3] modified the model including a failure criterion and were able to reproduce the typical cup–cone fracture. The yield function is defined by Equation (17), where f^* represents the voids volume fraction of the material, which introduces the failure criterion based on the void volume at the beginning of the coalescence mechanism and the void's volume when there is a total loss of strength; q_1, q_2 and q_3 are model parameters that, according to Tvergaard and Needleman, can be considered to be around 1.5, 1.0 and 2.25, respectively:

$$\phi = \left(\frac{\bar{\sigma}}{\sigma_y}\right)^2 + 2q_1 f^* \cosh\left(-q_2 \frac{3\sigma_H}{2\sigma_y}\right) - \left(1 + q_3 f^{*2}\right). \tag{17}$$

This model introduces the effect of triaxiality by means of the hydrostatic stress σ_H and the von Mises equivalent stress $\bar{\sigma}$. Since its appearance, it has been often used to study fracture in ductile materials and is still a referent model in the field. The model is able to reproduce ductile fracture in many materials [26,62,63], has been compared with other models proving to be very precise [64] and has even been used to study cases of ductile-brittle damage transition [65].

Many researchers have paid attention to this model and have adjusted its formulation to make it usable in other situations. For example, Hao and Brocks [66] proposed a variation that included the influence of temperature, Steglich et al. [67] combined the GTN and cohesive models to reproduce fracture in particle-reinforced metals, Zhang [38] modified the model formulation to reproduce coalescence in terms of the plastic limit by Thomason. In its original formulation, the GTN model is limited, since, in cases of null hydrostatic pressure, damage does not develop; to solve this problem, Nahson and Hutchinson [68] proposed a modification that took into account cases of pure shear and cases where the hydrostatic pressure was very low; nevertheless, it was later proved that this version overestimated damage in cases with high triaxialities [69]. Later versions of the model by several authors can be found [63,70–73], some even including the Lode angle in their formulation [74,75].

In spite of the wide acceptance and usage of this model, it has also been found to present some drawbacks, mainly two: the parameters that feed the model cannot be experimentally measured, since they refer to voids' volume fractions that cannot be quantified by experimental procedures, and a distinct set of calibrated parameters can be found to be valid for the same material [67].

3.2. Triaxiality-Dependent Cohesive Model

The cohesive crack concept was proposed in the seventies by Hillerborg [76], based on the work by Dugdale and Barenblatt [77], and was applied in the framework of Linear Elastic Fracture Mechanics (LEFM). Many models have appeared since then and have based the material damage on a softening function that is defined by only two parameters, the tensile strength f_t and the fracture energy G_f. The main advantages of these models are their simplicity and their easy calibration, which can be done by standardised experimental tests [78,79], ensuring an objective procedure of calibrating and modelling the fracture process. Besides this, they can be combined with any material model for the continuum, including those discussed in the previous sections. For further details on this numerical model, the reader is referred to [80,81].

This numerical approach has been used by many researchers since its appearance and has allowed implementing interface cohesive elements for quasi-brittle materials under mode I or under the combination of modes I, II and III loading [82–88], smeared crack models [89–93] and embedded crack models [94–100].

Although the cohesive zone model is in principle valid in quasi-brittle materials, such as concrete or cement-based materials, it has also been used in metals. Siegmund and Brocks [101] used a cohesive crack model to study the crack growth in elastic-plastic materials and Scheider and Brocks [5,102] used a cohesive model to simulate the cup–cone fracture and even made it dependent on triaxiality.

In [14], fracture pattern 2 was modelled by means of a cohesive zone model where the cohesive parameters f_t and G_f were dependent on the stress triaxiality. To do this, an interface element reproduced the fracture behaviour, which was fed by the stress triaxiality values of the adjacent element (see Figure 18). In this case, since the fracture of a metal is reproduced, a rectangular-shaped softening function is recommended [77]; given that, for numerical reasons, a rectangular shape induces many convergence problems, a parabolic shape that resembles a rectangular behaviour is used.

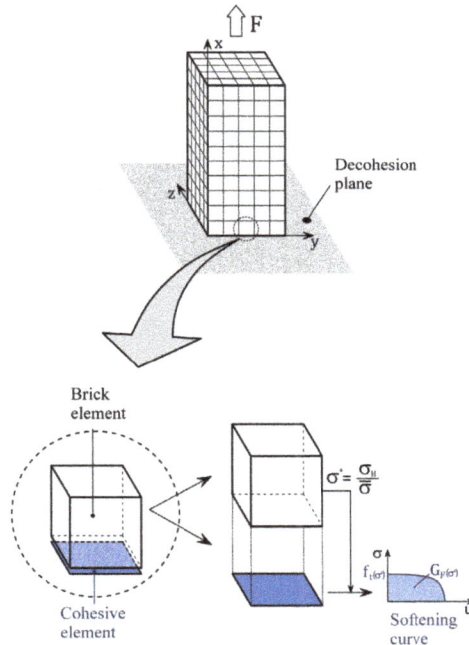

Figure 18. General graphical representation of the triaxiality-dependent cohesive zone interface element. Reproduced from [14], with permission from Elsevier, 2016.

This model is fed by f_t and G_f values that can be obtained by standardised tests [25,103] and reproduces reasonably well the decohesive process assumed in the center of the specimens that leads to the dark region that can be observed in the fracture surface. As can be observed in Figure 19, when this model is applied for modelling fracture in a cylindrical specimen of Material 1, fracture initiates in the center of the specimen, unloading that part of the eventual fracture plane and, therefore, transfering a stress increment to the surrounding region. As strain increases, this decohesive process propagates from inside to outside.

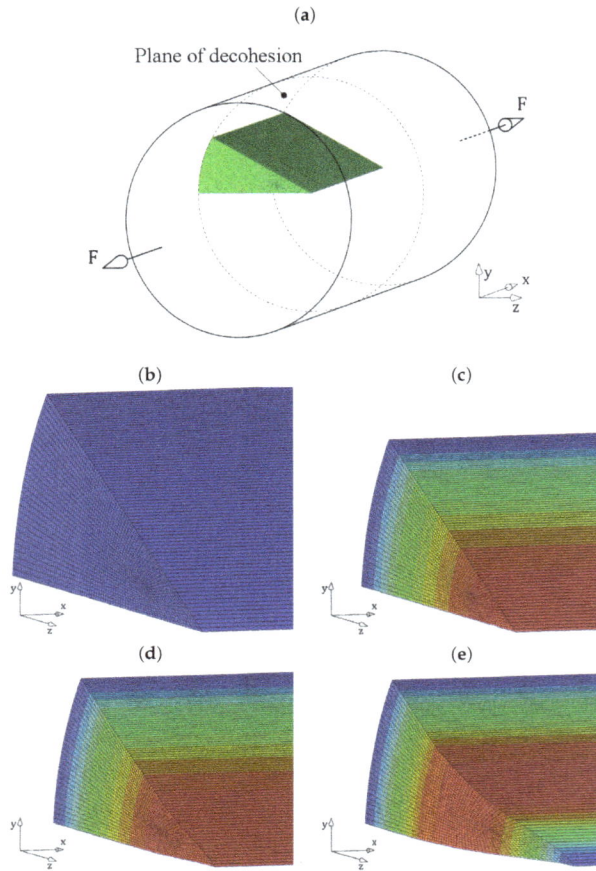

Figure 19. Evolution of decohesion at the fracture plane: (**a**) grid used to simulate the fracture process; (**b**) fracture plane before loading; (**c**) stress concentrates in the center as loading is applied; (**d**) decohesion begins; (**e**) the decohesive process propagates from inside to outside. Reproduced from [14], with permission from Elsevier, 2016.

Figure 20a,b show the load–strain curves experimentally obtained for Material 1 specimens of distinct diameters (3, 6 and 9 mm) with a fixed initial gauge length of 12.5 mm and a proportional to the diameter gauge length 1ϕ, respectively. These figures compare the experimental curves with those obtained numerically with the triaxiality-dependent model described above. All the numerical results were obtained with the same material parameters for each model and not individually adjusted. The fracture parameters were obtained by standardised tests, resulting in the values of Table 3. Figure 20a,b also provide the instants of decohesion initiation and failure, which are defined based on the radius of the decohesion zone. Finally, Figure 21a,b compare the experimental and numerical results in terms of the strain of fracture (ε_{ult}). It is interesting to observe that in the numerical models the decohesive process starts at the last moment of the test, close to the ultimate strain value, which is in good agreement with the experimental observations (see results on embrittled specimens by means of liquid nitrogen and XRCT images in Figure 15).

Table 3. Fracture parameters used for the numerical results of Figures 20 and 21.

f_t [MPa]	G_F [MPa·m]
1450	8.0×10^{-3}

(a)

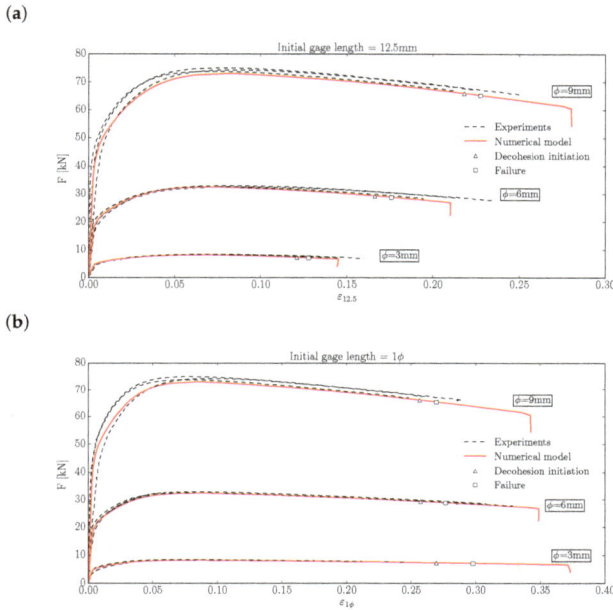

(b)

Figure 20. Load–strain curves obtained experimentally and numerically for each of the three considered diameters using an initial gauge length (**a**) 12.5 mm and (**b**) 1ϕ. Reproduced from [14], with permission from Elsevier, 2016.

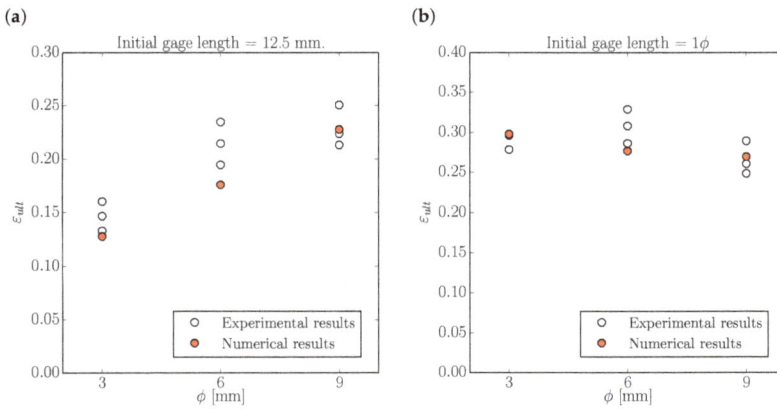

Figure 21. Ultimate strain obtained experimentally and numerically for each of the three considered diameters using an initial gauge length (**a**) 12.5 mm and (**b**) 1ϕ. Reproduced from [14], with permission from Elsevier, 2016.

This model provides reasonably good agreement with the experimental results, although it does not provide a fracture criterion for the ductile-brittle transition observed in this type of fracture (see Figure 1a).

The similar approach has been successfully employed later [104] with gray cast iron, but, instead of using interface elements, by means of crack-embedded elements.

4. Conclusions

In this paper, a review of the recent advances made on the study of two more usual fracture patterns observed in construction steels has been carried out, paying special attention to the material behaviour after maximum loading and the analysis of the fracture surfaces.

From the experimental point of view, although the ultimate strain values are usually neglected in practice, the experimental results carried out using a digital image correlation system have proved to be very repetitive and independent from the specimen radius and length. Hence, in principle, they could be used as reliable data if certain considerations are taken into account:

- If specimens of different radii are to be compared, a proportional-to-the-radius initial gauge length should be considered.
- The reference gauge length must be centered with the eventual fracture plane, otherwise the strain gradient would affect the measurements; this can be accomplished by using a digital image correlation system and not with conventional extensometers.

X-ray computed tomography allows identifying internal damage in steel, at least if thin enough specimens are used. This technique has helped to identify the internal damage evolution during a tensile test in the necking region. In the case of Material 1, which corresponds to an eutectoid steel used for manufacturing prestressing wires, the internal damage that eventually leads to a brittle fracture mechanism can be identified.

The study of the fracture surfaces' geometries suggests that the fracture mechanisms of both analysed materials differ notably, not only for their final shape (cup–cone pattern and flat surface with internal dark circular region), but also because, in the case of Material 1, this geometry proves not to be proportional to the specimen diameter, while, in the case of Material 2, this proportionality can be observed. Hence, both analysed materials show distinct fracture behaviours; in particular, it is interesting to note that Material 1 is an eutectoid steel and thus with a pearlitic structure that increases the material strength but reduces its ductility.

A brief overview on the most extended numerical models for ductile materials has confirmed that the triaxiality dependence must always be considered in this type of fracture and, in many cases, also the influence of the Lode angle. The use of a cohesive crack formulation has proved to provide reasonably good results even for metallic materials; this approach has been used by some researchers in the last years, including an application by the authors by using interface elements with the finite element method.

All the mentioned information can help to understand and support studying the behaviour of distinct steels beyond their maximum bearing capacity, which is usually neglected and considered unreliable. The works presented seem to point out in a different direction, since, if the experimental work is carefully carried out, results seem to be pretty repetitive and reliable. A better understanding of this issue can help to extend the usage of these materials beyond their current limits, which is considered of interest, since the last part of the stress–strain diagram, between maximum loading and failure, is mainly responsible for the energy absorbed by the material before collapsing. This could help to design better strategies in projected structures that could lead to higher structural safety conditions.

Therefore, the conclusions derived from this review work can be summarised as follows:

- If reliable values of the stress–strain curve beyond the maximum engineering stress are to be obtained, digital image correlation extensometry is preferred. The gauge length must be proportional to the specimen radius and placed so its midpoint is coincident with the fracture plane.

- X-ray computed tomography helps to identify internal damage in steel specimens, at least if thin enough specimens are used.
- Triaxiality must always be considered as a key factor when numerically reproducing fracture in steel. The Lode parameter can also be important in some cases.
- A triaxiality-dependent cohesive model is able to successfully reproduce fracture in eutectoid steel bars under tension.

Acknowledgments: The authors gratefully acknowledge the financial support provided for this research by the Ministry of Economy and Competitiveness of Spain by means of the Research Fund Project BIA 2016-78742-C2-2-R.

Author Contributions: Jaime C. Gálvez and José M. Atienza conceived and designed the experiments; Fernando Suárez performed the experiments; Fernando Suárez, David A. Cendón and Jaime C. Gálvez analyzed the data; Fernando Suárez and David A. Cendón carried out the numerical modelling; Fernando Suárez wrote the paper.

Conflicts of Interest: The authors declare no conflict of interest. The funding sponsors had no role in the design of the study; in the collection, analyses, or interpretation of data; in the writing of the manuscript, and in the decision to publish the results.

References

1. International Organization for Standardization. *E. 6892-1. Metallic Materials-Tensile Testing-Part 1: Method of Test at Room Temperature*; ISO: Geneva, Switzerland, 2009.
2. Bluhm, J.I.; Morrissey, R.J. *Fracture in a Tensile Specimen*; Defense Technical Information Center: Fort Belvoir, VA, USA, 1966.
3. Tvergaard, V.; Needleman, A. Analysis of the cup–cone fracture in a round tensile bar. *Acta Metall.* **1984**, *32*, 157–169.
4. Besson, J.; Steglich, D.; Brocks, W. Modeling of crack growth in round bars and plane strain specimens. *Int. J. Solids Struct.* **2001**, *38*, 8259–8284.
5. Scheider, I.; Brocks, W. Simulation of cup–cone fracture using the cohesive model. *Eng. Fract. Mech.* **2003**, *70*, 1943–1961.
6. Barsoum, I.; Faleskog, J. Rupture mechanisms in combined tension and shear—Micromechanics. *Int. J. Solids Struct.* **2007**, *44*, 5481–5498.
7. Barsoum, I.; Faleskog, J. Rupture mechanisms in combined tension and shear—Experiments. *Int. J. Solids Struct.* **2007**, *44*, 1768–1786.
8. Huespe, A.E.; Needleman, A.; Oliver, J.; Sánchez, P.J. A finite thickness band method for ductile fracture analysis. *Int. J. Plast.* **2009**, *25*, 2349–2365.
9. Huespe, A.E.; Needleman, A.; Oliver, J.; Sánchez, P.J. A finite strain, finite band method for modeling ductile fracture. *Int. J. Plast.* **2012**, *28*, 53–69.
10. Hutchinson, J.W.; Tvergaard, V. Shear band formation in plane strain. *Int. J. Solids Struct.* **1981**, *17*, 451–470.
11. Ayaso, J.; González, B.; Matos, J.C.; Vergara, D.; Lorenzo, M.; Toribio, J. Análisis fractográfico cuantitativo del comportamiento en fractura de aceros perlíticos progresivamente trefilados. *Anal. Mech. Fract.* **2005**, *22*, 128–133.
12. González, B.; Matos, J.C.; Toribio, J. Relación microestructura-propiedades mecánicas en acero perlítico progresivamente trefilado. *Anal. Mech. Fract.* **2009**, *1*, 142–147.
13. Rodríguez, R.; Toribio, J.; Ayaso, F.J. Defectos microestructurales que gobiernan la fractura anisótropa en aceros fuertemente trefilados. *Anal. Mech. Fract.* **2009**, *1*, 148–153.
14. Suárez, F.; Gálvez, J.C.; Cendón, D.A.; Atienza, J.M. Fracture of eutectoid steel bars under tensile loading: Experimental results and numerical simulation. *Eng. Fract. Mech.* **2016**, *158*, 87–105.
15. Suárez, F. Estudio de la Rotura en Barras de Acero: Aspectos Experimentales y Numéricos. Ph.D. Thesis, Universidad Politécnica de Madrid, Madrid, Spain, 2013.
16. Peters, W.; Ranson, W. Digital imaging techniques in experimental stress analysis. *Opt. Eng.* **1982**, *21*, 213427, doi:10.1117/12.7972925.

17. Chu, T.; Ranson, W.; Sutton, M.A. Applications of digital-image-correlation techniques to experimental mechanics. *Exp. Mech.* **1985**, *25*, 232–244.

18. Hung, P.C.; Voloshin, A. In-plane strain measurement by digital image correlation. *J. Braz. Soc. Mech. Sci. Eng.* **2003**, *25*, 215–221.

19. Sutton, M.A.; Orteu, J.J.; Schreier, H. *Image Correlation for Shape, Motion and Deformation Measurements: Basic Concepts, Theory and Applications*; Springer Science & Business Media: New York, NY, USA, 2009.

20. Pan, B.; Wang, Z.; Lu, Z. Genuine full-field deformation measurement of an object with complex shape using reliability-guided digital image correlation. *Opt. Express* **2010**, *18*, 1011–1023.

21. Pan, B.; Dafang, W.; Yong, X. Incremental calculation for large deformation measurement using reliability-guided digital image correlation. *Opt. Lasers Eng.* **2012**, *50*, 586–592.

22. Suárez, F.; Gálvez, J.C.; Cendón, D.A.; Atienza, J.M. Study of the last part of the stress-deformation curve of construction steels with distinct fracture patterns. *Eng. Fract. Mech.* **2016**, *166*, 43–59.

23. Guinea, G.; Rojo, F.; Elices, M. Stress intensity factors for internal circular cracks in fibers under tensile loading. *Eng. Fract. Mech.* **2004**, *71*, 365–377.

24. Rojo, F. Aplicación de la Mecánica de la Fractura a la Rotura Frágil de Fibras De sémola. Ph.D. Thesis, Universidad Politécnica de Madrid, Madrid, Spain, 2003.

25. ASTM. *E 399-90: Standard Test Method for Plane-Strain Fracture Toughness of Metallic Materials*; Annual Book of ASTM Standards: West Conshohocken, PA, USA, 1997; Volume 3, pp. 506–536.

26. Mirza, M.S.; Barton, D.C.; Church, P.; Sturges, J.L. Ductile Fracture of Pure Copper : An Experimental and Numerical Study. *J. Phys. IV France* **1997**, *7*, 891–896.

27. Toribio, J. A fracture criterion for high-strength steel notched bars. *Eng. Fract. Mech.* **1997**, *57*, 39–404.

28. Toribio, J.; Ayaso, F. Anisotropic fracture behaviour of cold drawn steel: A materials science approach. *Mater. Sci. Eng. A* **2003**, *343*, 265–272.

29. Toribio, J.; Vergara, D.; Lorenzo, M. Hydrogen effects in multiaxial fracture of cold-drawn pearlitic steel wires. *Eng. Fract. Mech.* **2017**, *174*, 243–252.

30. Gurson, A.L. *Continuum Theory of Ductile Rupture by Void Nucleation and Growth: Part I. Yield Criteria and Flow Rules for Porous Ductile Media*; Number Part 1 in Technical Report; Division of Engineering, Brown University: Providence, RI, USA, 1977.

31. Bridgman, P.W. *Studies in Large Plastic Flow and Fracture: With Special Emphasis on the Effects of Hydrostatic Pressure*; Metallurgy and Metallurgical Engineering Series; Harvard University Press: Cambridge, MA, USA, 1952.

32. Johnson, G.; Cook, W. A constitutive model and data for metals subjected to large strains, high strain rates and high temperatures. In Proceedings of the Seventh International Symposium on Ballistics, The Hague, The Netherlands, 19–21 April 1983.

33. Bao, Y. Prediction of Ductile Crack Formation in Uncracked Bodies. Ph.D. Thesis, Massachusetts Institute of Technology, Cambridge, MA, USA, 2003.

34. Bao, Y.; Wierzbicki, T. On fracture locus in the equivalent strain and stress triaxiality space. *Int. J. Mech. Sci.* **2004**, *46*, 81–98.

35. Bao, Y. Dependence of ductile crack formation in tensile tests on stress triaxiality, stress and strain ratios. *Eng. Fract. Mech.* **2005**, *72*, 505–522.

36. Borvik, T.; Langseth, M.; Hopperstad, O.; Malo, K. Ballistic penetration of steel plates. *Int. J. Impact Eng.* **1999**, *22*, 855–886.

37. Lode, W. Versuche über den Einfluß der mittleren Hauptspannung auf das Fließen der Metalle Eisen, Kupfer und Nickel. *Z. Phys. A Hadron. Nuclei* **1926**, *36*, 913–939.

38. Zhang, Z.L.; Thaulow, C.; Ødegård, J. A complete Gurson model approach for ductile fracture. *Eng. Fract. Mech.* **2000**, *67*, 155–168.

39. Xue, L.; Wierzbicki, T. Ductile fracture initiation and propagation modeling using a new fracture criterion. In Proceedings of the 9th European Mechanics of Materials Conference (EMMC9), Moret sur Loing, France, 9–12 May 2006; pp. 181–186.

40. Bai, Y.; Wierzbicki, T. A new model of metal plasticity and fracture with pressure and Lode dependence. *Int. J. Plast.* **2008**, *24*, 1071–1096.

41. Erice, B. Flow and Fracture Behaviour of High Performance Alloys. Ph.D. Thesis, Universidad Politécnica de Madrid, Madrid, Spain, 2012.

42. Erice, B.; Gálvez, F. A coupled elastoplastic-damage constitutive model with Lode angle dependent failure criterion. *Int. J. Solids Struct.* **2014**, *51*, 93–110.

43. Mirone, G.; Corallo, D. A local viewpoint for evaluating the influence of stress triaxiality and Lode angle on ductile failure and hardening. *Int. J. Plast.* **2010**, *26*, 348–371.

44. Barsoum, I.; Faleskog, J. Micromechanical analysis on the influence of the Lode parameter on void growth and coalescence. *Int. J. Solids Struct.* **2011**, *48*, 925–938.

45. Johnson, G.R.; Cook, W.H. Fracture characteristics of three metals subjected to various strains, strain rates, temperatures and pressures. *Eng. Fract. Mech.* **1985**, *21*, 31–48.

46. Sharma, P.; Chandel, P.; Bhardwaj, V.; Singh, M.; Mahajan, P. Ballistic impact response of high strength aluminium alloy 2014-T652 subjected to rigid and deformable projectiles. *Thin-Walled Struct.* **2017**, doi:10.1016/j.tws.2017.05.014.

47. Ouyang, Q.; Weng, G.; Soh, A.; Guo, X. Influences of nanotwin volume fraction on the ballistic performance of coarse-grained metals. *Theor. Appl. Mech. Lett.* **2017**, *7*, doi:10.1016/j.taml.2017.09.012.

48. Holmen, J.K.; Hopperstad, O.S.; Børvik, T. Influence of yield-surface shape in simulation of ballistic impact. *Int. J. Impact Eng.* **2017**, *108*, 136–146.

49. Sharma, P.; Chandel, P.; Mahajan, P.; Singh, M. Quasi-Brittle Fracture of Aluminium Alloy 2014 under Ballistic Impact. *Procedia Eng.* **2017**, *173*, 206–213.

50. Burley, M.; Campbell, J.; Dean, J.; Clyne, T. Johnson–Cook parameter evaluation from ballistic impact data via iterative FEM modelling. *Int. J. Impact Eng.* **2018**, *112*, 180–192.

51. Morales-Alonso, G.; Cendón, D.; Gálvez, F.; Sánchez-Gálvez, V. Influence of the softening curve in the fracture patterns of concrete slabs subjected to blast. *Eng. Fract. Mech.* **2015**, *140*, 1–16.

52. Imbalzano, G.; Tran, P.; Ngo, T.D.; Lee, P.V. A numerical study of auxetic composite panels under blast loadings. *Compos. Struct.* **2016**, *135*, 339–352.

53. Gambirasio, L.; Rizzi, E. An enhanced Johnson–Cook strength model for splitting strain rate and temperature effects on lower yield stress and plastic flow. *Comput. Mater. Sci.* **2016**, *113*, 231–265.

54. Liang, X.; Wang, Z.; Wang, R. Deformation model and performance optimization research of composite blast resistant wall subjected to blast loading. *J. Loss Prev. Process Ind.* **2017**, *49*, 326–341.

55. Wilkins, M.; Streit, R.; Reaugh, J. *Cumulative-Strain-Damage Model of Ductile Fracture: Simulation and Prediction of Engineering Fracture Tests*; Technical Report; Lawrence Livermore National Lab.: Livermore, CA, USA; Science Applications, Inc.: San Leandro, CA, USA, 1980.

56. Lemaitre, J. Coupled elasto-plasticity and damage constitutive equations. *Comput. Methods Appl. Mech. Eng.* **1985**, *51*, 31–49.

57. Lemaitre, J. A continuous damage mechanics model for ductile fracture. *Trans. ASME J. Eng. Mater. Technol.* **1985**, *107*, 83–89.

58. Xue, L. Damage accumulation and fracture initiation in uncracked ductile solids subject to triaxial loading. *Int. J. Solids Struct.* **2007**, *44*, 5163–5181.

59. Xue, L.; Wierzbicki, T. Ductile fracture initiation and propagation modeling using damage plasticity theory. *Eng. Fract. Mech.* **2008**, *75*, 3276–3293.

60. Børvik, T.; Hopperstad, O.; Berstad, T.; Langseth, M. A computational model of viscoplasticity and ductile damage for impact and penetration. *Eur. J. Mech.-A/Solids* **2001**, *20*, 685–712.

61. Rice, J.R.; Tracey, D.M. On the ductile enlargement of voids in triaxial stress fields. *J. Mech. Phys. Solids* **1969**, *17*, 201–217.

62. Nègre, P.; Steglich, D.; Brocks, W. Crack extension in aluminium welds: A numerical approach using the Gurson–Tvergaard–Needleman model. *Eng. Fract. Mech.* **2004**, *71*, 2365–2383.

63. Fei, H.; Yazzie, K.; Chawla, N.; Jiang, H. The effect of random voids in the modified Gurson model. *J. Electron. Mater.* **2012**, *41*, 177–183.

64. Li, H.; Fu, M.W.; Lu, J.; Yang, H. Ductile fracture: Experiments and computations. *Int. J. Plast.* **2011**, *27*, 147–180.

65. Needleman, A.; Tvergaard, V. Numerical modeling of the ductile-brittle transition. *Int. J. Fract.* **2000**, *101*, 73–97.

66. Hao, S.; Brocks, W. The Gurson–Tvergaard–Needleman-model for rate and temperature-dependent materials with isotropic and kinematic hardening. *Comput. Mech.* **1997**, *20*, 34–40.

67. Steglich, D.; Siegmund, T.; Brocks, W. Micromechanical modeling of damage due to particle cracking in reinforced metals. *Comput. Mater. Sci.* **1999**, *16*, 404–413.

68. Nahshon, K.; Hutchinson, J.W. Modification of the Gurson Model for shear failure. *Eur. J. Mech.- A/Solids* **2008**, *27*, 1–17.

69. Nielsen, K.L.; Tvergaard, V. Effect of a shear modified Gurson model on damage development in a FSW tensile specimen. *Int. J. Solids Struct.* **2009**, *46*, 587–601.

70. Nahshon, K.; Xue, Z. A modified Gurson model and its application to punch-out experiments. *Eng. Fract. Mech.* **2009**, *76*, 997–1009.

71. Jackiewicz, J. Use of a modified Gurson model approach for the simulation of ductile fracture by growth and coalescence of microvoids under low, medium and high stress triaxiality loadings. *Eng. Fract. Mech.* **2011**, *78*, 487–502.

72. Nielsen, K.L.; Tvergaard, V. Ductile shear failure or plug failure of spot welds modelled by modified Gurson model. *Eng. Fract. Mech.* **2010**, *77*, 1031–1047.

73. Xu, F.; Zhao, S.; Han, X. Use of a modified Gurson model for the failure behaviour of the clinched joint on Al6061 sheet. *Fatigue Fract. Eng. Mater. Struct.* **2014**, *37*, 335–348.

74. Morgeneyer, T.F.; Besson, J. Flat to slant ductile fracture transition: Tomography examination and simulations using shear-controlled void nucleation. *Scr. Mater.* **2011**, *65*, 1002–1005.

75. Vadillo, G.; Reboul, J.; Fernández-Sáez, J. A modified Gurson model to account for the influence of the Lode parameter at high triaxialities. *Eur. J. Mech.-A/Solids* **2016**, *56*, 31–44.

76. Hillerborg, A.; Modéer, M.; Petersson, P.E. Analysis of crack formation and crack growth in concrete by means of fracture mechanics and finite elements. *Cem. Concr. Res.* **1976**, *6*, 773–781.

77. Dugdale, D.S. Yielding of steel sheets containing slits. *J. Mech. Phys. Solids* **1960**, *8*, 100–104.

78. Vandewalle, L.; Dupont, D. *Bending Test and Interpretation*; RILEM Publication PRO: Bagneux, France, 2003; Volume 31, pp. 1–14.

79. RILEM-TCS. Determination of the fracture energy of mortar and concrete by means of three-point bend tests on notched beams. *Mater. Struct.* **1985**, *18*, 285–290.

80. Bažant, Z.P.; Planas, J. *Fracture and Size Effect in Concrete and Other Quasibrittle Materials*; New Directions in Civil Engineering; CRC Press: Boca Raton, FL, USA, 1997.

81. Bažant, Z.P.; Bittnar, Z.; Jirásek, M.; Mazars, J. *Fracture and Damage in Quasibrittle Structures: Experiment, Modeling and Computation*; CRC Press: Boca Raton, FL, USA, 2004.

82. García-Álvarez, V.O.; Carol, I.; Gettu, R. Numerical simulation of fracture in concrete using joint elements. *Anal. Mech. Fract.* **1994**, *11*, 75–80.

83. Xie, M.; Gerstle, W. Energy-based cohesive crack propagation modeling. *J. Eng. Mech.* **1995**, *121*, 1349–1358.

84. Carol, I.; Prat, P.C.; López, C.M. Normal/Shear Cracking Model: Application to Discrete Crack Analysis. *J. Eng. Mech.* **1997**, *123*, 765–773.

85. Cendón, D. Estudio de la Fractura en Modo Mixto de Hormigones y Morteros. Ph.D. Thesis, Universidad Politécnica de Madrid, Madrid, Spain, 2001.

86. Gálvez, J.C.; Cendón, D.A.; Planas, J.; Elices, M. Fractura en modo mixto de probetas de hormigón con doble entalla bajo solicitación de compresión: simulación numérica. *Anal. Mech. Fract.* **2001**, *18*, 219–225.

87. Gálvez, J.C.; Cervenka, J.; Cendón, D.A.; Saouma, V. A discrete crack approach to normal/shear cracking of concrete. *Cem. Concr. Res.* **2002**, *32*, 1567–1585.

88. Gálvez, J.C.; Cendón, D.A. Simulación de la fractura del hormigón en modo mixto. *Rev. Int. Metodos Numér.* **2002**, *18*, 31–58.

89. Rashid, Y. Ultimate strength analysis of prestressed concrete pressure vessels. *Nuclear Eng. Des.* **1968**, *7*, 334–344.

90. Suidan, M.; Schnobrich, W.C. Finite element analysis of reinforced concrete. *J. Struct. Div.* **1973**, *99*, 2109–2122.

91. Gupta, A.K.; Akbar, H. Cracking in reinforced concrete analysis. *J. Struct. Eng.* **1984**, *110*, 1735–1746.

92. De Borst, R.; Nauta, P. Non-orthogonal cracks in a smeared finite element model. *Eng. Comput.* **1985**, *2*, 35–46.

93. Jirásek, M.; Zimmermann, T. Rotating crack model with transition to scalar damage. *J. Eng. Mech.* **1998**, *124*, 277–284.

94. Simo, J.; Oliver, J. A new approach to the analysis and simulation of strain softening in solids. In *Proceedings of the Conference on Fracture and Damage in Quasibrittle Structures*; CRC Press: New York, NY, USA, 1994; pp. 25–39.

95. Larsson, R.; Runesson, K.; Sture, S. Embedded localization band in undrained soil based on regularized strong discontinuity–theory and FE-analysis. *Int. J. Solids Struct.* **1996**, *33*, 3081–3101.

96. Reyes, E. Rotura de la Fábrica de Ladrillo Bajo Solicitaciones de Tracción y Cortante. Ph.D. Thesis, Universidad de Castilla la Mancha, Ciudad Real, Spain, 2004.

97. Sancho, J.M.; Planas, J.; Cendón, D.A.; Reyes, E.; Gálvez, J.C. An embedded crack model for finite element analysis of concrete fracture. *Eng. Fract. Mech.* **2007**, *74*, 75–86.

98. Sancho, J.M.; Planas, J.; Fathy, A.M.; Gálvez, J.C.; Cendón, D.A. Three-dimensional simulation of concrete fracture using embedded crack elements without enforcing crack path continuity. *Int. J. Numer. Anal. Methods Geomech.* **2007**, *31*, 173–187.

99. Reyes, E.; Gálvez, J.C.; Casati, M.J.; Cendón, D.A.; Sancho, J.M.; Planas, J. An embedded cohesive crack model for finite element analysis of brickwork masonry fracture. *Eng. Fract. Mech.* **2009**, *76*, 1930–1944.

100. Gálvez, J.C.; Planas, J.; Sancho, J.M.; Reyes, E.; Cendón, D.A.; Casati, M.J. An embedded cohesive crack model for finite element analysis of quasi-brittle materials. *Eng. Fract. Mech.* **2012**, *109*, 369–386.

101. Siegmund, T.; Brocks, W. A numerical study on the correlation between the work of separation and the dissipation rate in ductile fracture. *Eng. Fract. Mech.* **2000**, *67*, 139–154.

102. Scheider, I. Derivation of separation laws for cohesive models in the course of ductile fracture. *Eng. Fract. Mech.* **2009**, *76*, 1450–1459.

103. ASTM. *ASTM E 1820-01. Standard Test Method for Measurement of Fracture Toughness*; Technical Report; ASTM: West Conshohocken, PA, USA, 2001.

104. Cendón, D.; Jin, N.; Liu, Y.; Berto, F.; Elices, M. Numerical Assessment of Gray Cast Iron Notched Specimens by Using a Triaxiality-Dependent Cohesive Zone Model. *Theor. Appl. Fract. Mech.* **2017**, *90*, 259–267.

MDPI

St. Alban-Anlage 66

4052 Basel

Switzerland

Tel. +41 61 683 77 34

Fax +41 61 302 89 18

www.mdpi.com

Metals Editorial Office

E-mail: metals@mdpi.com

www.mdpi.com/journal/metals